ELECTROCHEMICAL CELLS – NEW ADVANCES IN FUNDAMENTAL RESEARCHES AND APPLICATIONS

Edited by **Yan Shao**

Electrochemical Cells – New Advances in Fundamental Researches and Applications
Edited by Yan Shao

Published by InTech

Janeza Trdine 9, 51000 Rijeka, Croatia

Publishing Process Manager Oliver Kurelic

Technical Editor Teodora Smiljanic

Cover Designer InTech Design Team

Additional hard copies can be obtained from orders@intechopen.com

Electrochemical Cells – New Advances in Fundamental Researches and Applications,
Edited by Yan Shao
p. cm.
ISBN 978-953-51-0032-4

For EU product safety concerns:
IN TECH d.o.o., Prolaz Marije Krucifikse Kozulić 3, 51000 Rijeka, Croatia,
info@intechopen.com or visit our website at intechopen.com.

Meet the editor

Dr Yan Shao's academic background covers Chemistry, Physics, and Materials Science & Engineering. Currently, Dr Shao holds a PhD degree (2001) in Materials Science & Engineering from University of California, Los Angeles (UCLA), and BS (1998) and MS degrees (2001) in Chemistry, both from Tsinghua University. He also has postdoctoral research experience with Nobel laureate professor Alan J. Heeger in University of California, Santa Barbara (UCSB). Besides technical experience in material preparation, characterization, identification, and various electrical and electrochemical measurement methods, Dr Shao also has extensive project management experiences for developing new materials and material systems in both academic and industrial environment. He has already published 20 peer-reviewed journal papers, 8 Invention Patents, 1 book chapter with 1 editor highlight on Science, and multiple presentations in conferences.

Contents

Preface

According to the definition, electrochemical cells are the devices transferring electrical energy from chemical reactions into electricity, or helping chemical processes through the introduction of electrical energy or electrical field. A common example in this category is battery, which has evolved into a big family and is currently used in all kinds of applications.

As a relatively old scientific topic, the concept and application of electrochemical cells has been invented and utilized by human society has a long history. With the advance of modern science and technology, the research field of electrochemical cells has been more and more involved in the emerging areas of nano-technology, bio-technology, and novel energy storage and conversion systems, especially when serious attentions have been paid to energy, health, and environmental issues for modern industrial society.

It has become useful and necessary to summarize the new advances in both fundamental and practical researches in this field when we want to review the new development in recent years. In this book, parallel efforts have been put on both new advances in fundamental research and recent developments for applications of electrochemical cells, which include electrode design, electrochemical probe, liquid electrolytes, fuel cells, electrochemical detectors for health and environment consideration. Of course, electrochemical cells have been developed into a relatively large research category, which also means this book can only cover a corner of these topics.

Yan Shao,
Center for Polymers and Organic Solids,
Mitsubishi Chemical Center for Advanced Materials,
University of California, Santa Barbara,
USA

Part 1

New Advances in Fundamental Research in Electrochemical Cells

1

A Review of Non-Cottrellian Diffusion Towards Micro- and Nano-Structured Electrodes

Katarína Gmucová
Institute of Physics, Slovak Academy of Sciences
Slovak Republic

1. Introduction

The past few decades have seen a massive and continued interest in studying electrochemical processes at artificially structured electrodes. As is well known, the rate of redox reactions taking place at an electrode depends on both the mass transport towards the electrode surface and kinetics of electron transfer at the electrode surface. Three modes of mass transport can be considered in electrochemical cells: diffusion, migration and convection. The diffusional mass transport is the movement of molecules along a concentration gradient, from an area of high concentration to an area of low concentration. The migrational mass transport is observed only in the case of ions and occurs in the presence of a potential gradient. Convectional mass transport occurs in flowing solutions at rotating disk electrodes or at the dropping mercury electrode.

In 1902 Cottrell derived his landmark equation describing the diffusion current, I, flowing to a planar, uniformly accessible and smooth electrode of surface area, A, large enough not to be seriously affected by the edge effect, in contact with a semi-infinite layer of electrolyte solution containing a uniform concentration, c_O, of reagent reacting reversibily and being present as a minor component with an excess supporting electrolyte under unstirred conditions, during the potential-step experiment (Cottrell, 1902)

$$I = nFAc_O\sqrt{\frac{D}{\pi t}}, \tag{1}$$

where n is the number of electrons entering the redox reaction, F is the Faraday constant, D is the diffusion coefficient, and t is time.

It has long been known that the geometry, surface structure and choice of substrate material of an electrode have profound effects on the electrochemical response obtained. It is also understood that the electrochemical response of an electrode is strongly dependent on its size, and that the mass transport in electrochemical cell is affected by the electrode surface roughness which is generally irregular in both the atomic and geometric scales. Moreover, the instant rapid development in nanotechnology stimulates novel approaches in the preparation of artificially structured electrodes. This review seeks to condense information on the reasons giving rise or contributing to the non-Cottrellian diffusion towards micro- and nano structured electrodes.

2. Electrode geometry

Cottrell equation, derived for a planar electrode, can be applied to electrodes of other simple geometries, provided that the temporal and spatial conditions are such that the semi-infinite diffusion to the surface of the electrode is approximately planar. However, in both the research and application spheres various electrode geometries are applied depending on the problem or task to be solved. Most electrodes are impaired by an "edge effect" of some sort and therefore do not exhibit uniform accessibility towards diffusing solutes. Only the well defined electrode geometry allows the data collected at the working electrode to be reliably interpreted. The diffusion limited phenomena at a wide variety of different electrode geometries have been frequently studied by several research teams. Aoki and Osteryoung have derived the rigorous expressions for diffusion-controlled currents at a stationary finite disk electrode through use of the Wiener-Hopf technique (Aoki & Osteryoung, 1981). The chronoamperometric curve they have obtained varies smoothly from a curve represented by the Cottrell equation and can be expressed as the Cottrell term multiplied by a power series in the parameter $\sqrt{Dt/r}$, where r is the electrode radius. Later, a theoretical basis for understanding the microelectrodes with size comparable with the thickness of the diffusion layer, providing a general solution for the relation between current and potential in the case of a reversible reaction was given by the same authors (Aoki & Osteryoung, 1984). A user-friendly version of the equations for describing diffusion-controlled current at a disk electrode resulting from any potential perturbation was derived by Mahon and Oldham (Mahon & Oldham, 2005). Myland and Oldahm have proposed a method that permits the derivation of Cottrell's equation without explicitly solving Fick's second law (Myland & Oldham, 2004). The procedure, based on combining two techniques – the Green's Function technique and the Method of Images, has been shown to successfully treat several electrochemical situations. Being dependent on strict geometric conditions being met, it may provide a vehicle for a novel approach to electrochemical simulation involving diffusion in nonstandard geometries. In the same year Oldham reported an exact method used to find the diffusion-controlled faradaic current for certain electrode geometries that incorporate edges and vertices, which is based on Green's equation (Oldham, 2004). Gmucová and co-workers described the real electrochemical response of neurotransmitter dopamine on a carbon fiber microelectrode as a power function, i.e., $\propto t^{\beta}$ (Gmucová et al., 2004). That power function expanded to the polynomial terms can be, in conformity with (Aoki & Osteryoung, 1981; Mahon & Oldham, 2005), regarded as a Cottrell term, multiplied by a series of polynomial terms used to involve corrections to the Cottrell equation.

The variation of the diffusion layer thicknesses at planar, cylindrical, and spherical electrodes of any size was quantified from explicit equations for the cases of normal pulse voltammetry, staircase voltammetry, and linear sweep voltammetry by Molina and co-workers (Molina et al., 2010a). Important limiting behaviours for the linear sweep voltammetry current-potential curves were reported in all the geometries considered. These results are of special physical relevance in the case of disk and band electrodes which possess non-uniform current densities since general analytical solutions were derived for the above-mentioned geometries for the first time. Explicit analytical expressions for diffusion layer thickness of disk and band electrodes of any size under transient conditions were reported by Molina and co-workers (Molina et al., 2011b). Here, the evolution of the mass transport from linear (high sizes) to radial (microelectrodes) was characterized, and the conditions required to attain a stationary state were discussed. The use of differential pulse voltammetry at spherical electrodes and microelectrodes for the study of the kinetic of charge transfer processes was analyzed and an analytical solution was presented by Molina and co-workers (Molina et al., 2010b). The repored expressions are valid for any value of the electrode radius, the heterogeneous rate constant and the transfer coefficient. The anomalous shape of differential pulse voltammetry curves for quasi-reversible processes with small values of the transfer coefficient was reported, too. Moreover, general working curves were given for the determination of kinetic parameters from the position and height of differential pulse voltammetry peak. Sophisticated methods based on graphic programming units have been used by Cuttress and Compton to facilitate digital electrochemical simulation of processes at elliptical discs, square, rectangular, and microband electrodes (Cuttress & Compton, 2010a; Cuttress & Compton, 2010b).

A general, explicit analytical solution for any multipotential waveform valid for an electrochemically reversible system at an electrode of any geometry is continually in the centre of interest. This problem has been solved many times (e.g., Aoki et al., 1986; Cope & Tallman, 1991; Molina et al., 1995; Serna & Molina, 1999). A general theory for an arbitrary potential sweep voltammetry on an arbitrary topography (fractal or nonfractal) of an electrode operating under diffusion-limited or reversible charge-transfer conditions was developed by Kant (Kant, 2010). This theory provides a possibility to make clear various anomalies in measured electrochemical responses. Recently, analytical explicit expressions applicable to the transient *I-E* response of a reversible charge transfer reaction when both species are initially present in the solution at microelectrodes of different geometries (spheres, disks, bands, and cylinders) have been deduced (Molina et al., 2011a).

3. Electrochemical cells with bulk resistance

Mathematical modeling of kinetics and mass-transfer in electrochemical events and related electroanalytical experiments, generally consists of dealing with various physico-chemical parameters, as well as complicated mathematical problems, even in their simplest statement.

An analysis of the transient response in potential controlled experiments is a standard procedure which can yield information about many electrochemical processes and several kinetic parameters. However, a resistance in series (i.e., solution resitance, electrode coating resistance, sample resitance in solid state electrochemistry) can have a serious effect on electrochemical measurements. Thus, the presence of migration leads to essential deviations from the Cottrellian behaviour. Electrochemical systems that exhibit bulk ohmic resistances cannot be characterized accurately using the Cottrell equation. Electrochemical experiments in solution without added supporting electrolyte, i.e., without suppressed migration, became possible with the progress of microelectrodes. The expressions for current vs. time responses to applied voltage steps across the whole system, and corresponding concentration profiles within the cell or membrane were derived by Nahir and Buck and compared with experimental results (Nahir & Buck, 1992). Voltammetry in solutions of low ionic strength has been reviewed by Ciszkowska and Stojek (Ciszkowska & Stojek, 1999). A mathematical model of migration and diffusion coupled with a fast preceding reaction at a microelectrode was developed by Jaworski and co-workers (Jaworski et al., 1999). Myland and Oldham have shown that on macroelectrodes the Cottrellian dependence can be preserved even when supporting electrolyte is absent. The limiting current, however, was shown to depart in magnitude from the Cottrellian prediction by a factor (greater or less than unity) that depends on the charge numbers of the salt's ions and that of the electroproduct (Myland & Oldham, 1999). A generalized theory of the steady-state voltammetric response of a microelectrode in the absence of supporting electrolyte and for any values of diffusion coefficients of the substrate and the product of an electrode process was presented by Hyk and Stojek (Hyk & Stojek, 2002).

The influence of supporting electrolyte on the drugs detection was studied and data obtained using cyclic voltammetry, steady-state voltammetry and voltcoulometry on the same analyte were compared to each other by Orlický and co-workers. Under unsupported conditions different detection limits of the above mentioned methods were observed. Some species were easily observed by the kinetics-sensitive voltcoulometry even for concentrations near or under the sensitivity limit of voltammetric methods (Orlický et al., 2003). Thus, systems obeing deviations from Cottrell behaviour should find their application in sensorics. Later, it has been revealed that the dopamine diffusion current towards a carbon fiber microelectrode fulfills, within experimental errors and for concentration similar to those in a rat striatum, the behaviour theoretically predicted by the Cottrell equation. Nevertheless, under unsupported or weakly supported conditions non-Cottrellian responses were observed. Moreover, markedly non-Cottrellian responses were observed for dopamine concentrations lower or higher than the physiological ones in the rat striatum. It has been also shown, that the non-Cottrellian behaviour of diffusion current involves the nonlinearity of the dopamine calibration curve obtained by kinetics-sensitive voltcoulometry, while voltammetric calibration curve remains linear (Gmucová et al., 2004). Similarly, Caban and co-workers analysed the contribution of migration to the transport of polyoxometallates in the gels by methods of different sensitivity to migration (Caban et al., 2006).

Mathematical models of the ion transport regarded as the superposition of diffusion and migration in a potential field were analyzed by Hasanov and Hasanoglu (Hasanov & Hasanoglu, 2008). Based on the Nernst-Planck equation the authors have derived explicit analytical formulae for the concentration of the reduced species and the current response in the case of pure diffusive as well as diffusion - migration model, for various concentrations at initial conditions. The proposed approach can predict an influence of ionic diffusivities, valences, and initial and boundary concentrations to the behaviour of non-Cottrellian current response. In addition to these, the analytical formulae obtained can also be used for numerical and digital simulation methods for Nernst-Planck equations. The mathematical model of the nonlinear ion transport problem, which includes both the diffusion and migration, was solved by the same authors (Hasanov & Hasanoglu, 2009). They proposed a numerical iteration algorithm for solving the nonlocal identification problem related to nonlinear ion transport. The presented computational results are consistent with experimental results obtained on real systems.

The quantitative understanding of generalized Cottrellian response of moderately supported electrolytic solution at rough electrode/electrolyte interface was enabled with the Srivastav's and Kant's work (Srivastav & Kant, 2010). Here, the effect of the uncompensated solution resistance on the reversible charge transfer at an arbitrary rough electrode was studied and the significant deviation from the classical Cottrellian behavior was explained as arising from the resistivity of the solution and geometric irregularity of the interface. In the short time domain it was found to be dependent primarily on the resistance of the electrolytic solution and the real area of the surface. Results obtained for various electrode roughness models were reported. In the absence of the surface roughness, the current crossover to classical Cottrell response as the diffusion length exceeds the diffusion-ohmic length, but in the presence of roughness, there is formation of anomalous intermediate region followed by classical Cottrell region. Later, the theoretical results elucidating the influence of an uncompensated solution resistance on the anomalous Warburg's impedance in case of rough surfaces has been published by the same authors (Srivastav & Kant, 2011).

4. Modified electrodes

Modified electrodes include electrodes where the surface was deliberately altered to impart functionality distinct from the base electrode. During last decades a large number of different strategies for physical and chemical electrode modification have been developed, aimed at the enhancement in the detection of species under interest. Particularly in biosciences and environmental sciences such electrodes became of great importance. One of the issues raised in the research of redox processes taking place at modified electrodes has been the analysis of changes in the diffusion towards their altered surfaces.

Historically, liquid and solid electrochemistry grew apart and developed separately for a long time. Appearance of novel materials and methods of thin films preparation lead to massive development of chemically modified electrodes (Alkire et al., 2009). Such electrodes represent relatively modern approach to electrode systems with thin film of a selected chemical bonded or coated onto the electrode surface. A wide spectrum of their possible applications turned the spotlight of electrochemical research towards the design of electrochemical devices for applications in sensing, energy conversion and storage, molecular electronics etc. Only several examples of possible electrode coatings are mentioned in this chapter, all of them in close contact with the study of the electron transfer kinetic on them.

4.1 Micro- and nanoparticle modified electrodes

Marked deviations from Cottrellian behaviour were encountered in the theoretical study (Thompson et al., 2006) describing the diffusion of charge over the surface of a microsphere resting on an electrode at a point, in the limit of reversible electrode kinetics. A realistic physical problem of truncated spheres on the electrode surface was modelled in the above mentioned work, and the effect of truncation angle on chronoamperometry and voltammetry was explored. It has been shown that the most Cottrell-like behaviour is observed for the case of a hemispherical particle resting on the surface, but only at short times is the diffusion approximated well by a planar diffusion model. Concurrently, Thompson and Compton have developed a model for the voltammetric response due to surface charge injection at a single point on the surface of a microsphere on whose surface the electro-active material is confined. The cyclic voltammetric response of such system was investigated, the Fickian diffusion constrained on spherical surfaces showed strong deviations from the responses expected for planar diffusion. The Butler–Volmer condition was imposed for the electron transfer kinetics. It was found that the peak-to-peak separations differ from those expected for the planar-diffusion model, as well as the peak currents and the asymmetry of the voltammetric wave at higher sweep rates indicate the heterogeneous kinetics. The wave shape was explained by the competing processes of divergent and convergent diffusion (Thompson & Compton, 2006). Later, the electrochemical catalytic mechanism at a regularly distributed array of hemispherical particles on a planar surface was studied using simulated cyclic voltammetry (Ward et al., 2011). As is known, a high second-order rate constant can lead to voltammetry with a split wave. The conditions under which anomalous 'split-wave' phenomenon in cyclic voltammogram is observed were elucidated in the above-mentioned work.

In recent years significant attention is paid to the use of nanoparticles in many areas of electrochemistry. Underlying this endeavour is an expectation that the changed morphology and electronic structure between the macro- and nanoscales can lead to usefully altered electrode reactions and mechanisms. Thus, the use of nanoparticles in electroanalysis became an area of research which is continually expanding. Within both the trend towards the miniaturisation of electrodes and the ever-increasing progress in preparation and using nanomaterials, a profound development in electroanalysis has been connected with the design and characterisation of electrodes which have at least one dimension on the nano-scale.

In a nanostructured electrode, a larger portion of atoms is located at the electrode surface as compared to a planar electrode. Nanoparticle modified electrodes possess various advantages over macroelectrodes when used for electroanalysis, e.g., electrocatalysis, higher effective surface area, enhancement of mass transport and control over electrode microenvironment. An overview of the investigations carried out in the field of nanoparticles in electroanalytical chemistry was given in two successive papers (Welch & Compton, 2006; Campbell & Compton, 2010). Particular attention was paid to examples of the advantages and disadvantages nanoparticles show when compared to macroelectrodes and the advantages of one nanoparticle modification over another. From the works detailed in these reviews, it is clear that metallic nanoparticles have much to offer in electroanalysis due to the unique properties of nanoparticulate materials (e.g., enhanced mass transport, high surface area, improved signal-to-noise ratio). The unique properties of nanoparticulate materials can be exploited to enhance the response of electroanalytical techniques. However, according to the authors, at present, much of the work is empirical in nature. Belding and co-workers have compared the behaviour of nanoparticle-modified electrodes with that of conventional unmodified macroelectrodes (Belding et al., 2010). Here, a conclusion has been made that the voltammetric response from a nanoparticle-modified electrode is substantially different from that expected from a macroelectrode.

The first measurement of comparative electrode kinetics between the nano- and macroscales has been recently reported by Campbell and co-workers. The electrode kinetics and mechanism displayed by the nanoparticle arrays were found to be qualitatively and quantitatively different from those of a silver macrodisk. As was argued by Campbell and co-workers, the electrochemical behaviour of nanoparticles can differ from that of macroelectrodes for a variety of reasons. The most significant among them is that the size of the diffusion layer and the diffuse double layer at the nanoscale can be similar and hence diffusion and migration are strongly coupled. By comparison of the extracted electrode kinetics the authors stated that for the nanoparticle arrays, the mechanism is likely to be a rate-determining electron transfer followed by a chemical step. As the kinetics displayed by the nanoparticle arrays show changed kinetics from that of a silver macrodisk, they have inferred a change in the mechanism of the rate-determining step for the reduction of 4-nitrophenol in acidic media between the macro- and nanoscales (Campbell et al., 2010). Zhou and co-workers have found the shape and size of voltammograms obtained on silver nanoparticle modified electrodes to be extremely sensitive to the nanoparticle coverage, reflecting the transition from convergent to planar diffusion with increased coverage (Zhou et al., 2010). A system of iron oxide nanoparticles with mixed valencies deposited on photovoltaic amorphous hydrogenated silicon was studied by the kinetic sensitive voltcoulometry by Gmucová and co-workers. This study was motivated by the previously observed orientation ordering in similar system of nanoparticles involved by a laser irradiation under the applied electric field (Gmucová et al., 2008a). A significant dependence of the kinetic of the redox reactions, in particular oxidation reaction of ferrous ions, was observed as a consequence of the changes in the charged deep states density in amorphous hydrogenated silicon (Gmucová et al., 2008b).

4.2 Carbon nanotubes modified electrodes

Both the preparation and application of carbon nanotubes modified electrodes have been reviewed by Merkoçi, and by Wildgoose and co-workers (Merkoçi, 2006; Wildgoose et al., 2006). The comparative study of electrochemical behaviour of multiwalled carbon nanotubes and carbon black (Obradović et al., 2009) has revealed that although the electrochemical characteristics of properly activated carbon black approaches the characteristics of the carbon nanotubes, carbon nanotubes are superior, especially regarding the electron-transfer properties of the nanotubes with corrugated walls. The kinetics of electron-transfer reactions depends on the morphology of the samples and is faster on the bamboo-like structures, than on the nanotubes with smooth walls. Different oxidation properties of coenzyme NADH on carbon fibre microelectrode and carbon fibre microelectrode modified with branching carbon nanotubes have been reported by Zhao and co-workers (Zhao et al., 2010).

4.3 Thin film or membrane modified electrodes

Thin-layer cells, thin films and membrane systems show theoretical *I-t* responses that deviate from Cottrell behaviour. Although the diffusion was often assumed to be the only transport mechanism of the electroactive species towards polymer coated electrodes, the migration can contribute significantly. The bulk resistance of film corresponds to a resistance in series with finite diffusional element(s) and leads to ohmic *I-t* curves at short times. Subsequently, this resistance and the interacting depletion regions give rise to the non-Cottrellian behaviour of thin systems. According to Aoki, when an electrode is coated with a conducting polymer, the Nernst equation in a stochastic process is defined (Aoki, 1991). In such a case the electrode potential is determined by the ratio of the number of conductive (oxidized) species to that of the insulating (reduced) species experienced at the interface which is formed by electric percolation of the conductive domain to the substrate electrode. Examples of evaluating the potential for the case where the film has a random distribution of the conductive and insulating species were presented for three models: a one-dimensional model, a seven-cube model and a cubic lattice model.

Lange and Doblhofer solved the transport equations by digital simulation techniques with boundary conditions appropriate for the system electrode/membrane-type polymer coating (Lange & Doblhofer, 1987). They have concluded that the current transients follow Cottrell equation, however, the observed "effective" diffusion coefficients are different from the tabular ones. In the 90s an important effort has been devoted to examination of the nature of the diffusion processes of membrane-covered Clark-type oxygen sensors by solving the axially symmetric two-dimensional diffusion equation. Gavaghan and co-workers have presented a numerical solution of 2D equations governing the diffusion of oxygen to a circular disc cathode protected from poisoning by the medium to be measured by a tightly stretched plastic membrane which is permeable to oxygen (Gavaghan et al., 1992).

The current-time behaviour of membrane-covered microdisc clinical sensors was examined with the aim to explain their poor performance when pulsed (Sutton et al., 1996). It has been shown by Sutton and co-workers that the Cottrellian hypothesis is not applicable to this type of sensor and it is not possible to predict this behaviour from an analytical expression, as might be the case for membrane-covered macrodisc sensors and unshielded microdisc electrodes.

Gmucová and co-workers have shown that changes in kinetic of a redox reaction manifested as a deviation from the Cottrellian behaviour can be utilized in the preparation of ion selective electrodes. The electroactive hydrophobic end of a molecule used for the Langmuir-Blodgett film modification of a working electrode can induce a change in the kinetic of redox reactions. Ion selective properties of the poly(3-pentylmethoxythiophene) Langmuir–Blodgett film modified carbon-fiber microelectrode have been proved using a model system, mixture of copper and dopamine ions. While in case of the typical steady-state voltammetry the electrode remains sensitive to both the copper and dopamine, the kinetic-sensitive properties of voltcoulometry disable the observation of dopamine (Gmucová et al., 2007).

Recently, a sensing protocol based on the anomalous non-Cottrellian diffusion towards nanostructured surfaces was reported by Gmucová and co-workers (Gmucová et al., 2011). The potassium ferrocyanide oxidation on a gold disc electrode covered with a system of partially decoupled iron oxides nanoparticle membranes was investigated using the kinetic-sensitive voltcoulometry. Kinetic changes were induced by the altered electrode surface morphology, i.e., micro-sized superparamagnetic nanoparticle membranes were curved and partially damaged under the influence of the applied magnetic field. Thus, the targeted changes in the non-Cottrellian diffusion towards the working electrode surface resulted in a marked amplification of the measured voltcoulometric signal. Moreover, the observed effect depends on the membrane elasticity and fragility, which may, according to the authors, give rise to the construction of sensors based on the influence of various physical, chemical or biological external agents on the superparamagnetic nanoparticle membrane Young's moduli.

4.4 Spatially heterogeneous electrodes

Porous electrodes, partially blocked electrodes, microelectrode arrays, electrodes made of composite materials, some modified electrodes and electrodes with adsorbed species are spatially heterogeneous in the electrochemical sense. The simulation of non-Cottrellian electrode responses at such surfaces is challenging both because of the surface variation and because of the often random distribution of the zones of different electrode activity. The Cottrell equation becomes invalid even if the electrode reaction causes motion of the electrolyte/electrode boundary. Thereby it was modified by Oldham and Raleigh to take account of this effect, as well as to the data published on the inter-diffusion of silver and gold (Oldham & Raleigh, 1971).

Davies and co-workers have shown that by use of the concept of a ''diffusion domain'' computationally expensive three-dimensional simulations may be reduced to tractable two-dimensional equivalents which gives results in excellent agreement with experiment (Davies et al., 2005). Their approach predicts the voltammetric behaviour of electrochemically heterogeneous electrodes, e.g., composites whose different spatial zones display contrasting electrochemical behaviour toward the same redox couple. Four categories of response on spatially heterogeneous electrode have been defined by the authors depending on the blocked and unblocked electrode surface zones dimensions. In the performed analysis of partially blocked electrodes the difference between "macro" and ''micro'' was shown to be critical. The question how to specify whether the dimensions of the electro-active or inert zones of heterogeneous electrodes fall into one category or another one can be answered using the Einstein equation, which indicates that the approximate distance, δ, diffused by a species with a diffusion coefficient, D, in a time, t, is $\delta = \sqrt{2Dt}$. The work carried out in the Compton group on methods of fabricating and characterising arrays of nanoelectrodes, including multi-metal nanoparticle arrays for combinatorial electrochemistry, and on numerical simulating and modelling of the electrochemical processes was reviewed in the frontiers article written by Compton (Compton et al., 2008).

An improved sensitivity of voltammetric measurements as a consequence of either electrode or voltammetric cell exposure to low frequency sound was reported by Mikkelsen and Schrøder (Mikkelsen & Schrøder, 1999; Mikkelsen & Schrøder, 2000). According to the authors the longitudinal waves of sound applied during measurements make standing regions with different pressures and densities, which make streaming effects in the boundary layer at least comparable to the conventional stirring. As an alternative explanation of the marked sensitivity enhancement the authors suggested a possible change in the electrical double layer structure. Later, a study of the dopamine redox reactions on the carbon fiber microelectrode by the kinetics-sensitive voltcoulometry (Gmucová et al., 2002) revealed an impressive shift towards the ideal kinetic described by Cottrell equation, achieved by an electrochemical pretreatment of the electrode accompanied by its simultaneous exposure to the low frequency sound.

The diffusion equation including the delay of a concentration flux from the formation of a concentration gradient, called diffusion with memory, was formulated by Aoki and solved under chronoamperometric conditions (Aoki, 2006). A slower decay than predicted by the Cottrell equation was obtained.

A theoretical study of the current–time relationship aimed at the explanation of anomalous response in differential pulse polarography was reported by Lovrić and Zelić. The effect was explained by the adsorption of reactant at the electrode surface (Lovrić & Zelić, 2008). The situation connected with the formation of metal preconcentration at the electrode surface, followed by electrodissolution was modelled by Cutters and Compton. The theory to explore the electrochemical signals in such a case at a microelectrode or ultramicroelectrode arrays was derived (Cutress & Compton, 2009).

5. Fractal concepts

A possible cause of the deviation of measured signals from the ideal Cottrellian one is of geometric origin. The irregular (rough, porous or partially active) electrode geometry can and does cause current density inhomogeneities which in turn yield deviations from ideal behaviour. Kinetic processes at non-idealised, irregular surfaces often show non-conventional behaviour, and fractals offer an efficient way to handle irregularity in general terms. Rough and partially active electrodes are frequently modelled using fractal concepts; their surface roughness of limited length scales irregularities is often characterized as self-affine fractal. Fractal geometry is an efficient tool for characterizing irregular surfaces in very general terms. An introduction to the methods of fractal analysis can be found in the work (Le Mehaute & Crepy, 1983). Electrochemistry at fractal interfaces has been reviewed by Pajkossy (Pajkossy, 1991). Diffusion-limited processes on such interfaces show anomalous behavior of the reaction flux.

Pajkossy and co-workers have published an interesting series of papers devoted to the electrochemistry on fractal surfaces (Nyikos & Pajkossy, 1986; Pajkossy & Nyikos, 1989a; Pajkossy & Nyikos, 1989b; Nyikos et al., 1990; Borosy et al., 1991). Diffusion to rough surfaces plays an important role in diverse fields, e.g., in catalysis, enzyme kinetics, fluorescence quenching and spin relaxation. Nyikos and Pajkossy have shown that, as a consequence of fractal electrode surface, the diffusion current is dependent on time as $i \propto t^{-(D_f-1)/2}$, where D_f is the fractal dimension (Nyikos & Pajkossy, 1986). For a smooth, two-dimensional interface ($D_f = 2$) the Cottrell behaviour $i \propto t^{-1/2}$ is obtained. In electrochemical terms this corresponds to a generalized Cottrell equation (or Warburg impedance) and can be used to describe the frequency dispersion caused by surface roughness effects. Later, the verification of the predicted behaviour for fractal surfaces with $D_f > 2$ (rough interface), and $D_f < 2$ (partially blocked surface or active islands on inactive support) was reported (Pajkossy & Nyikos, 1989a). The fractal decay kinetics has been shown to be valid for both contiguous and non-contiguous surfaces, rough or partially active surfaces. Using computer simulation, a mathematical model, and direct experiments on well defined fractal electrodes the fractal decay law has been confirmed for different surfaces. According to the authors, this fractal diffusion model has a feature which deserves some emphasis: this being its generality. It is based on a very general assumption, i.e., self-similarity of the irregular interface, and nothing specific concerning the electrode material, diffusing substance, etc. is assumed. Based on the generalized Cottrell equation, the calculation and experimental verification of linear sweep and cyclic voltammograms on fractal electrodes have been performed (Pajkossy & Nyikos, 1989). The generalized model has been shown to be valid for non-linear potential sweeps as well. Its experimental verification on an electrode with a well defined fractal geometry $D_f = 1.585$ was presented for a rotating disc electrode of fractal surface (Nyikos et al., 1990). The fractal approximation has been shown to be useful for describing the geometrical aspects of diffusion processes at realistic rough or irregular-interfaces (Borosy et al., 1991). The authors have concluded that diffusion towards a self-affine fractal surface with much smaller vertical irregularity than horizontal irregularity leads to the conventional Cottrell relation between current and time of the Euclidean object, not the generalised Cottrell relation including fractal dimension.

The most important conclusions, as outlined in (Pajkossy, 1991), are as follows. If a capacitive electrode is of fractal geometry, then the electrode impedance will be of the constant phase element form (i.e., the impedance, Z, depends on the frequency, ω, as $Z \propto (i\omega)^{\beta}$ with $0 < \beta < 1$). However, no unique relation between fractal dimension D_f and constant phase element exponent can be established. Assume that a real surface is irregular from the geometrical point of view and that the diffusion-limited current can be measured on it, the surface irregularities can be characterized by a single number, the fractal dimension. The time dependence of the diffusion limited flux to a fractal surface is a power-law function of time, and there is a unique relation ($\beta = (D_f - 1)/2$) between fractal dimension and the exponent β. This equation provides a possibility for the experimental determination of the fractal dimension.

The determination of fractal dimension of a realistic surface has been reported by Ocon and co-workers (Ocon et al., 1991). The thin columnar gold electrodeposits (surface roughness factor 50-100) grown on gold wire cathodes by electroreducing hydrous gold oxide layers have been used for this purpose, the fractal dimension has been determined by measuring the diffusion controlled current of the $Fe(CN)^{4-}/Fe(CN)^{3-}$ reaction. Several examples of diffusion controlled electrochemical reactions on irregular metal electrodeposits type of electrodes were described in (Arvia & Salvarezza, 1994). Using the fractal geometry relevant information about the degree of surface disorder and the surface growth mechanism was obtained and the kinetic of electrochemical reactions at these surfaces was predicted.

Kant has discussed rigorously the anomalous current transient behaviour of self-affine fractal surface in terms of power spectral density of the surface (Kant, 1997). The non-universality and dependence of intermediate time behaviour on the strength of fractality of the interface has been reported, the exact result for the low roughness and the asymptotic results for the intermediate and large roughness of self-affine fractal surfaces have been derived. The intermediate time behaviour of the reaction flux for the small roughness interface has been shown to be proportional to $t^{-1/2}$ + const $t^{-3/2+H}$, however, for the large roughness interfaces the dependence $\sim t^{-1+H/2}$, where H is Hurst's exponent, was found. For an intermediate roughness a more complicated form has been obtained.

Shin and co-workers investigated the diffusion toward self-affine fractal interfaces by using diffusion-limited current transient combined with morphological analysis of the electrode surface (Shin et al., 2002). Here, the current transients from the electrodes with increasing morphological amplitude (roughness factor) were roughly characterised by the two-stage power dependence before temporal outer cut-off of fractality. Moreover, the authors suggested a method to interpret the anomalous current transient from the self-affine fractal electrodes with various amplitudes. This method, describing the anomalous current transient behaviour of self-affine electrodes, includes the determination of the apparent self-similar scaling properties of the self-affine fractal structure by the triangulation method.

A general transport phenomenon in the intercalation electrode with a fractal surface under the constraint of diffusion mixed with interfacial charge transfer has been modelled by using the kinetic Monte Carlo method based upon random walk approach (Lee & Pyun, 2005). Go and Pyun (Go & Pyun, 2007) reviewed anomalous diffusion towards and from fractal interface. They have explained both the diffusion-controlled and non-diffusion-controlled transfer processes. For the diffusion coupled with facile charge-transfer reaction the electrochemical responses at fractal interface were treated with the help of the analytical solutions to the generalised diffusion equation. In order to provide a guideline in analysing anomalous diffusion coupled with sluggish charge-transfer reaction at fractal interface, i.e., non-diffusion-controlled transfer process across fractal interface, this review covered the recent results concerned to the effect of surface roughness on non-diffusion-controlled transfer process within the intercalation electrodes. It has been shown, that the numerical analysis of diffusion towards and from fractal interface can be used as a powerful tool to elucidate the transport phenomena of mass (ion for electrolyte and atom for intercalation electrode) across fractal interface whatever controls the overall transfer process.

A theoretical method based on limited scale power law form of the interfacial roughness power spectrum and the solution of diffusion equation under the diffusion-limited boundary conditions on rough interfaces was developed by Kant and Jha (Kant & Jha, 2007). The results were compared with experimentally obtained currents for nano- and micro-scales of roughness and are applicable for all time scales and roughness factors. Moreover, this work unravels the connection between the anomalous intermediate power law regime exponent and the morphological parameters of limited scales of fractality.

Kinetic response of surfaces defined by finite fractals has been addressed in the context of interaction of finite time independent fractals with a time-dependent diffusion field by a novel approach of Cantor Transform that provides simple closed form solutions and smooth transitions to asymptotic limits (Nair & Alam, 2010). In order to enable automatic simulation of electrochemical transient experiments performed under conditions of anomalous diffusion in the framework of the formalism of integral equations, the adaptive Huber method has been extended onto integral transformation kernel representing fractional diffusion (Bieniasz, 2011).

The fractal dimension can be simply estimated using the kinetics-sensitive voltcoulometry introduced by Thurzo and co-workers (Thurzo et al., 1999). On the basis of the multipoint analysis principles the transient charge is sampled at three different events in the interval between subsequent excitation pulses and the sampled values are combined according the appropriate filtering scheme. The third sampling event chosen at the end of measuring period and slow potential scans make the observation of non-Cottrellian responses easier. The parameter β that enters the power-law time dependence of the transient charge, as well as the fractal dimension can be simply determined from two voltcoulograms obtained for two distinct sets of sampling events (Gmucová et al., 2002).

6. Conclusion

The electrode surface attributes have a profound influence on the kinetic of electron transfer. The continued progress in material research has induced the marked progress in the preparation of electrochemical electrodes with enhanced sensitivity or selectivity. If such a sophisticated electrode with microstructured, nanostructured or electroactive surface is used a special attention should be paid to a careful examination of changes initiated in the diffusion towards its surface. Newly designed types of electrochemical electrodes often result in more or less marked deviations from the ideal Cottrell behaviour. Various modifications of the relationship (Equation (1)) have been investigated to describe the processes in real electrochemical cells. A raising awareness of the importance of a detailed knowledge on the kinetic of charge transfer during the studied redox reaction has lead to a significant number of theoretical, computational, phenomenological and, last but not least, experimental studies. Based on them one can conclude: nowadays, an un-usual behaviour is the Cottrellian one.

7. Acknowledgment

This work was supported by the ASFEU project Centre for Applied Research of Nanoparticles, Activity 4.2, ITMS code 26240220011, supported by the Research & Development Operational Programme funded by the ERDF and by Slovak grant agency VEGA contract No.: 2/0093/10.

8. References

Alkire, R.C.; Kolb, D.M.; Lipkowsky & J. Ross, P.N. (Eds.). (2009). *Advances in Electrochemical Science and Engineering, Volume 11, Chemically modified electrodes,* WILEY-VCH Verlag GmbH & Co. KGaA, ISBN 978-3-527-31420-1, Weinheim Germany

Aoki, K. & Osteryoung, J. (1981) Diffusion-Controlled Current at the Stationary Finite Disk Electrode. *J. Electroanal. Chem.* Vol.122, No.1, (May 1981), pp. 19-35, ISSN 1572-6657

Aoki, K. & Osteryoung, J. (1984) Formulation of the Diffusion-Controlled Current at Very Small Stationary Disk Electrodes. *J. Electroanal. Chem.* Vol.160, No.1-2 (January 1984), pp. 335–339, ISSN 1572-6657

Aoki, K.; Tokuda, K.; Matsuda, H. & Osteryoung, J. (1986) Reversible Square-Wave Voltammograms Independence of Electrode Geometry. *J. Electroanal. Chem.* Vol.207, No.1-2, (July 1986), pp. 335–339, ISSN 1572-6657

Aoki, K. (1991) Nernst Equation Complicated by Electric Random Percolation at Conducting Polymer-Coated Electrodes. *J. Electroanal. Chem.* Vol.310, No.1-2, (July 1991), pp. 1–12, ISSN 1572-6657

Aoki, K. (2006) Diffusion-Controlled Current with Memory. *J. Electroanal. Chem.* Vol.592, No.1, (July 2006), pp. 31-36, ISSN 1572-6657

Arvia, A.J. & Salvarezza, R.C. (1994). Progress in the Knowledge of Irregular Solid Electrode Surfaces. *Electrochimica Acta* Vol.39, No.11-12, (August 1994), pp. 1481–1494, ISSN 0013-4686

Belding, S.R.; Campbell, F.W.; Dickinson, E.J.F. & R.G. Compton, R.G. (2010) Nanoparticle-Modified Electrodes. *Physical Chemistry Chemical Physics* Vol.12, No.37, (October 2010), pp. 11208–11221, ISSN 1463-9084

Bieniasz, L.K. (2011). Extension of the Adaptive Huber Method for Solving Integral Equations Occurring in Electroanalysis, onto Kernel Function Representing Fractional Diffusion. *Electroanalysis,* Vol.23, No.6, (June 2011), pp. 1506-1511, ISSN 1521-4109

Borosy, A. P.; Nyikos, L. & Pajkossy, T. (1991). Diffusion to Fractal Surfaces-V. Quasi-Random Interfaces. *Electrochimica Acta.* Vol.36, No.1, (1991), pp. 163–165, ISSN 0013-4686

Buck, R.P.; Mahir, T.M.; Mäckel, R. & Liess, H.-D. (1992) Unusual, Non-Cottrell Behavior of Ionic Transport in Thin Cells and in Films. *J. Electrochem. Soc.* Vol.139, No.6, (June 1992), pp. 1611–1618, ISSN 0013-4651

Caban, K.; Lewera, A.; Zukowska, G.Z.; Kulesza, P.J.; Stojek, Z. & Jeffrey, K.R. (2006) Analysis of Charge Transport in Gels Containing Polyoxometallates Using Methods of Different Sensitivity to Migration. *Analytica Chimica Acta.* Vol.575, No.1, (August 2006), pp. 144–150, ISSN 0003-2670

Campbell, F.W. & Compton, R.G. (2010) The Use of Nanoparticles in Electroanalysis: an Updated Review. *J. Phys. Chem. C.* Vol.396, No.1, (January 2010), pp. 241–259, ISSN 1618-2650

Campbell, F.W.; Belding, S.R. & Compton, R.G. (2010). A Changed Electrode Reaction Mechanism between the Nano- and Macroscales. *Chem. Phys. Chem.,* Vol. 1, No.13, (September 2010), pp. 2820-2824, ISSN 1439-7641

Ciszkowska, M. & Stojek, Z. (1999) Voltammetry in Solutions of Low Ionic Strength. Electrochemical and Analytical Aspects. *J. Electroanal. Chem.* Vol.466, No.2, (May 1999), pp. 129-143, ISSN 1572-6657

Compton, R.G.; Wildgoose, G.G.; Rees, N.V.; Streeter, I. & Baron, R. (2008) Design, Fabrication, Characterisation and Application of Nanoelectrode Arrays. *Chem. Phys. Letters.* Vol.459, No.1-6, (June 2008), pp. 1–17, ISSN 0009-2614

Cope, D.K. & Tallman, D.E. (1991) Representation of Reversible Current for Arbitrary Electrode Geometries and Arbitrary Potential Modulation. *J. Electroanal. Chem.* Vol.303, No.1-2, (March 1991), pp. 1-15, ISSN 1572-6657

Cottrel, F.G. (1902) Der Reststrom bei galvanischer Polarisation, betrachtet als ein Diffusionsproblem. *Z. Phys. Chem.* Vol.42 (1902), pp. 385-431, ISSN 0942-9352

Cutress, I.J. & Compton, R.G. (2009) Stripping Voltammetry at Microdisk Electrode Arrays: Theory. *Electroanalysis* Vol.21, No.24, (December 2009), pp. 2617-2625, ISSN 1521-4109

Cutress, I.J. & Compton, R.G. (2010a) Using Graphics Processors to Facilitate Explicit Digital Electrochemical Simulation: Theory of Elliptical Disc Electrodes. *J. Electroanal. Chem.* Vol.643, No.1-2, (May 2010), pp. 102-109, ISSN 1572-6657

Cutress, I.J. & Compton, R.G. (2010b) Theory of Square, Rectangular, and Microband Electrodes through Explicit GPU Simulation. *J. Electroanal. Chem.* Vol.645, No.2, (July 2010), pp. 159-166, ISSN 1572-6657

Davies, T.J.; Banks, C.E. & Compton, R.G. (2005). Voltammetry at Spatially Heterogeneous Electrodes. *J. Solid State Electrochem.,* Vol.9, No.12, (December 2005), pp. 797-808, ISSN 1433-0768

Gavaghan, D.J., Rollet, J.S. & Hahn, C.E.W. (1992). Numerical Simulation of the Time-Dependent Current to Membrane-Covered Oxygen Sensors. Part I. The Switch-on Transient. *J. Electroanal. Chem.* Vol.325, No.1-2, (March 1992), pp. 23–44, ISSN 1572-6657

Gmucová, K.; Thurzo, I.; Orlický, J. & Pavlásek, J. (2002). Sensitivity Enhancement in Double-Step Voltcoulometry as a Consequence of the Changes in Redox Kinetics on the Microelectrode Exposed to Low Frequency Sound. *Electroanalysis,* Vol.14, No.13, (July 2002), pp. 943-948, ISSN 1521-4109

Gmucová, K.; Orlický, J. & Pavlásek, J. (2004). Non-Cottrell Behaviour of the Dopamine Redox Reaction Observed on the Carbon Fibre Microelectrode by the Double-Step Voltcoulometry. *Collect. Czech. Chem. Commun.,* Vol.69, No.2, (February 2004), pp. 419-425, ISSN 0010-0765

Gmucová, K.; Weis, M.; Barančok, D.; Cirák, J.; Tomčík, P. & Pavlásek, J. (2007). Ion Selectivity of a Poly(3-pentylmethoxythiophene) LB-Layer Modified Carbon-Fiber Microelectrode as a Consequence of the Second Order Filtering in Voltcoulometry. *J. Biochem. Biophys. Medthods,* Vol.70, No.3, (April 2007), pp. 385-390, ISSN 0165-022X

Gmucová, K.; Weis, M.; Nádaždy. V.; & Majková, E. (2008a). Orientation Ordering of Nanoparticle Ag/Co Cores Controlled by Electric and Magnetic Fields. *ChemPhysChem.,* Vol.9, No.7, (May 2008), pp. 1036–1039, ISSN 1439-7641

Gmucová, K.; Weis, M.; Nádaždy. V.; Capek, I.; Šatka, A.; Chitu, L. Cirák, J. & Majková, E. (2008b). Effect of Charged Deep States in Hydrogenated Amorphous Silicon on the Behaviour of Iron Oxides Nanoparticles Deposited on its Surface. *Applied Surface Science,* Vol.254, No.21, (August 2008), pp. 7008-7013, ISSN 0169-4332

Gmucová, K.; Weis, M.; Benkovičová, M.; Šatka, A. & Majková, E. (2011). Microstructured Nanoparticle Membrane Sensor Based on Non-Cottrellian Diffusion. *J. Electroanal. Chem.,* Vol.659, No.1 (August 2011), pp. 58-62, ISSN 1572-6657

Go, J.Y. & Pyun S.I. (2007). A Review of Anomalous Diffusion Phenomena at Fractal Interface for Diffusion-Controlled and Non-Diffusion-Controlled Transfer processes. *J. Solid State Electrochem.,* Vol.11, No.2, (February 2007), pp. 323–334, ISSN 1433-0768

Hasanov, A. & Hasanoglu, Ş. (2008). Analytical Formulaes and Comparative Analysis for Linear Models in Chronoamperometry Under Conditions of Diffusion and Migration. *J. Math. Chem.,* Vol.44, No.1, (July 2008), pp. 133–141, ISSN 1572-8897

Hasanov, A. & Hasanoglu, Ş. (2009). An Analysis of Nonlinear Ion Transport Model Including Diffusion and Migration. *J. Math. Chem.* Vol.46, No.4, (November 2009), pp. 1188–1202, ISSN 1572-8897

Hyk, W. & Stojek, Z. (2002). Generalized Theory of Steady-State Voltammetry Without a Supporting Electrolyte. Effect of Product and Substrate Diffusion Coefficient Diversity. *Anal. Chem.* Vol. 74, No.18, (September 2002), pp. 4805–4813, ISSN 1520-6882

Jaworski, A.; Donten, M.; Stojek, Z. & Osteryoung, J.G. (1999). Migration and Diffusion Coupled with a Fast Preceding Reaction. Voltammetry at a Microelectrode. *Anal. Chem.* Vol.71, No.1, (January 1999), pp. 167-173, ISSN 0003-2700

Kant, R. (1997). Diffusion-Limited Reaction Rates on Self-Affine Fractals. *J. Phys. Chem B.* Vol.101, No.19, (May 1997), pp. 3781–3787, ISSN 1520-5207

Kant, R. & Jha, Sh.K. (2007). Theory of Anomalous Diffusive Reaction Rates on Realistic Self-affine Fractals. *J. Phys. Chem C.* Vol.111, No.38, (September 2007), pp. 14040–14044, ISSN 1932-7447

Kant, R. (2010) General Theory of Arbitrary Potential Sweep Methods on an Arbitrary Topography Electrode and Its Application to Random Surface Roughness. *J. Phys. Chem C.* Vol.114, No.24, (June 2010), pp. 10894–10900, ISSN 1932-7447

Lange, R. & Doblhofer, K. (1987) The Transient Response of Electrodes Coated with Membrane-Type Polymer Films under Conditions of Diffusion and Migration of the Redox Ions. *J. Electroanal. Chem.* Vol.237, No.1, (November 1987), pp. 13–26, ISSN 1572-6657

Lee, J.-W. & Pyun, S.-I. (2005). A Study on the Potentiostatic Current Transient and Linear Sweep Voltammogram Simulated from Fractal Intercalation Electrode: Diffusion

Coupled with Interfacial Charge Transfer. *Electrochimica Acta.* Vol.50, No.9, (March 2005), pp. 1947–1955, ISSN 0013-4686

Le Mehaute, A. & Crepy, G. (1983). Introduction to Transfer and Motion in Fractal Media: the Geometry of Kinetics. *Solid St. Ionics.* Vol.9-10, No.1, (December 1983), pp. 17–30, ISSN 0167-2738

Lovrić, M. & Zelić, M. (2008) Non-Cottrell Current–Time Relationship, Caused by Reactant Adsorption in Differential Pulse Polarography. *J. Electroanal. Chem.* Vol.624, No.1-2, (December 2008), pp. 174–178, ISSN 1572-6657

Mahon, P.J. & Oldham, K.B. (2005). Diffusion-Controlled Chronoamperometry at a Disk Electrode. *Anal. Chem.* Vol.77, No.19, (October 2005), pp. 6100–6101, ISSN 1520-6882

Mikkelsen, Ø. & Schrøder, K.H. (1999). Sensitivity Enhancements in Stripping Voltammetry from Exposure to Low Frequency Sound. *Electroanalysis,* Vol.11, No.6, (May 1999), pp. 401-405, ISSN 1521-4109

Mikkelsen, Ø. & Schrøder, K.H. (2000). Low Frequency Sound for Effective Sensitivity Enhancement in Staircase Voltammetry. *Anal. Lett,* Vol.33, No.7, (April 2000), pp. 1309-1326, ISSN 0003-2719

Merkoçi, A. (2006). Carbon Nanotubes in Analytical Sciences. *Microchimica Acta,* Vol.152, No.3-4, (January 2006), pp. 157-174, ISSN 1436-5073

Molina, A.; Serna, C. & Camacho, L.J. (1995) Conditions of Applicability of the Superposition Principle in Potential Multipulse Techniques: Implications in the Study of Microelectrodes. *J. Electroanal. Chem.* Vol.394, No.1-2, (September 1995), pp. 1–6, ISSN 1572-6657

Molina, A.; Gonzáles, J.; Martínez-Ortiz, F. & Compton, R.G. (2010a) Geometrical Insights of Transient Diffusion Layers. *J. Phys. Chem. C.* Vol.114, No.9, (March 2010), pp. 4093–4099, ISSN 1932-7447

Molina, A.; Martínez-Ortiz, F.; Laborda, E. & Compton, R.G. (2010b) Characterization of Slow Charge Transfer Processes in Differential Pulse Voltammetry at Spherical Electrodes and Microelectrodes. *Electrochimica Acta* Vol.55, No.18, (Jul 2010), pp. 5163–5172, ISSN 0013-4686

Molina, A.; Gonzalez, J.; Henstridge, M.C. & Compton, R.G. (2011a) Voltammetry of Electrochemically Reversible Systems at Electrodes of any Geometry: A General, Explicit Analytical Characterization. *J. Phys. Chem. C.* Vol.115, No.10, (March 2011), pp. 4054–4062, ISSN 1932-7447

Molina, A.; Gonzalez, J.; Henstridge, M.C. & Compton, R.G. (2011b) Analytical Expressions for Transient Diffusion Layer Thicknesses at Non Uniformly Accessible Electrodes. *Electrochimica Acta* Vol.56, No.12, (April 2011), pp. 4589–4594, ISSN 0013-4686

Myland, J.C. & Oldham, K.B. (1999). Limiting Currents in Potentiostatic Voltammetry without Supporting Electrolyte. *Electrochemistry Communications.* Vol.1, No.10, (October 1999), pp. 467–471, ISSN 1388-2481

Myland, J.C. & Oldham, K.B. (2004). Cottrell's Equation Revisited: An Intuitive, but Unreliable, Novel Approach to the Tracking of Electrochemical Diffusion. *Electrochemistry Communications.* Vol.6, No.4, (February 2004), pp. 344–350, ISSN 1388-2481

Nahir, T.M. & Buck, R.P. (1992) Modified Cottrell Behavior in Thin Layers: Applied Voltage Steps under Diffusion Control for Constant-Resistance Systems. *J. Electroanal. Chem.* Vol.341, No.1-2, (December 1992), pp. 1–14, ISSN 1572-6657

Nair, P.R. & Alam, M.A. (2010). Kinetic Response of Surfaces Defined by Finite Fractals. *Fractals.* Vol.18, No.4, (December 2010), pp. 461–476, ISSN 1793-6543

Nyikos, L. & Pajkossy, T. (1986). Diffusion to Fractal Surfaces. *Electrochimica Acta.* Vol.31, No.10, (October 1986), pp. 1347–1350, ISSN 0013-4686

Nyikos, L.; Pajkossy, T.; Borosy, A.P. & Martemyanov, S.A. (1990). Diffusion to Fractal Surfaces-IV. The Case of the Rotating Disc Electrode of Fractal Surfaces. *Electrochimica Acta.* Vol.35, No.9, (September 1990), pp. 1423–1424, ISSN 0013-4686

Obradović, M.D.; Vuković. G.D.; Stevanović, S.I.; Panić, V.V.; Uskoković, P.S.; Kowal, A. & Gojković, S.Lj. (2009). A Comparative Study of the Electrochemical Properties of Carbon Nanotubes and Carbon Black. *J. Electroanal. Chem.* Vol.634, No.2, (July 2009), pp. 22–30, ISSN 1572-6657

Ocon, P.; Herrasti, P.; Vázquez, L.; Salvarezza, R.C.; Vara, J.M. & Arvia. A.J. (1991) Fractal Characterisation of Electrodispersed Electrodes. *J. Electroanal. Chem.* Vol.319, No.1-2, (December 1991), pp. 101–110, ISSN 1572-6657

Oldham, K.B. & Raleigh, D.O. (1971) Modification of the Cottrell Equation to Account for Electrode Growth; Application to Diffusion Data in the Ag-Au System. *J. Electrochem. Soc.* Vol.118, No.2, (February 1971), pp. 252–255, ISSN 0013-4651

Oldham, K.B. (2004). Electrode "Edge Effects" Analyzed by the Green Function Method. *J. Electroanal. Chem.* Vol.570, No.2, (Jun 2004), pp. 163–170, ISSN 1572-6657

Orlický, J.; Gmucová, K.; Thurzo, I. & Pavlásek, J. (2003). Monitoring of Oxidation Steps of Ascorbic Acid Redox Reaction by Kinetics-Sensitive Voltcoulometry in Unsupported and Supported Aqueous Solutions and Real Samples. *Analytical Sciences,* Vol.19, No.4, (April 2003), pp. 505-509, ISSN 0910-6340

Pajkossy, T. & Nyikos, L. (1989a). Diffusion to Fractal Surfaces-II. Verification of Theory. *Electrochimica Acta.* Vol.34, No.2, (1989), pp. 171–179, ISSN 0013-4686

Pajkossy, T. & Nyikos, L. (1989b). Diffusion to Fractal Surfaces-III. Linear Sweep and Cyclic Voltammograms. *Electrochimica Acta.* Vol.34, No.2, (1989), pp. 181–186, ISSN 0013-4686

Pajkossy, T. (1991). Electrochemistry of Fractal Interfaces. *J. Electroanal. Chem.* Vol.300, No.1-2, (February 1991), pp. 1–11, ISSN 1572-6657

Serna, C. & Molina, A. (1999) General Solutions for the I/t Response for Reversible Processes in the Presence of Product in a Multipotential Step Experiment at Planar and Spherical Electrodes whose Areas Increase with any Power of Time. *J. Electroanal. Chem.* Vol.466, No.1, (May 1999), pp. 8-14, ISSN 1572-6657

Shin, H.-Ch.; Pyun, S.-I. & Go, J.-Y. (2002). A Study on the Simulated Diffusion-Limited Current Transient of a Self-Affine Fractal Electrode Based upon the Scaling Property. *J. Electroanal. Chem.* Vol.531, No.2, (August 2002), pp. 101–109, ISSN 1572-6657

Srivastav, S. & Kant, R. (2010) Theory of Generalized Cottrellian Current at Rough Electrode with Solution Resistance Effects. *J. Phys. Chem C.* Vol.114, No.24, (June 2010), pp. 10066–10076, ISSN 1932-7447

Srivastav, S. & Kant, R. (2011) Anomalous Warburg Impedance: Influence of Uncompensated Solution Resistance. *J. Phys. Chem C.* Vol.115, No.24, (June 2011), pp. 12232–12242, ISSN 1932-7447

Sutton, L.; Gavaghan, D.J. & Hahn, C.E.W. (1996). Numerical Simulation of the Time-Dependent Current to Membrane-Covered Oxygen Sensors. Part IV. Experimental Verification that the Switch-on Transient is Non-Cottrellian for Microdisc Electrodes. *J. Electroanal. Chem.* Vol.408, No.1-2, (May 1996), pp. 21–31, ISSN 1572-6657

Thompson, M.; Wildgoose, G.G. & Compton, R.G. (2006). The Theory of Non-Cottrellian Diffusion on the Surface of a Sphere or Truncated Sphere. *ChemPhysChem.* Vol.7, No.6, (June 2006), pp. 1328–1336, ISSN 1439-7641

Thompson, M. & Compton R.G. (2006). Fickian Diffusion Constrained on Spherical Surfaces: Voltammetry. *ChemPhysChem.* Vol.7, No.9, (September 2006), pp. 1964–1970, ISSN 1439-7641

Thurzo, I.; Gmucová, K.; Orlický, J. & Pavlásek, J. (1999). Introduction to a Kinetics-Sensitive Double-Step Voltcoulometry. *Rev. Sci. Instrum,* Vol.70, No.9, (September 1999), pp. 3723-3734, ISSN 0034-6748

Ward, K.R.; Lawrence, N.S.; Hartshorne, R.S. & Compton, R.G. (2011) Cyclic Voltammetry of the EC′ Mechanism at Hemispherical Particles and Their Arrays: The Split Wave. *J. Phys. Chem. C.* Vol.115, No.22, (June 2011), pp. 11204–11215, ISSN 1932-7447

Welch, Ch. M. & Compton, R.G. (2006) The Use of Nanoparticles in Electroanalysis: A Review. *Anal. Bioanal. Chem.* Vol.384, No.3, (February 2006), pp. 601–619, ISSN 1618-2650

Wildgoose, G.G.; Banks, C.E.; Leventis, H.C. & Compton, R.G. (2006). Chemically Modified Carbon Nanotubes for Use in Electroanalysis. *Microchimica Acta,* Vol.152, No.3-4, (January 2006), pp. 187-214, ISSN 1436-5073

Zhao, X.; Lu, X.; Tze, W.T.I. & Wang P. (2010). A Single Carbon Fiber Microelectrode with Branching Carbon Nanotubes for Bioelectrochemical Processes. *Biosensors and Bioelectronics* Vol.25, No.10, (July 2010), pp. 2343–2350, ISSN 09565663

Zhou, Y.-G.; Campbell, F.W.; Belding, S.R. & Compton, R.G. (2010) Nanoparticle Modified Electrodes: Surface Coverage Effects in Voltammetry Showing the Transition from Convergent to Linear Diffusion. The Reduction of Aqueous Chromium (III) at Silver Nanoparticle Modified Electrodes. *Chem. Phys. Letters.* Vol.497, No.4-6, (September 2010), pp. 200–204, ISSN 0009-2614

2

Modeling and Quantification of Electrochemical Reactions in RDE (Rotating Disk Electrode) and IRDE (Inverted Rotating Disk Electrode) Based Reactors

Lucía Fernández Macía, Heidi Van Parys, Tom Breugelmans, Els Tourwé and Annick Hubin
Electrochemical and Surface Engineering Group, Vrije Universiteit Brussel Belgium

1. Introduction

Whether it is for the design of a new electrochemical reactor or the optimization of an existing electrochemical process, it is of primordial importance to have the possibility to predict the behavior of a system. For example, the process of electrogalvanization of steel on an industrial scale would not be possible without knowing the main and side reactions taking place during the deposition of zinc on the metallic surface. However, understanding their mechanism is only the first point. The characteristic parameters of the reaction need to be identified and quantified in order to obtain a correct reactor design and to achieve the optimal operating conditions. Nowadays, part of the technological know-how still relies on best practice guidance, most often gained from years of experience with trial and error. Condensing those findings into empirical models may help to control some of the process parameters and make predictions of the system behavior within a small operation window. The problem, however, is that such a model acts as a black box and that a profound comprehension of the physical and electrochemical phenomena will fail to come.

The aim of a kinetic study is the determination of the mechanism of the electrochemical reaction and the quantification of its characteristic parameters: charge transfer parameters (rate constants and transfer coefficients) and mass transfer parameters (diffusion coefficients). Nevertheless, determining kinetic parameters accurately from the experimental results remains complex.

Linear sweep voltammetry (LSV) in combination with a rotating disk electrode (RDE) is a widely used technique to study electrode kinetics. Different methods exist to extract the values of the process parameters from polarization curves. The Koutecky-Levich graphical method is frequently used to determine the mass transfer parameters (Diard et al., 1996) : the slope of a plot of the inverse of the limiting current versus the inverse of the square root of the rotation speed of the rotating disk electrode is proportional to the diffusion coefficient. If more than one diffusing species is present, this method provides the mean diffusion coefficient of all species. The charge transfer current density is determined from the inverse of the intercept. In practical situations, however, the experimental observation of a limiting current can sometimes be masked by other reactions, e.g., in (Gattrell et al., 2004), and in that case Koutecky-Levich method cannot be used.

Also, to calculate the charge transfer parameters, a plot of the natural logarithm of the charge transfer current density as a function of potential, known as a Tafel plot, is often constructed. In the linear region of this curve, the transfer coefficient can be deduced from the slope and the rate constants from the intercept. The Tafel method is well established for simple reaction mechanisms (Bamford & Compton, 1986; Diard et al., 1996), but it becomes much more complicated for complex mechanisms (Gattrell et al., 2004; Wang et al., 2004). When there are significant diffusional or ohmic effects in the electrolyte, or additional electrode reactions, the Tafel plot deviates from linearity (Yeum & Devereux, 1989) and the charge transfer parameters cannot be determined.

Besides these well-known graphical methods, some authors suggest other methods to extract the kinetic parameters from an LSV experiment. They usually involve the fitting of theoretical expressions to the experimental data. In (Rocchini, 1992) the charge transfer parameters are estimated by fitting experimental polarization curves with exponential polynomials. Obviously, this method is only valid if the reaction rate is determined by charge transfer alone. Caster et al. fit convolution potential sweep voltammetry experiments with equations for a reversible charge transfer reaction with only one reactant present initially and under conditions of planar diffusion (Caster et al., 1983). No other steps are allowed to occur either before or after the electrode reaction. Yeum and Devereux propose an iterative method for fitting complex electrode polarization curves (Yeum & Devereux, 1989). They split up the total current density into contributions from the partial reactions and use simplified expressions for the current-potential relations. With these expressions they try to find the parameters that optimize the correlation between model and experimental data by minimizing a least squares cost function. This optimization is done by trial-and-error. In (Rusling, 1984) tabulated dimensionless current functions are fitted to linear sweep voltammograms. Therefore, a least squares cost function is minimized; however, no details on the minimizing algorithm are given.

In a series of papers, Harrison describes a hardware/software system for the complete automation of electrode kinetic measurements (Aslam et al., 1980; Cowan & Harrison, 1980a;b; Denton et al., 1980; Harrison, 1982a;b; Harrison & Small, 1980a;b). This involves the fitting of the data using a library of reaction schemes to determine the model parameter values. A quasi-Newton method is used to minimize the modulus of the differences between experiment and theory or the sum of the weighted squares of the differences. Although it is emphasized that care has to be taken in weighting the observations, no information on the determination of the weighting factors is given. Moreover, no criterion to decide whether the fitting is acceptable or not is discussed.

In (Bortels et al., 1997; Van den Bossche et al., 1995; 2002; Van Parys et al., 2010) a numerical approach is developed in order to define the underlying reaction mechanism . By using the MITReM (Multiple Ion Transport and Reaction Model) model, mass transport by convection, diffusion and migration but also the presence of homogeneous reactions in the electrolyte, are accounted for. The related model parameters such as diffusion coefficients, rate constants and transfer coefficients are adjusted in order to improve the agreement between experimental and simulated polarization curves. Thus, the best parameter values, corresponding to the best simulated curve, are selected by a *chi-by-eye* approach, without a statistical evaluation.

Although the reaction and transport models are defined more precisely, the lack of a fitting tool does not allow a reliable determination of the model parameters.

A quantitative, accurate and statistically founded modeling approach of electrochemical reactions has been the focus of an extensive work in our research group (Aerts et al., 2011; Tourwé et al., 2007; 2006; Van Parys et al., 2008). It is a generally applicable method to model an electrochemical reaction and to determine its mass and charge transfer parameters quantitatively. The reliability of the model parameters and the accuracy of the parameter fitting are key-elements of the method. A plausible reaction mechanism and the characteristic parameters of the electrochemical reaction are extracted from LSV experiments with a rotating disk electrode. Compared to others, this method offers the advantage that it uses one integrated expression that accounts for mass and charge transfer steps, and this without simplifying their mathematical expressions. The whole polarization curve is considered, rather than just some part in which only mass or charge transfer are supposed to be rate determining.

In this paper, we explain throughly this modeling methodology for the rotating disk electrode (RDE) and the inverted rotating disk electrode (IRDE) configurations. The modeling and quantification of the electrochemical parameters are applied to redox reactions with one electron transfer mechanism: the ferri/ferrocyanide system and the hexaammineruthenium (III)/(II) system.

2. Linear sweep voltammetry in combination with a rotating disk electrode

Linear sweep voltammetry with a rotating disk electrode (LSV/RDE) is a powerful technique for providing information on the mechanism and kinetics of an electrochemical reaction. Since the current density is a measure for the rate of an electrochemical reaction, LSV provides a stationary method to measure the rate as a function of the potential. In other words, the technique is used to distinguish between the elementary reactions taking place at the electrode as a function of the applied potential. Different elementary steps are often coupled, however, the overall current is determined by the slowest process (rate determining step). As a steady state technique, linear sweep voltammetry can only give mechanistic information about rate determining elementary reactions.

To determine a quantitative model for an electrochemical process, first a plausible reaction model is proposed and afterwards combined with a transport model. The combination of both models enables the formulation of the mass balances of the species and the conservation laws, which results in a set of non-linear partial differential equations, where the electrochemical reactions constitute a boundary condition at the electrode. While the reaction model is proper to the reaction under study, the transport model is merely determined by the mass transport of the species in the electrochemical reactor. As a result, it is possible to direct an electrochemical investigation in an adapted experimental reactor (electrochemical cell) under conditions for which the description of the transport phenomena can be simplified, without a loss of precision.

For controlling the mass transport contribution to the overall electrochemical kinetics, a rotating disk electrode possesses favorable features. The RDE configuration provides analytical equations to describe the mass transport and hydrodynamics in the electrochemical cell. It is known that a simplified transport model can be used if an RDE and diluted solutions are used in the experimental set-up. The hydrodynamic equations and the convective-diffusion equation for a rotating disk electrode have been solved rigorously for the steady state (Levich, 1962; Slichting, 1979). The axial symmetry of the configuration of the RDE reactor and the uniform current distribution allow a one-dimensional description. Moreover, at sufficient flow rate (when natural convection can be ignored), the hydrodynamics in diluted solution are not influenced by changes in concentrations due to electrochemical reactions. The mathematical problem can thus be solved more easily. Levich reduced the equation of convection transport to an ordinary differential equation (Albery & Hitchman, 1971; Levich, 1962; Slichting, 1979).

To model an electrochemical reaction and determine its mass and charge transfer parameters quantitatively, an electrochemical data fitting tool has been developed in our research group. From an analytical approach, it is designed to extract a quantitative reaction mechanism from polarization curves.

3. Analytical fitting of electrochemical parameters

The proposed analytical modeling of electrochemical reactions is founded on four building blocks. Figure 1 illustrates the structure of the modeling methodology.

The results of the experimental study These are the current-potential couples defining the polarization curve. The mean of multiple experiments that are performed under identical conditions is the experimental data for the modeling. The standard deviation of the experiments is used in the fitting procedure.

The mathematical expression for the proposed model The proposed model is based on well-considered reaction and transport models for the studied reaction. The mathematical expression of the reaction-transport model is derived from the basic equations that describe what happens during an electrochemical reaction. It has the following form: *current = function (potential, experimental parameters, model parameters)*, where the *experimental parameters* describe the experimental conditions, like e.g. temperature, rotation speed of the RDE, concentration, ..., and the *model parameters* are the unknown parameters that need to be quantitatively determined, like e.g. rate constants, transfer coefficients, ...

The fitting procedure In this block the differences between experimental and theoretical data are minimized. A weighted least squares cost function is formulated. The Gauss-Newton and Levenberg-Marquardt method are implemented to minimize this cost function and eventually provide the parameter values which best describe the data. Moreover, the standard deviations of the estimated parameters are also calculated.

A statistical evaluation If a statistical evaluation of the fitting results demonstrate a good description of the experiment by the model, a quantitative reaction mechanism is obtained. If, on the other hand, no good agreement between experiment and model is achieved, a new mechanism has to be proposed and the previous steps should be repeated.

3.1 The mathematical expression for the proposed reaction and transport models

Once the reaction model is defined, the transport model must be included. Using an RDE as the working electrode for the LSV experiments, the transport equations can be simplified. In addition, assuming that the electrolyte is a diluted solution, the migration term in the transport model can be neglected. This section provides the basic equations for mass and

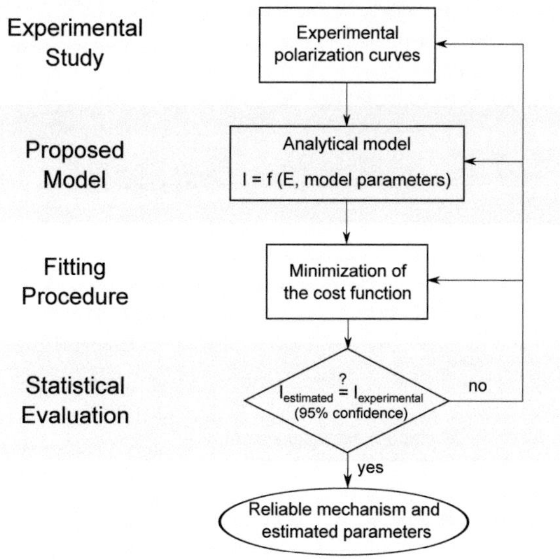

Fig. 1. The four building blocks of the modeling methodology.

charge transfer, which can be found in numerous textbooks (Bamford & Compton, 1986; Diard et al., 1996; Newman, 1973; Pletcher, 1991; Thirsk & Harrison, 1972; Vetter, 1967).

Consider a uniformly accessible planar electrode, immersed in an electrolyte that contains electroactive species and an excess of inert supporting electrolyte. At the surface an electrochemical reaction is taking place, which has P partial heterogeneous electrochemical or chemical reactions with N_v electroactive species in the electrolyte or in the electrode material and N_s electroactive species present in an adsorbed phase on the electrode surface. N is the total number of electroactive species involved in the reaction: $N = N_v + N_s$. The jth step of the reaction can be written as:

$$\sum_{i=1}^{N} r_{ij} X_i \underset{K'_j}{\overset{K_j}{\leftrightarrow}} \sum_{i=1}^{N} p_{ij} X_i \pm n_j e \tag{1}$$

with r_{ij} and p_{ij} the stoichiometric coefficients, the index i refers to the considered species, the index j to the partial reaction. K_j and K'_j are the rate constants. For an electrochemical reaction, they depend on the electrode potential. n_j is the number of electrons exchanged in the jth partial reaction. For an electrochemical reaction n_j is preceded by a plus sign if the reaction is written in the sense of the oxidation and by a minus sign if written in the sense of the reduction. For a chemical reaction n_j equals zero.

The global reaction is described by the relations that connect the electrode potential E to the faradaic current density i_f, the interfacial concentrations of the volume species $X_i(0)$ and the surface concentrations X_i of the adsorbed species[1]. Under steady-state conditions and using an RDE, the general system of equations describing this electrochemical systems is given by:

- the rate (v_j) expressions for each partial reaction:

$$v_j = K_j \prod_{i=1}^{N} X_i^{r_{ij}} - K'_j \prod_{i=1}^{N} X_i^{p_{ij}} \tag{2}$$

X_i is the concentration of the electroactive species. In electrode kinetics, concentrations are usually employed rather than activities and thus the rate constants include the product of activity coefficients. K_j and K'_j are the potential dependent rate constants of the partial reaction, given by:

$$K_j \text{ or } K'_j = k_{ox,j} \exp\left(\frac{\alpha_{ox,j} n_j F E(t)}{RT}\right) \tag{3}$$
$$K_j \text{ or } K'_j = k_{red,j} \exp\left(\frac{-\alpha_{red,j} n_j F E(t)}{RT}\right)$$

with $\alpha_{ox,j} + \alpha_{red,j} = 1$. F is the Faraday constant (96485 C/mol), R is the ideal gas constant (8.32 J/mol K) and T is the absolute temperature (K). $k_{ox,j}$ is the rate constant of the partial reaction j in the sense of the oxidation and $k_{red,j}$ is the one in the sense of the reduction. $\alpha_{ox,j}$ and $\alpha_{red,j}$ are the transfer coefficients in the sense of the oxidation and reduction respectively. They are supposed to be independent of the electrode potential (Diard et al., 1996). If adsorbed species are involved in the electrochemical reaction the rate constants K_j and K'_j may depend on the coverage.

- the relations that connect the faradaic current density to the rates of the partial electrochemical reactions:

$$i_f = F \sum_{j=1}^{P} s_{ej} v_j \tag{4}$$

where $s_{ej} = \pm n_j$. The + or − sign is fixed by the following convention: an oxidation current is counted positive, a reduction current negative.

- the relations expressing the transformation rate of the electroactive species at the interface electrode/electrolyte, and the continuity relation which expresses that this transformation rate is equal to the interfacial mass transport flux of the species X_i:

$$\begin{aligned} v_{X_i} &= J_{X_i} \\ &= -m_{X_i}[X_i^* - X_i(0)] \\ &= \sum_{j-1}^{P} s_{ij} v_j \\ i &= 1, \dots, N_v \end{aligned} \tag{5}$$

[1] To simplify the notation the symbol X_i is used to describe the chemical species in a reaction and its concentration $X_i = [X_i]$

or

$$v_{X_i}(t) = 0 \qquad (6)$$
$$i = N_v + 1, \ldots, N$$

were J_{X_i} is the molecular flux (expressed in mol/m^2 s) of X_i, equal to the number of moles of X_i going per unit of time across a unit plane, perpendicularly oriented to the flow of the species. X_i^* is the bulk concentration of species X_i. m_{X_i} is the mass transport rate constant for diffusion and convection of species X_i, given by:

$$m_{X_i} = 0.620 D_{X_i}^{2/3} \nu^{-1/6} \omega^{1/2} \qquad (7)$$

with D_{X_i} the diffusion coefficient of X_i, ν the kinematic viscosity and ω the rotation speed of the RDE.

At the maximum diffusion rate, the concentration at the electrode is zero and the limiting current is achieved. The Levich equation provides an expression of the limiting current as a function of the mass transport rate constant and the bulk concentration:

$$I_{lim} = 0.620 nFSD_{X_i}^{2/3} \nu^{-1/6} \omega^{1/2} X_i^* \qquad (8)$$

with S the electrode surface and n the number of electrons exchanged in the reaction. This equation thus applies to the totally mass-transfer-limited condition at the RDE.

3.2 The fitting procedure

Many comprehensive textbooks about parameter estimation and minimization algorithms are available. The development of the fitting procedure for the analytical modeling of electrochemical reactions is founded on a few of them (Fletcher, 1980; Kelley, 1999; Norton, 1986; Pintelon & Schoukens, 2001; Press et al., 1988; Sorenson, 1980).

Given a set of observations, one often wants to condense and summarize the data by fitting it to a 'model' that depends on adjustable parameters. In this work the 'model' is the current-potential relation, describing the polarization curve, which is derived from the basic laws for mass and charge transfer (given in section 3.1). The fitting of this mathematical expression provides the values of characteristic parameters (rate constants, transfer coefficients, diffusion coefficients), resulting in a quantitative reaction mechanism for the electrochemical reaction.

The basic approach is usually the same: a *cost function* that measures the agreement between the data and the model with a particular choice of parameters is designed. The cost function usually defines a distance between the experimental data and the model and is conventionally arranged so that small values represent close agreement. The parameters of the model are then adjusted to achieve a minimum in the cost function. This is schematically illustrated in Figure 2. It shows an imaginary experiment, modeled with the well-known Butler-Volmer equation, which describes the rate of an electrochemical reaction under charge transport control. By changing the values of the transfer coefficients and the rate constants, the distance between model and experiment varies for each data point. The cost function takes all these distances into account. The *estimates* or *best-fit-parameters* are the arguments that minimize the cost function.

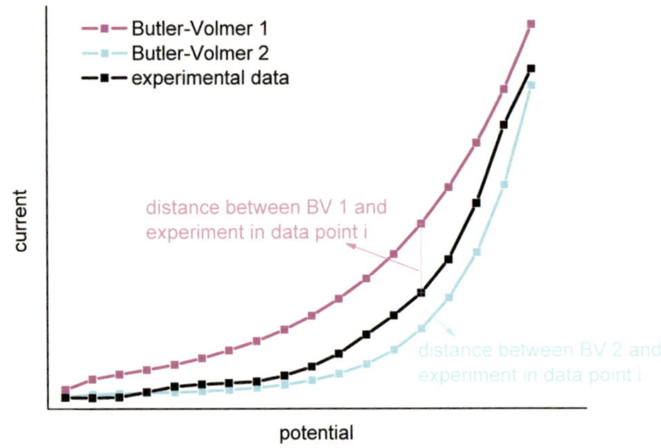

Fig. 2. The cost function as a measure of the agreement between the experiment and the model.

In most data fitting problems the match between the model and the measurements is quantified by a least squares cost function. Consider a multiple input, single output system modeled by $y_0(l) = F(u_0(l), \theta)$ with l the measurement index, $y(l) \in \mathbb{R}, u(l) \in \mathbb{R}^{1\times M}$, and $\theta \in \mathbb{R}^{n_\theta \times 1}$ the parameter vector. The aim is to estimate the parameters from noisy observations of the output of the system: $y(l) = y_0(l) + n_y(l)$. This is done by minimizing the sum of the squared errors, $e(l) = y(l) - y_0(l)$:

$$\hat{\theta}_{LS} = \underset{\theta}{arg\ min}\, V_{LS}(\theta) \text{ with } V_{LS} = \frac{1}{2}\sum_{l=1}^{N_{dp}} e^2(l) \tag{9}$$

with N_{dp} the number of data points. $arg\ min_\theta$ stands for: "find the argument θ that minimizes ...".

In this work polarization curves will be modeled. The current response (= output y) to a linearly changing potential (= input u_0) is measured in N_{dp} different points (with index l) and the transfer coefficients, rate constants and diffusion coefficients (= θ) need to be estimated. The function F is a theoretical expression which relates the current to the potential and the model parameters.

The Gauss-Newton algorithm is very well suited to deal with the least squares minimization problem. The numerical solution is found by applying the following iterative process:

1. Solve $J(\theta^{(k)})^T J(\theta^{(k)})\delta^{(k)} = -J(\theta^{(k)})^T e(\theta^{(k)})$
2. Set $\theta^{(k+1)} = \theta^{(k)} + \delta^{(k)}$

with $J(\theta)$ the $N_{dp} \times n_\theta$ Jacobian matrix. This is the matrix of all first-order partial derivatives of the error e. To improve the numerical conditioning of the expression 1 of the iterative process, the Jacobian can be scaled by multiplying each column with the corresponding parameter value from the previous iteration. As the Gauss-Newton method is an iterative method, it generates a sequence of points, noted as $\theta^{(1)}, \theta^{(2)}, \theta^{(3)}, \ldots$, or $\{\theta^{(k)}\}$. The iteration mechanism stops when a convergence test is satisfied, and the following criterion is used here:

$$\left| \frac{\theta_i^{(k+1)} - \theta_i^{(k)}}{\theta_i^{(k+1)}} \right| \leq \epsilon_i \;\; \forall i \tag{10}$$

where ϵ_i is the relative error in the minimization algorithm, which is defined by the user. That means that the fitting stops when the parameter values from one iteration to another do not change more than a low ϵ_i value.

Using singular value decomposition (SVD) techniques, the Gauss-Newton algorithm can be solved without forming the product $J(\theta^{(k)})^T J(\theta^{(k)})$ so that more complex problems can be solved because the numerical errors are significantly reduced. By SVD the matrix J is transformed into the product $J = U\Sigma V^T$ with U and V orthogonal matrices: $U^T U = I$ and $V^T V = V V^T = I$. Σ is a diagonal matrix with the singular values on the diagonal. This leads to the following expression for δ:

$$\delta = -V\Sigma^{-1}U^T e \tag{11}$$

Under quite general assumptions on the noise, some regularity conditions on the model $F(u_0(l), \theta)$ and the excitation (choice of $u_0(l)$), consistency [2] of the least squares estimator is proven. Asymptotically (for the number of data points going to infinity) the covariance matrix C_{LS} of the estimated model parameters is given by:

$$C_{LS} = (J(\theta^{(k)})^T J(\theta^{(k)}))^{-1} J(\theta^{(k)})^T C_y J(\theta^{(k)}) (J(\theta^{(k)})^T J(\theta^{(k)}))^{-1} \tag{12}$$

with $C_y = cov\{n_y\}$ the covariance matrix of the noise.

The Levenberg-Marquardt algorithm is a popular alternative to the Gauss-Newton method. This method is often considered to be the best type of method for non-linear least squares problems, but the rate of convergence can be slow. The iterative process is given by:

1. Solve $(J(\theta^{(k)})^T J(\theta^{(k)}) + \lambda^2 I)\delta^{(k)} = -J(\theta^{(k)})^T e(\theta^{(k)})$
2. Set $\theta^{(k+1)} = \theta^{(k)} + \delta^{(k)}$

or using SVD:

$$\delta = -V \begin{pmatrix} \frac{s_1}{s_1^2+\lambda^2} & 0 & 0 \\ 0 & \frac{s_2}{s_2^2+\lambda^2} & 0 \\ & & \\ 0 & 0 & \ddots \end{pmatrix} U^T e \tag{13}$$

with s the singular values and λ is called the Levenberg-Marquardt factor. As starting value $\lambda = \frac{s_1}{100}$ is chosen. If the value of the cost function decreases after performing an iteration, a new iteration is performed with $\lambda_{new} = \lambda_{old} * 0.4$. If the cost function increases, $\lambda_{new} = \lambda_{old} * 10$ is chosen and the old value of θ is maintained.

Some numerical aspects of these methods are also worth mentioning. Whether the Gauss-Newton or the Levenberg-Marquardt algorithm is used, the expression for equation to be solved is always written as:

[2] An estimator $\hat{\theta}$ is strongly consistent if it converges almost surely to θ_0: a.s. $\lim_{N_{dp}\to\infty} \hat{\theta}(N_{dp}) = \theta_0$, with θ_0 the true (unknown) value θ.

$$X\delta = e \tag{14}$$

Consequently, a change Δe of e, also causes δ to change by $\Delta\delta$. The problem is said to be *well conditioned* if a small change of e results in a small change of δ. If not, the problem is *ill-conditioned*. It can be shown that

$$\frac{\| \Delta\delta \|_2}{\| \delta \|_2} \leq \frac{s_{max}}{s_{min}} \frac{\| \Delta e \|_2}{\| e \|_2} \tag{15}$$

with $\| \ \|_2$ the 2-norm given by:

$$\| x \|_2 = \sum_i x_i^2 \tag{16}$$

and $\kappa = \frac{s_{max}}{s_{min}}$, the ratio of the largest singular value to the smallest, is called the condition number of X. If κ is large, the problem is ill-conditioned.

In Eq. (9) all measurements are equally weighted. In many problems it is desirable to put more emphasis on one measurement with respect to the other. This is done to make the difference between measurement and model smaller in some regions. If the covariance matrix of the noise is known, then it seems logical to suppress measurements with high uncertainty and to emphasize those with low uncertainty. In practice it is not always clear what weighting should be used. If it is, for example, known that model errors are present, then the user may prefer to put in a dedicated weighting in order to keep the model errors small in some specific operation regions instead of using the weighting dictated by the covariance matrix.

In general, the weighted least squares estimator $\hat{\theta}_{WLS}$ is:

$$\hat{\theta}_{WLS} = \underset{\theta}{arg\ min}\, V_{WLS}(\theta) \text{ with } V_{WLS} = \frac{1}{2} e(\theta)^T W e(\theta) \tag{17}$$

where $W \in \mathbb{R}^{N \times N}$ is a symmetric positive definite weighting matrix. All the remarks on the numerical aspects of the least squares estimator are also valid for the weighted least squares. This can be easily understood by applying the following transformation: $\epsilon = Se$ with $S^T S = W$ so that $V_{WLS} = \epsilon^T \epsilon$, which is a least squares estimator in the transformed variables. This also leads to the following Gauss-Newton algorithm to minimize the cost function:

$$\theta^{(k+1)} = \theta^{(k)} + \delta^{(k)}, \text{with } J(\theta^{(k)})^T W J(\theta^{(k)}) \delta^{(k)} = -J(\theta^{(k)})^T W e(\theta^{(k)}) \tag{18}$$

Eq. (12) is generalized to (noticing that $W^T = W$):

$$C_{WLS} = (J(\theta^{(k)})^T W J(\theta^{(k)}))^{-1} J(\theta^{(k)})^T W C_y W J(\theta^{(k)}) (J(\theta^{(k)})^T W J(\theta^{(k)}))^{-1} \tag{19}$$

By choosing $W = C_y^{-1}$ the expression simplifies to:

$$C_{WLS} = (J(\theta^{(k)})^T C_y^{-1} J(\theta^{(k)}))^{-1} \tag{20}$$

It can be shown that, among all possible positive definite choices for W, the best one is $W = C_y^{-1}$ since this minimizes the covariance matrix (Pintelon & Schoukens, 2001). These expressions depend on the covariance matrix of the noise C_y. In practice this knowledge should be obtained from measurements. In this work a sample covariance matrix obtained from repeated measurements is used. It can be shown that in this case the covariance matrix of the estimates is given by (Schoukens et al., 1997):

$$C_{WLS} = \frac{M-1}{M-5}(J(\theta^{(k)})^T C_y^{-1} J(\theta^{(k)}))^{-1} \tag{21}$$

where M is the number of polarization curves that are measured. At least 11 experiments have to be performed to use this expression (Schoukens et al., 1997).

After SVD of $C_y^{-1}J$ this equation becomes:

$$C_{WLS} = V\Sigma^{-2}V^T\frac{M-1}{M-5} \tag{22}$$

In this work the Gauss-Newton and Levenberg-Marquardt methods are implemented. The fitting starts with the Gauss-Newton method, but when the cost function is no longer decreasing after an iteration, it switches to Levenberg-Marquardt. The reason for starting with the Gauss-Newton method is that this method usually converges more rapidly than the Levenberg-Marquardt algorithm.

3.3 A statistical evaluation

After the determination of the best-fit parameter values, the validity of the selected model should be tested: does this model describe the available data properly or are there still indications that a part of the data is not explained by the model, indicating remaining model errors? We need the means to assess whether or not the model is appropriate, that is, we need to test the *goodness-of-fit* against some useful statistical standard.

With the best-fit-parameters, the modeled data is calculated and compared with the experimental data. Also, the difference between the experimental and calculated data is plotted. If the model describes the experiment appropriately, this difference should lie in the confidence band. This interval is defined by $\pm$ two times the experimental standard deviation of the current, calculated from a series of repeated experiments. When performing multiple experiments, 95 % of the experiments are expected to fall in this interval.

If the statistical evaluation does not show a good agreement between model and experiments, the model cannot be accepted indicating that any of the modeling steps is not well designed. In that case, we have to go back to the previous modeling steps: the experimental study, the proposed model (reaction or transport model) or/and the fitting procedure, and check their performance and the assumptions and simplifications that have been made to develop every step of the modeling methodology. Adjustments or possible alternatives must then be proposed.

An important message is that the fitting of parameters is not the end-all of parameter estimation. To be genuinely useful, a fitting procedure should provide (i) parameters, (ii) error estimates on the parameters, and (iii) a statistical measure of goodness-of-fit. When the third item suggests that the model is an unlikely match to the data, then items (i) and (ii) are probably worthless. Unfortunately, many practitioners of parameter estimation never proceed beyond item (i). They deem a fit acceptable if a graph of data and model "look good". This approach is known as *chi-by-eye* and should definitively be avoided.

4. Analytical modeling of LSV/RDE experiments

4.1 The experimental procedure

4.1.1 Composition of the electrolytes

The chemicals used for the study of the hexaammineruthenium (III)/(II) redox reaction are: $[Ru(NH_3)_6]Cl_3$ (Sigma-Aldrich, 98%), $Na_2HPO_4 \cdot 2H_2O$ and $NaH_2PO_4 \cdot 12H_2O$ (both ProLabo A.R.). Solutions are made with once-distilled and deionized water. A 0.1 M phosphate buffer pH 7 solution is used as the supporting electrolyte and the concentration of the electroactive species hexaammineruthenium (III) is 0.001 M. The following chemicals are used for the ferri/ferrocyanide reaction: $K_4[Fe(CN)_6] \cdot 3H_2O$, $K_3[Fe(CN)_6]$ and KCl (all Merck P.A.). The supporting electrolyte is a 1 M KCl solution and the concentrations of the electroactive components ferri/ferrocyanide are 0.005 M. In that way, a negligible migration flux, constant activity and diffusion coefficients of the electroactive species, a low electrolyte resistance and a uniform current distribution are assumed for both electrochemical systems.

4.1.2 Experimental set-up

A typical three electrode set-up is used for the LSV/RDE experiments (Bamford & Compton, 1986; Diard et al., 1996). The electrochemical cell contains a Ag/AgCl reference electrode (Schott-Geräte), a rotating disk working electrode and a platinum grid as counter electrode. A platinum RDE is used for the ferri/ferrocyanide system and a gold RDE for the hexaamineruthenium (III)/(II) system. The RDE electrodes are fabricated by embedding a 4 mm diameter polycrystalline platinum or gold rod in an insulating mantle of polyvinylidenefluoride.

The electrode is rotated by an RDE control system of Autolab. For the hexaammineruthenium (III)/(II) reaction, the rotation speed is set to 500 and 1000 rpm. In the experiments with the ferri/ferrocyanide system, the rotation speeds are 300, 1000, 1500 and 2000 rpm. The voltammograms are measured using a high resolution galvanostat/potentiostat PGSTAT100 (Autolab Instruments) of Ecochemie, controlled by the GPES 4.8 or the Nova 1.5 softwares. In the LSV experiments, the potential is swept from 0.3 to -0.4 V vs NHE for the $Ru(NH_3)_6^{+3}$/ $Ru(NH_3)_6^{+2}$ reaction and from 0.8 to 0.2 V vs NHE for the $Fe(CN)_6^{-3}/Fe(CN)_6^{-4}$ reaction. The scan rate is taken constant at 1 mV/s. The step potential is set to 0.00015 V, in this way a maximum number of data points is measured.

All measurements are performed in a 200 ml glass electrolytic cell, thermostatted at $25 \pm 0.5\,^{\circ}C$ using a water jacket connected to a thermostat bath (Lauda RE304). Prior to the measurements, the electrolyte is deoxygenated by bubbling with nitrogen gas (Air Liquide) for 10 min, while during the experiment a nitrogen blanket is maintained over the cell. This results in a substantial flattening of the reduction plateau of both the ferricyanide and the hexaamineruthenium (III).

4.1.3 Electrode pretreatment

The reproducibility of the measurements is strongly increased by means of applying the following pretreatment of the working electrode surface: 1) mechanical polishing of the electrode on a rotating disk (Struers DP10, on cloth), successively using diamond paste of 7 μm and 1 μm grain size (Struers) for the Pt electrode, and 9 μm, 3 μm diamond paste and

0.04 μm Al_2O_3 paste (Struers) for the Au electrode; 2) ultrasonic rinsing with deionized water followed by degreasing with chloroform, also in an ultrasonic bath (Elma model T470/H); 3) four cyclic voltammograms are performed before each experiment in order to remove oxide and trace contaminants from the metallic surface (Robertson et al., 1988). During the cyclic voltammetry measurements, the potential is swept between -0.35 V to 1.45 V vs Ag/AgCl at a scan velocity of 50 mV/s on the Au electrode (hexaammineruthenium (III)/(II) system) and between +0.55 and -0.45 V vs Ag/AgCl at a scan velocity of 10 mV/s, on the Pt electrode (ferri/ferrocyanide system).

4.2 Modeling the ferri/ferrocyanide redox reaction

The simplest electrochemical reactions, which can be found among the different kinds of electrode processes, are those where electrons are exchanged across the interface by flipping oxidation states of transition metal ions in the electrolyte adjacent to the electrode surface (Bamford & Compton, 1986), i.e. an ET (electron transfer) mechanism. The electrode acts as the source or sink of electrons for the redox reaction and is supposed to be inert. The reduction of ferricyanide to ferrocyanide (Angell & Dickinson, 1972; Bamford & Compton, 1986; Bruce et al., 1994; Iwasita et al., 1983) is an example of such a mechanism:

$$Fe(CN)_6^{-3} + e^- \rightleftharpoons Fe(CN)_6^{-4} \qquad (23)$$

4.2.1 Results of the experimental study

A set of equivalent polarization curves of the ferri/ferrocyanide reaction is used as the experimental data, which is the input to the model methodology. As advised in (Tourwé et al., 2006), 11 voltammograms are measured under identical conditions for every rotation speed. The resulting voltammograms for the $Fe(CN)_6^{-3}/Fe(CN)_6^{-3}$ system are shown in Figure 3.

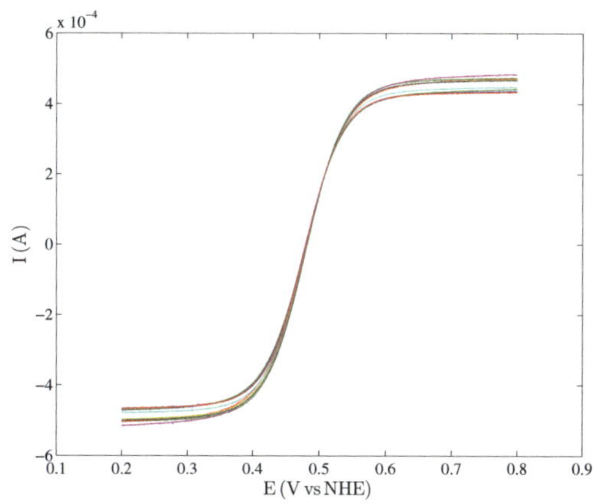

Fig. 3. Voltammograms of the reduction/oxidation of 0.005 M ferri/ferrocyanide in 1 M KCl, at 2000 rpm.

4.2.2 The analytical expression for the current as a function of the potential

The analytical expression is derived for the reduction/oxidation of ferri/ferrocyanide (reaction (23)). The basic equations that describe this mechanism, when studied under steady-state conditions on a rotating disk electrode, are:

$$v = K_{ox}C_{red}(0) - K_{red}C_{ox}(0) \tag{24}$$

$$I = iS = i_f S = FSv \tag{25}$$

$$v = v_{ox/red} = -m_{ox}[C^*_{ox} - C_{ox}(0)] = m_{red}[C^*_{red} - C_{red}(0)] \tag{26}$$

$$i_{lim,ox/red} = \pm F m_{red/ox} C^*_{red/ox} = \pm 0.620 F D^{2/3}_{red/ox} \nu^{-1/6} \omega^{1/2} C^*_{red/ox} \tag{27}$$

$$K_{ox/red} = k_{ox/red} exp \frac{\pm \alpha_{ox/red} nFE}{RT} \tag{28}$$

From this set of equations an expression for the current as a function of the potential, the experimental parameters and the model parameters can be derived. Considering $\alpha_{ox} + \alpha_{red} = 1$, the model parameters are D_{red}, D_{ox}, k_{ox}, k_{red} and α_{ox}. The experimental parameters are C^*_{red}, C^*_{ox}, S, n and T, which are all known. ν and ω are eliminated in the expression for the current, by using the limiting current density as a model parameter, rather than the diffusion coefficient. Solving this set of equations leads to the following current-potential relation (Tourwé et al., 2007):

$$I = \frac{nFS\left(k_{ox} exp\left(\frac{\alpha_{ox} nFE}{RT}\right) C^*_{red} - k_{red} exp\left(\frac{-\alpha_{red} nFE}{RT}\right) C^*_{ox}\right)}{1 + nFS\left(\frac{k_{ox} exp\left(\frac{\alpha_{ox} nFE}{RT}\right) C^*_{red}}{i_{lim,ox}} - \frac{k_{red} exp\left(\frac{-\alpha_{red} nFE}{RT}\right) C^*_{ox}}{i_{lim,red}}\right)} \tag{29}$$

4.2.3 The fitting results and their statistical evaluation

For the ferri/ferrocyanide system, the analytical model is based on 5 parameters: the rate constant of the reduction reaction, k_{red}, the rate constant of the oxidation reaction, k_{ox}, the charge transfer coefficient, α_{ox}, the reduction limiting current density, $i_{lim,red}$ and the oxidation limiting current density, $i_{lim,ox}$. The mathematical expression for the current as a function of the potential (Eq. (29)) is fitted to the mean of the equivalent experimental voltammograms, using the method described previously.

Four rotation speeds are considered for the study of the ferri/ferrocyanide redox reaction: 300, 1000, 1500 and 2000 rpm. Figure 4.a presents a comparison between the mean experimental voltammogram at 2000 rpm and the modeled voltammogram, calculated with the best-fit-parameters, and the 95% confidence interval $\pm 2\sigma$, with σ the standard deviation of the current in the set of experimental curves. The difference between the experimental and the modeled curve (Figure 4.b) lies in the 95% confidence band. This means that the model is able to describe the experiments appropiately.

In order to check the validity of the analytical modeling to estimate the model parameters, an evaluation of the parameter values determined at different rotation speeds is carried out. In Figure 5 the 95% confidence intervals for the best fit-parameters at four rotation speeds are compared. It can be seen that for all parameters the 95% confidence intervals obtained at the different rotation speeds overlap with the interval for the lowest rotation speed (300 rpm). The latter is, however, much larger than those obtained for the higher rotation speeds. This is due to the fact that at lower rotation speeds mass transfer rapidly becomes the rate determining step. The potential region where charge transfer influences the reaction rate is smaller than for the other rotation speeds and, as a consequence, the charge transfer

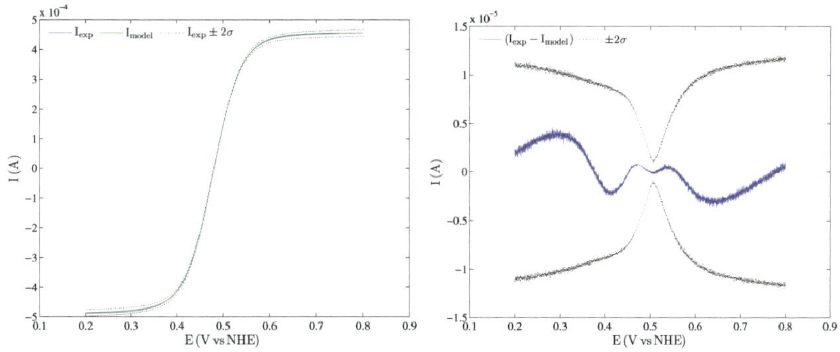

(a) (b)

Fig. 4. Modeling the reduction/oxidation of ferri/ferrocyanide: (a) Comparison of the modeled curve and the mean experimental curve, at 2000 rpm. (b) Difference between the experiments and the model within the 95% confidence interval.

parameters are determined less accurately. This does not imply that the results at this lowest rotation speed have to be rejected. When calculating the best estimates of the parameters, the higher uncertainty is taken into account. The 95% confidence intervals for k_{ox}, k_{red} and α_{ox} obtained for the intermediate rotation speeds (1000, 1500 and 2000 rpm) overlap, indicating that the results are independent of the rotation speed.

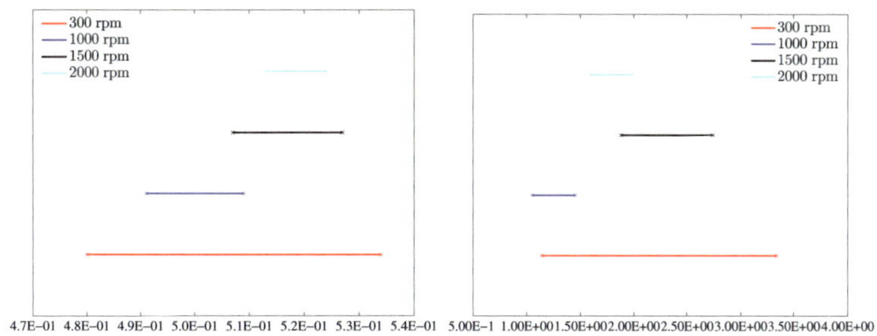

(a) α_{ox} (b) k_{red} (m/s)

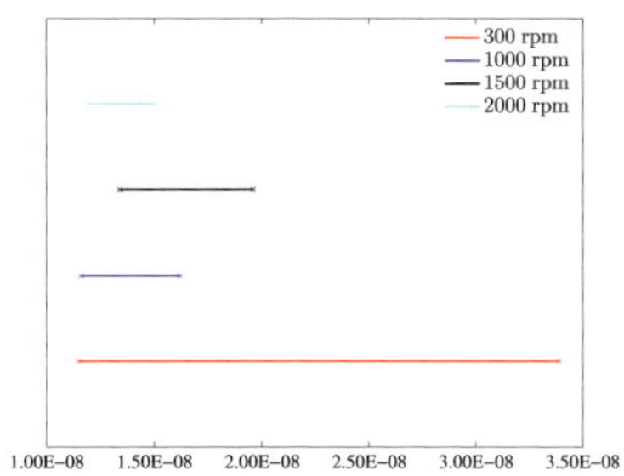

(c) k_{ox} (m/s)

Fig. 5. Comparison of the 95 % confidence intervals of the charge transfer parameters obtained at 300 rpm, 1000 rpm, 1500 rpm and 2000 rpm.

The best-fit estimates of the charge transfer parameters are calculated as the weighted mean of the values obtained at the four rotation speeds taken into consideration. Introducing weighting factors helps to adjust the impact of one result on the mean with regard to their level of uncertainty. The maximum likelihood estimate of normally distributed variables x_i with different variances σ_i^2 and the same mean value μ is given by (Sorenson, 1980):

$$\mu = \frac{\Sigma_i \frac{x_i}{\sigma_i^2}}{\Sigma_i \frac{1}{\sigma_i^2}} \tag{30}$$

This results in the following values for the charge transfer coefficient and the rate constants: α_{ox} = 5.14E-01 $\pm$ 1.28E-02, k_{ox} = 1.41E-08 m/s $\pm$ 4.50E-09 m/s and k_{red} = 1.61E+00 m/s $\pm$ 4.82E-01 m/s.

The diffusion coefficients can be calculated from the Levich equation (Eq. 8). Using the values of $i_{lim,red}$ and $i_{lim,ox}$, determined for 300, 1000, 1500 and 2000 rpm, this results in D_{red} = 8.07E-10 m^2/s $\pm$ 2.23E-17 m^2/s and D_{ox} = 8.31E-10 m^2/s $\pm$ 2.75E-17 m^2/s. The obtained parameter values compare well with the values presented in literature (Angell & Dickinson, 1972; Beriet & Pletcher, 1993; Bruce et al., 1994; Jahn & Vielstich, 1962).

4.3 The analytical modeling of the hexaammineruthenium (III)/(II) redox reaction

The reduction of hexaammineruthenium (III) to hexaammineruthenium (II) is extensively described in literature (Beriet & Pletcher, 1994; Deakin et al., 1985; Elson et al., 1975; Khoshtariya et al., 2003; Marken et al., 1995; Muzikar & Fawcett, 2006) as a one electron transfer reaction, i.e.,

$$Ru(NH_3)_6^{+3} + e^- \rightleftharpoons Ru(NH_3)_6^{+2} \tag{31}$$

Due to the fact that the hexaammineruthenium (II) complex is not stable in solution, the electrochemical reaction (31) is studied here only in the direction of the reduction, with the hexaammineruthenium (III) being the only electroactive species initially present in the electrolyte.

For the reduction of $Ru(NH_3)_6^{+3}$ to $Ru(NH_3)_6^{+2}$, the mathematical expression describing the reaction mechanism is obtained, leaving out the oxidation parameters. The current-potential relation is formulated with the following expression:

$$I = \frac{-nFSk_{red}exp\left(\frac{-\alpha_{red}nFE}{RT}\right) C_{ox}^*}{i_{lim,red} - nFSk_{red}exp(\frac{-\alpha_{red}nFE}{RT})C_{ox}^*} \tag{32}$$

For the analytical modeling of the reaction, also 11 identical experimental curves are performed, at 500 and 1000 rpm. In the reduction of $Ru(NH_3)_6^{+3}$, an unexpected variation of the current values in the region of the limiting current is observed. This behavior differs from the characteristic curve of an ET mechanism and might be due to additional reactions in the supporting electrolyte. This contribution of the supporting electrolyte is measured and subtracted from the experimental polarization curves.

In Figure 6 it can be seen that the difference between the experimental curve and the modeled curve, at 500 rpm, lies in the 95% confidence band. At this point we could accept that the model is appropiate to describe the experiments, but an evaluation of the estimated values

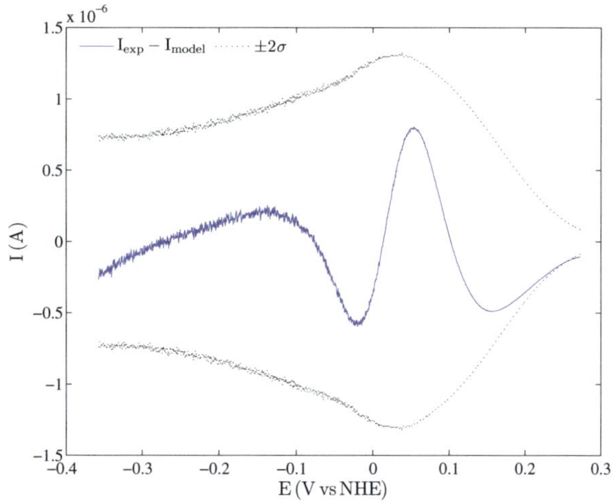

Fig. 6. Difference between the mean experimental and the modeled curves, at 500 rpm, within the 95% confidence band.

of the model parameters must be done. The model parameters are fitted for the experimental data obtained at two different rotation speeds. The best-fit values and their standard deviation obtained for 500 rpm and 1000 rpm are shown in Table 1. The values of the charge transfer coefficient are slightly different. In the case of the rate constant, the best-fit values at different rotation speeds present significant differences. Thus, the estimated values are not consistent with the results expected for the proposed model: while the limiting current density is proportional to the square root of ω, the values of the charge transfer parameters must be independent of the rotation speed.

The different results of the estimated parameters at the two rotation speeds could be due to interfering reactions in the supporting electrolyte or, more likely, to the oxidation of hexaamineruthenium (II) to hexaamineruthenium (III), occurring at the initial potentials of the polarization curve. The oxidation half-reaction is not taken in consideration in the formulation of the analytical current-potential expression; this could have an influence on the parameter estimation. According to our modeling strategy, a better estimation of the model parameters needs further investigation on the treatment of the experimental data and/or the formulation of the reaction mechanism. Present research is devoted to this topic.

Model parameters	500 rpm		1000 rpm	
	Estimate	Std	Estimate	Std
α_{red}	7.67E-01	2.95E-03	7.49E-01	2.15E-03
k_{red} (m/s)	6.17E-05	3.36E-07	7.14E-05	3.46E-07
$i_{lim,red}$ (A/m^2)	-3.29E+00	1.51E-03	-4.54E+00	2.20E-03

Table 1. Best-fit parameters and their standard deviation for 500 rpm and 1000 rpm.

5. The inverted rotating disk electrode (IRDE) reactor

The combination of linear sweep voltammetry with an RDE is a powerful tool to study electrochemical reactions. One limitation to the technique is the presence of gas in some electrochemical systems. The inverted rotating disk electrode (IRDE) is designed to tackle this problem. It allows an investigation of gas evolving reactions, preventing the presence of gas from being an obstacle to a quantitative modeling.

When the RDE set-up is used for the mechanistic study of gas evolution reactions, the formed gas bubbles tend to stick to the downward facing electrode surface and, hence, shield the active electrode surface. In contrast to the classical RDE configuration where the rotating disk is downward facing and positioned at the top of the electrochemical cell, in the IRDE configuration the rotating disk is now placed at the bottom of the cell, facing upwards. As a result, the generated bubbles rise freely and are no longer shielding the electrode surface. The bubbles are detached from the electrode by buoyancy or swept away by the rotational movement of the electrode.

Many industrial electrochemical systems deal with gas evolution, such as, for example, the water electrolysis, the chlorine and chlorate production or the side reactions during the electrowinning or electrodeposition of metals.

5.1 Design and construction of the IRDE reactor

The electrochemical cell consists of a cylindrical vessel of 74 mm diameter and 200 mm height, with a square water jacket around it. A schematic illustration of the IRDE design is presented in Figure 7 and a picture of the IRDE set-up is shown in Figure 8. The height of the working electrode protruding into the electrolyte is 30 mm. The counter and the reference electrode are located in the upper part of the cell, far away from the rotating electrode in order not to disturb the flow field near the working electrode. Because of the positioning of the reference electrode vs the working electrode, the current-potential data needs to be corrected for the ohmic drop. After emptying the cell, the vessel itself can be removed permitting an easy exchange of the working electrode. The cell is entirely made of plexiglas (polymethylmethacrylate or PMMA) to ensure full optical access to the cell. In this way, the gas bubbles can be characterized in-situ (e.g. bubble size, rise velocity, etc.) by means of optical imaging techniques.

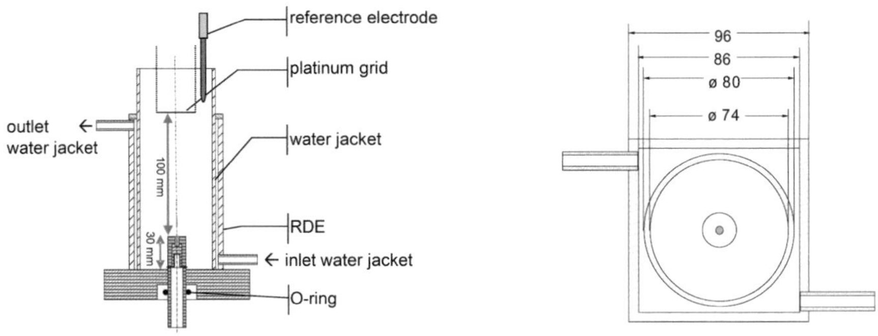

(a) Front view (b) Top view

Fig. 7. Schematic illustration of the IRDE cell.

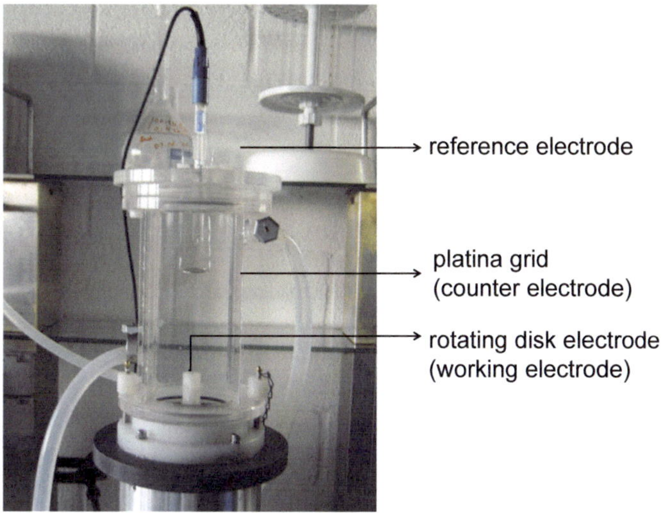

Fig. 8. Picture of the IRDE cell.

5.2 Evaluation of the IRDE reactor for kinetic studies of electrochemical reactions

Although the change in position of the working electrode from the top to the bottom of the cell seems to be a minor modification, it has to be examined whether the characteristics of mass transfer governing the RDE reactor are also valid for the IRDE reactor. To this purpose, the ferri/ferrocyanide redox system is used. Since theoretically charge transfer parameters are independent of the hydrodynamics, the same parameter values must be found in both the RDE and IRDE configuration. The equations of mass transfer, used for kinetic studies in an RDE configuration, are based on an analytical expression of the velocity field with the assumption of an infinite large electrode disk and bath dimensions. In practice, however, we deal with a finite sized rotating disk and a confined electrochemical cell. Although it is generally accepted that the analytical solution of the flow field derived by Levich (Newman, 1973) describes the velocity field near the electrode surface well, it is nevertheless not evident that the velocity field is described with the same accuracy for the IRDE configuration as for the RDE configuration.

The analytical modeling utilized for the RDE configuration is applied to the IRDE reactor. The experimental procedure described in section 4.1 is used for the LSV experiments of the ferri/ferrocyanide redox reaction on the IRDE. Eq. 29 is fitted to the mean polarization curves, obtained at 300, 1000 and 1500 rpm. The mean polarization curve is calculated from 11 independently recorded polarization curves. The 95 % confidence intervals of k_{ox}, k_{red} and α_{ox} obtained at 300, 1000 and 1500 rpm perfectly overlap (not shown). This means that the values of the charge transfer parameters, k_{ox}, k_{red} and α_{ox}, are independent of the rotation speed, as expected theoretically. The calculated weighted mean values of the model parameters for charge transfer, k_{ox}, k_{red} and α_{ox}, are respectively 1.03E-08 m/s $\pm$ 1.31E-09 m/s, 9.12E-01 $\pm$ 1.25E-01 m/s and 5.39E-01 $\pm$ 7.06E-03.

In Figure 9, the weighted mean values of the charge transfer parameters and their 95 % confidence interval, obtained in the IRDE and the RDE reactors, are compared. It can be seen that the confidence intervals of k_{ox} perfectly overlap, while a small deviation between the confidence intervals is observed for k_{red} and α_{ox}. However, the deviation between the confidence interval is smaller than 1% for α_{ox} and smaller than 5% for k_{red}. The deviation between the confidence intervals is determined by the difference between the maximum and the minimum values of the respective confidence intervals obtained in the RDE and the IRDE configurations, normalized to the mean value of the respective model parameter obtained in the RDE configuration. The values lie sufficiently close that it is reasonable to assume that the mass transfer characteristics of the RDE are also valid for the IRDE. The estimation of the charge transfer parameters does not depend on the hydrodynamics and mass transport within the electrochemical cell, which in its turn demonstrates that the IRDE is a suitable tool for the mechanistic study of the electrochemical reactions. Moreover, the differences between the values of the rate constants estimated for the RDE and the IRDE reactors narrow the window of values reported in literature (Angell & Dickinson, 1972; Beriet & Pletcher, 1993; Bruce et al., 1994; Jahn & Vielstich, 1962).

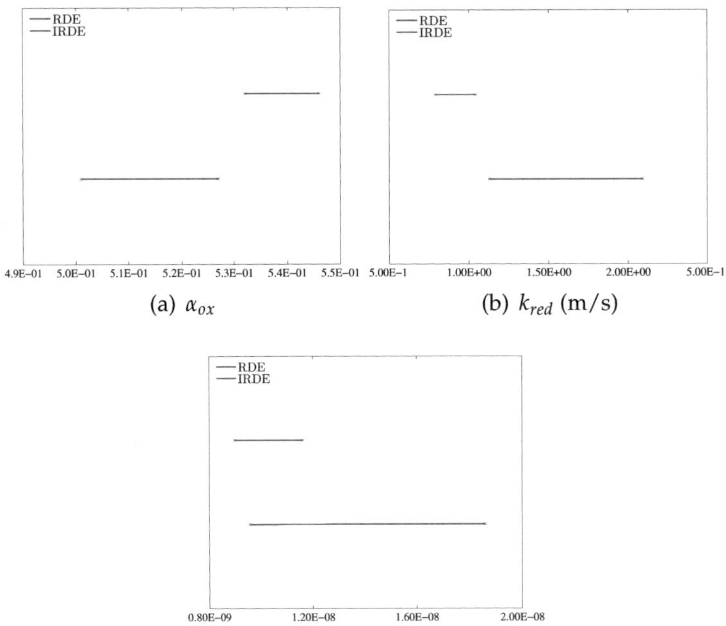

(a) α_{ox}

(b) k_{red} (m/s)

(c) k_{ox} (m/s)

Fig. 9. Comparison of the 95 % confidence intervals of the weighted mean of the charge transfer parameters obtained in the IRDE and the RDE configurations.

6. Towards the identification and quantification of characteristic parameters of complex electrochemical systems

In general, an accurate and fully statistical founded solution is the aim of all kinetic and mechanistic studies of electrochemical systems. The proposed analytical fitting model is applied to one electron transfer reactions. Yet the fitting can be extended to other reaction steps, such as chemical or adsorption reactions, or to other measurement techniques, such as, for example, electrochemical impedance spectroscopy, provided that they can be analytically formulated.

However, in many cases the electrochemical systems present more complicated reaction mechanisms, including multiple electrochemical or chemical reactions. The reaction model is thus too complex to be translated into analytical equations. Besides, when gas bubbles are involved in the electrochemical process, a new transport model describing the two-phase mass transport is designed (Maciel et al., 2009; Nierhaus et al., 2009; Van Damme et al., 2010). Modeling these complex systems requires a numerical approach. In our group, focus is put on the development of numerical models for such complex systems. Nevertheless, in order to achieve a statistical and accurate parameter estimation, it is clear that also a numerical fitting procedure has to be introduced. This aspect is under development at present in our group.

7. Conclusions

The strength of the analytical fitting model to quantify the kinetic parameters of an electrochemical reaction is shown. The coupling of statistically founded parameter estimation techniques with LSV/RDE experiments is an important innovative point of the modeling strategy.

The fitting methodology requires the proposition of an appropriate mechanism for the studied reaction and its mathematical translation into an expression that analytically describes the voltammogram. This expression depends on the mass and charge transfer parameters of the reaction (rate constants, transfer coefficients and diffusion coefficients). Powerful parameter estimation algorithms are used in the data fitting tool to adjust the values of the model parameters in order to obtain a good agreement between experimental and modeled data. The values of the model parameters that give rise to the best match, characterize the system quantitatively. Moreover, this method provides error estimates of the obtained parameter values. However, it is only after a statistical evaluation of the obtained results, that it is decided whether the model is able to describe the experiments.

The application of the analytical modeling for the study of the ferri/ferrocyanide reaction with LSV/RDE experiments demonstrate that the modeling methodology is valid to extract the quantitative mechanism of an electrochemical reaction. In the case of the hexaammineruthenium (III)/(II) reaction, however, the results of the analytical modeling point out the importance of a correct formulation of the reaction mechanism.

The IRDE reactor is built to facilitate the study of electrochemical gas evolution reactions. It offers the advantages of the classical RDE set-up, such as well-defined hydrodynamics and mass transport over a wide range of rotation speeds, while the gas bubbles can rise freely and do not shield the electrode surface. It is demonstrated that the IRDE configuration is valid for kinetic and mechanistic investigations of electrochemical reactions.

It has to be emphasized that for the existing fitting procedure the proposed reaction model describing the reactions taking place must be translated into an analytical equation. It is clear that in the presence of, for example, chemical reactions or gas bubbles, an analytical solution does not exist anymore. Therefore, for further modeling studies a fitting tool that makes use of numerical calculation procedures needs to be developed.

8. Acknowledgments

The authors thank the Flemish Institute for support of Scientific-Technological Research in Industry (IWT) and Vrije Universiteit Brussel for the financial support.

9. References

Aerts, T., Tourwe, E., Pintelon, R., De Graeve, I. & Terryn, H. (2011). Modelling of the porous anodizing of aluminium: Generation of experimental input data and optimization of the considered model, *Surface & Coatings Technology* 205(19): 4388–4396.

Albery, W. & Hitchman, M. (1971). *Ring-Disc Electrodes*, Clarendon Press.

Angell, D. & Dickinson, T. (1972). Kinetics of ferrous/ferric and ferro/ferricyanide reactions at platinum and gold electrodes. 1. Kinetics at bare-metal surfaces, *Journal of Electroanalytical Chemistry* 35: 55–&.

Aslam, M., Harrison, J. & Small, C. (1980). The automation of electrode kinetics. 2: The application to metal-deposition, *Electrochimica Acta* 25(5): 657–668.

Bamford, C. & Compton, R. (1986). *Electrode Kinetics: Principles and Methodology. Comprehensive Chemical Kinetics*, Vol. 26, Elsevier Science Publishers.

Beriet, C. & Pletcher, D. (1993). A microelectrode study of the mechanism and kinetics of the ferro ferricyanide couple in aqueous media - the influence of the electrolyte and its concentration, *Journal of Electroanalytical Chemistry* 361: 93–101.

Beriet, C. & Pletcher, D. (1994). A further microelectrode study of the influence of electrolyte concentration on the kinetics of redox couples, *Journal of Electroanalytical Chemistry* 375(1-2): 213–218.

Bortels, L., Van den Bossche, B. & Deconinck, J. (1997). Analytical solution for the steady-state diffusion and migration. Application to the identification of Butler-Volmer electrode reaction parameters, *Journal of Electroanalytical Chemistry* 422(1-2): 161–167.

Bruce, P., Lisowskaoleksiak, A., Los, P. & Vincent, C. (1994). Electrochemical impedance spectroscopy at an ultramicroelectrode, *Journal of Electroanalytical Chemistry* 367(1-2): 279–283.

Caster, D., Toman, J. & Brown, S. (1983). Curve fitting of semiderivative linear scan voltammetric responses - Effect of reaction reversibility, *Analytical Chemistry* 55(13): 2143–2147.

Cowan, K. & Harrison, J. (1980a). The automation of electrode kinetics. 3: The dissolution of Mg in Cl^-, F^- and OH^- containing aqueous solutions, *Electrochimica Acta* 25(7): 899–912.

Cowan, K. & Harrison, J. (1980b). Automation of electrode kinetics. 5: The dissolution of Al in Cl^- and F^- containing aqueous solutions, *Electrochimica Acta* 25(9): 1153–1163.

Deakin, M., Stutts, K. & Wightman, R. (1985). The effect of pH on some outer-sphere electrode-reactions at carbon electrodes, *Journal of Electroanalytical Chemistry* 182(1): 113–122.

Denton, D., Harrison, J. & Knowles, R. (1980). Automation of electrode kinetics. 4: The chlorine evolution reaction on a $RuO_2 - TiO_2$ plate electrode, *Electrochimica Acta* 25(9): 1147–1152.

Diard, J.-P., Le Gorrec, B. & Montella, C. (1996). *Cinétique électrochimique*, Hermann.

Elson, C., Itzkovitch, I., McKenney, J. & Page, J. (1975). Electrochemistry of ruthenium ammine complexes, *Canadian Journal of Chemistry-revue Canadienne De Chimie* 53(19): 2922–2929.

Fletcher, R. (1980). *Practical Methods of Optimization*, Vol. 1, John Wiley and Sons.

Gattrell, M., Park, J., MacDougall, B., Apte, J., McCarthy, S. & Wu, C. (2004). Study of the mechanism of the vanadium 4+/5+ redox reaction in acidic solutions, *Journal of the Electrochemical Society* 151(1): A123–A130.

Harrison, J. (1982a). Automation of electrode kinetic measurements. 7, *Electrochimica Acta* 27(8): 1113–1122.

Harrison, J. (1982b). Automation of electrode kinetic measurements. 8, *Electrochimica Acta* 27(8): 1123–1128.

Harrison, J. & Small, C. (1980a). The automation of electrode kinetic measurements. 1: The instrumentation and the fitting of the data using a library of reaction schemes, *Electrochimica Acta* 25(4): 447–452.

Harrison, J. & Small, C. (1980b). Automation of electrode kinetics. 6: Dissolution of Pb in H_2SO_4, *Electrochimica Acta* 25(9): 1165–1172.

Iwasita, T., Schmickler, W., Herrmann, J. & Vogel, U. (1983). The kinetic-parameters of the $Fe(CN)_6^{3-/4-}$ redox system - New results with the ring electrode in turbulent pipe-flow, *Journal of the Electrochemical Society* 130(10): 2026–2032.

Jahn, D. & Vielstich, W. (1962). Rates of electrode processes by the rotating disk method, *Journal of Electrochemical Society* 109: 849–852.

Kelley, C. (1999). *Iterative Methods for Optimization*, SIAM.

Khoshtariya, D., Dolidze, T., Vertova, A., Longhi, M. & Rondinini, S. (2003). The solvent friction mechanism for outer-sphere electron exchange at bare metal electrodes. the case of au/$Ru(NH_3)_6^{3+/2+}$ redox system, *Electrochemistry Communications* 5(3): 241–245.

Levich, V. (1962). *Physicochemical Hydrodynamics*, Prentice-Hall.

Maciel, P., Nierhaus, T., Van Damme, S., Van Parys, H., Deconinck, J. & Hubin, A. (2009). New model for gas evolving electrodes based on supersaturation, *Electrochemistry Communications* 11(4): 875–877.

Marken, F., Eklund, J. & Compton, R. G. (1995). Voltammetry in the presence of ultrasound - can ultrasound modify heterogeneous electron-transfer kinetics?, *Journal of Electroanalytical Chemistry* 395(1-2): 335–339.

Muzikar, M. & Fawcett, W. (2006). Medium effects for very fast electron transfer reactions at electrodes: the $[Ru(NH_3)_6]^{3+/2+}$ system in water, *Journal of Physical Chemistry B* 110(6): 2710–2714.

Newman, J. (1973). *Electrochemical Systems*, Englewood Cliffs, Prentice-Hall.

Nierhaus, T., Van Parys, H., Dehaeck, S., van Beeck, J., Deconinck, H., Deconinck, J. & Hubin, A. (2009). Simulation of the two-phase flow hydrodynamics in an IRDE reactor, *Journal of the Electrochemical Society* 156(9): P139–P148.

Norton, J. (1986). *An Introduction to Identification*, Academic Press.

Pintelon, R. & Schoukens, J. (2001). *System Identification - A Frequency Domain Approach*, IEEE Press.

Pletcher, D. (1991). *A First Course in Electrode Processes*, Electrochemical Consultancy.

Press, H., Teukolsky, S., Vetterling, W. & Flannery, B. (1988). *Numerical Recipes in C - The Art of Scientific Computing*, Cambridge University Press.

Robertson, B., Tribollet, B. & Deslouis, C. (1988). Measurement of diffusion-coefficients by DC and EHD electrochemical methods, *Journal of the Electrochemical Society* 135(9): 2279–2284.

Rocchini, G. (1992). The determination of the electrochemical parameters by the best-fitting with exponential polynomials, *Corrosion Science* 33(11): 1773–1788.

Rusling, J. (1984). Fitting tabulated current functions to linear-sweep voltammograms, *Analytica Chimica Acta* 162: 393–398.

Schoukens, J., Pintelon, R. & Rolain, Y. (1997). *Recent Advances in Total Least Squares Techniques and Errors-In-Variables Modeling*, SIAM, chapter Maximum Likelihood Estimation of Errors-In-Variables Models Using a Sample Covariance Matrix Obtained from Small Data Sets, pp. 59–68.

Slichting, H. (1979). *Boundary Layer Theory*, McGraw-Hill.

Sorenson, H. (1980). *Parameter Estimation - Principles and Problems*, Marcel Dekker.

Thirsk, H. & Harrison, J. (1972). *A Guide to the Study of Electrode Kinetics*, Academic Press.

Tourwé, E., Breugelmans, T., Pintelon, R. & Hubin, A. (2007). Extraction of a quantitative reaction mechanism from linear sweep voltammograms obtained on a rotating disk electrode. Part II: Application to the redoxcouple $Fe(CN)_6^{-3}/Fe(CN)_6^{-4}$, *Journal of Electroanalytical Chemistry* 609(1): 1–7.

Tourwé, E., Pintelon, R. & Hubin, A. (2006). Extraction of a quantitative reaction mechanism from linear sweep voltammograms obtained on a rotating disk electrode. Part I: Theory and validation, *Journal of Electroanalytical Chemistry* 594(1): 50–58.

Van Damme, S., Maciel, P., Van Parys, H., Deconinck, J., Hubin, A. & Deconinck, H. (2010). Bubble nucleation algorithm for the simulation of gas evolving electrodes, *Electrochemistry Communications* 12(5): 664–667.

Van den Bossche, B., Bortels, L., Deconinck, J., Vandeputte, S. & Hubin, A. (1995). Quasi-one-dimensional steady-state analysis of multi-ion electrochemical systems at a rotating disc electrode controlled by diffusion, migration, convection and homogeneous reactions, *Journal of Electroanalytical Chemistry* 397(1-2): 35–44.

Van den Bossche, B., Floridor, G., Deconinck, J., Van Den Winkel, P. & Hubin, A. (2002). Steady-state and pulsed current multi-ion simulations for a thallium electrodeposition process, *Journal of Electroanalytical Chemistry* 531(1): 61–70.

Van Parys, H., Telias, G., Nedashkivskyi, V., Mollay, B., Vandendael, I., Van Damme, S., Deconinck, J. & Hubin, A. (2010). On the modeling of electrochemical systems with simultaneous gas evolution. Case study: The zinc deposition mechanism, *Electrochimica Acta* 55(20): 5709–5718.

Van Parys, H., Tourwé, E., Breugelmans, T., Depauw, M., Deconinck, J. & Hubin, A. (2008). Modeling of mass and charge transfer in an inverted rotating disk electrode (IRDE) reactor, *Journal of Electroanalytical Chemistry* 622(1): 44–50.

Vetter, K. (1967). *Electrochemical Kinetics: Theoretical and Experimental Aspects*, Academic Press.

Wang, J., Markovic, N. & Adzic, R. (2004). Kinetic analysis of oxygen reduction on Pt(111) in acid solutions: Intrinsic kinetic parameters and anion adsorption effects, *Journal of Physical Chemistry B* 108(13): 4127–4133.

Yeum, K. & Devereux, O. (1989). An iterative method for fitting complex electrode polarization curves, *Corrosion* 45(6): 478–487.

3

Electrochemical Probe for Frictional Force and Bubble Measurements in Gas-Liquid-Solid Contactors and Innovative Electrochemical Reactors for Electrocoagulation/Electroflotation

Abdel Hafid Essadki
Ecole Supérieure de Technologie de Casablanca, Oasis, Casablanca
Hassan II Aïn Chock University
Morocco

1. Introduction

Electrochemistry constitutes an important discipline that involves many phenomena as mass transfer, migration because of the presence of electric field, and hydrodynamic especially in reactor with large scale.

In fact, electrochemical methods and electrochemical reactors have to be developed in order to resolve many problems in different area (mining, waste water treatment etc).

For measurement by electrochemical methods, probes have to be developed. For example, knowledge of the magnitude of the frictional forces between the solids and the gas-liquid mixture is very important for design of bioreactors. The growth of biomass on solid surfaces may be sensitive to shear stress. In fluidized bed bioreactors the suspended carriers are used for microbial immobilization or enzyme encapsulation. The knowledge of shear stress is important because some micro-organisms and cells attached to microcarriers are sensitive to excessive friction. The aim is to develop and verify a fast and inexpensive method for measuring the frictional forces in multiphase reactors beds.

On the other hand the performance of a multiphase reactor – for example, a bubble column- depends on the knowledge the bubble swarm properties as bubble shape, velocity allowing to determine the gas liquid mass transfer and liquid solid mass transfer.

To establish direct local information and precise bubble dimensions, various probes were used. The electrical resistivity probe was developed to measure the velocity and diameter of a bubble in a conducting liquid medium. In a non-conducting medium, the optical probe is more appropriate.

To ensure easy measurement of bubble sizes in a non transparent conducting medium and to establish direct local information in a reactor, the present work shows the possible use of electrochemical probe.

Paragraph two develops this aspect, the electrochemical probe developed serves in the same time as a probe able to measure the velocity gradient at the wall of a spherical sphere and to measure the volume bubble in a bubble column.

For the process aspect as the water and waste water treatment, the need of purifying water for human consumption is more and more required. Cleaning wastewater from industrial effluents before discharging is also a challenging work. In fact, innovative, cheap and effective techniques have to be developed.

Electrocoagulation (EC) is an electrochemical method for treating polluted water which has been successfully applied for treatment of soluble or colloidal pollutants, such as wastewater containing heavy metals, emulsions, suspensions, etc., but also drinking water for lead or fluoride removal. A typical EC unit includes therefore an EC cell/reactor, a separator for settling or flotation, and often a filtration step. Indeed, the benefits of EC include simplicity, efficiency, environmental compatibility, safety, selectivity, flexibility and cost effectiveness. In particular, the main points involve the reduction of sludge generation, the minimization of the addition of chemicals and little space requirements due to shorter residence time. The main deficiency is the lack of dominant reactor design and modeling procedures. The literature reveals any systematic approach for design and scale-up purpose. The most papers use laboratory-scale EC cells in which magnetic stirring is adjusted experimentally and the separation step by floatation/sedimentation is not studied.

That's why an innovative reactor is developed in order to optimize the cost of this process. This is the object of paragraph three.

2. Electrochemical method for measurements of frictional force and bubble size

This part involves the measurement character of electrochemical methods that lead to measure the velocity gradient and frictional force on a particle in a bubble column and in a fluidized bed. The possibility of measuring bubble volume is also discussed. Thus, an electrochemical probe may be proposed to measure in the same time the frictional force and the bubble volume. The application of such electrochemical method is very important in biochemical engineering.

2.1 Velocity gradient and frictional force measurement

Reiss and Hanratty (1963) developed a model to obtain the velocity gradient S_w, in the vicinity of a wall of rectangular or cylindrical shape. By using a microelectrode mounted flush with a tube wall, the apparent mass transfer coefficient K_a could be obtained by electrochemical technique. This model was established in the cases where the thickness of the concentration boundary layer (δ_c) was less than that of the hydrodynamic boundary layer (δ_v) (figure 1) and was verified in cases of high Schmidt number.

The mass balance in the vicinity of the microelectrode can be developed as follow:

$$\frac{\partial C}{\partial t} + S_w\, y \frac{\partial C}{\partial x} = D \frac{\partial^2 C}{\partial y^2} \tag{1}$$

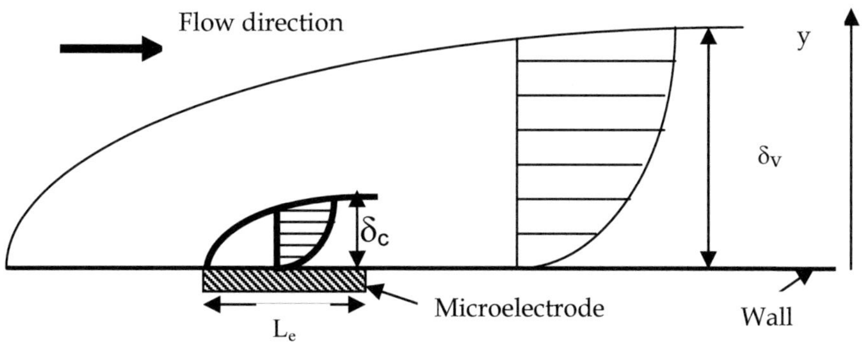

Fig. 1. Concentration and velocity profiles over the electrode surface.

The instantaneous rate of mass transfer is proportional to the concentration gradient at the surface averaged over the electrode:

$$N = D < \frac{\partial C}{\partial y} > \tag{2}$$

The mass transfer coefficient defined as : $K_a = N/(C\text{-}C_w)$, C_w is the wall concentration.

The analytical solution of (1) allows to determine the local mass transfer coefficient::

$$\overline{K_a} = \alpha \left(\frac{D^2 \overline{S_w}}{L_e} \right)^{1/3} \tag{3}$$

C : concentration, $mol.m^{-3}$
D : coefficient diffusion, $m^2.s^{-1}$,

The method can be extended to spherical walls (sphere particle). A sphere is equipped with an inside channel, bent through 90°, in which a gold thread of 1 mm diameter was introduced, cut flush with the surface. A rigid tube serves as support. The microelectrode can be directed relative to the average direction of the liquid by rotating the support (figure 2).

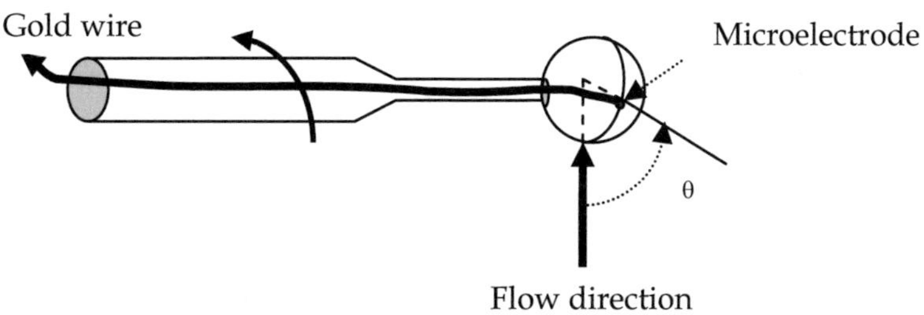

Fig. 2. The electrochemical electrode.

The electrochemical technique adopted consisted of the determination of the apparent local transfer coefficient (and subsequently the velocity gradient at the wall), based on measurements of the diffusion - limiting current during the reduction of an electroactive species.

The relationship between the apparent mass transfer coefficient and the velocity gradient is then deduced:

$$\overline{K_a} = 0.862\left(\frac{D^2\,\overline{S_w}}{d_e}\right)^{1/3} \tag{4}$$

d_e : diameter of microelectrode, m

In order to test the validity of this model to predict the frictional force in sphere particles in bubble column and fluidized bed, the experimental apparatus is then designed (figure 3).

It is composed of a plexiglass fluidization column, 157 cm height (distance between the perforated stainless steel plate serving as liquid distributor and the point of liquid overflow) and 9.4 cm diameter. A column filled with glass sphere was placed underneath the liquid distributor; this serves as a homogenization section to avoid any preferential passage phenomena. A liquid reservoir is equipped with a temperature controller and a liquid cooling coil. Rotameters are used to meausure the flow rates of liquid and gas (nitrogen) which are introduced via a gas injector consisting of two concentric circular tubes with 90 regularly spaced holes (0.4 mm diameter). An auxiliary electrode (nickel ring), a reference electrode furnished with a Luggin's capillary and a working electrode are connected to a potentiometer which is linked to microcomputer performing the acquisition of electric current signals and computations of the average and fluctuating current intensities.

The method depends on creating a diffusional limitation at the transfer surface studied (working electrode) and simultaneously avoids such limitations at the counter electrode surface (i.e the counter electrode potential remained almost constant). This is achieved by using a counter electrode with a large surface area compared with the working electrode.

The intensity of the limiting current of the working electrode is obtained by the graph presenting the current intensity versus potential. In order to achieve this, the reference electrode (saturated calomel), has to be placed as closed as possible to the working electrode, to minimize the ohmic resistance of the solution. For this purpose, a capillary of about 1 mm diameter is introduced between the working and reference electrodes.

Several possible reactions for electrochemical determination of mass transfer coefficients are available, but the most frequently used system is potassium ferricyanide to potassium ferrocyanide. Normally the reaction at the working electrode is the reduction of ferricyanide, whereas the oxidation of ferrocyanide takes place at the counter electrode:

Cathode: $Fe(CN)_6^{3-} + e^- \rightarrow Fe(CN)_6^{4-}$

Anode: $Fe(CN)_6^{4-} \rightarrow Fe(CN)_6^{4-} + e^-$

The ferricyanide concentration should thus in principle remain constant during the experiment. This is verified by using a turning disc electrode, as explained later. The

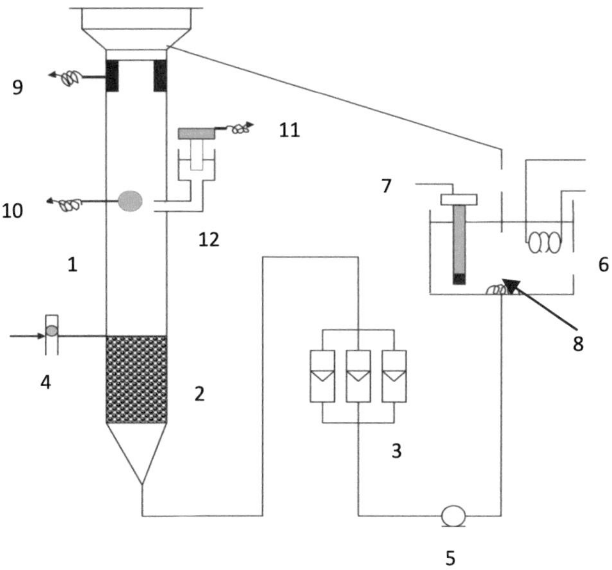

1- Column
2- Homogenization section
3- Rotameter
4- Rotameters
5- Pump
6- Liquid reservoir
7- Temperature controller
8- Cooling coil
9- Auxiliary electrode
10- Working electrode
11- Reference electrode
12- Capillary

Fig. 3. Experimental apparatus for the study of friction on a solid sphere.

potassium ferri-ferrocyanide system chosen consisted of a 0.5 mol.dm^{-3} sodium hydroxide solution with 5 x 10^{-3} mol.dm^{-3} each of potassium ferricyanide and potassium ferrocyanide. Sodium hydroxide is added as electrolyte support to minimize the migrational effects.

The physic-chemical properties of the solution are given in Table 1. The density of the solution was determined by pycnometry and the viscosity measured by a viscosmeter (Rheomat 15T-FC). The diffusion coefficient of the potassium ferricyanide is measured by means of a turning disc electrode, computed from the intensity of the limiting current on the disc using the relationship (Levart & Schumann, 1974):

$$J = D\, C_0\, Sc^{1/3}.(\omega/\nu)^{1/2}.\ 1/k \qquad (5)$$

$$k = 1.61173 + 0.480306\, Sc^{-1/3} + 0.23393\, Sc^{-2/3} + 0.113151\, Sc^{-1}$$

D: coefficient diffusion, m^2.s^{-1}
J : molar flux species, mol.m^{-2} s^{-1}
C_0 : electroactive species concentration, mol.m^{-3}
ω : angular velocity, rd.s^{-1}
ν : kinematic velocityof liquid, m^2.s^{-1}
Sc : Schmidt number defined by ν/D

This relationship is also used to determine the concentration of the solution once the diffusion coefficient is known (assumed to be concentration independent).

Temperature (°C)	10^{-3} ρ_l (kg.m^{-3})	10^3 η_l (Pa.s)	10^{10} D (m^2.s^{-1})
25	1.02	1.17	5.17

Table 1. Physico-chemical properties of the solution used in the study.

The apparent mass transfer coefficient at any microelectrode position θ, is given by:

$$K_a = \frac{I_l}{n.F.A_e.C_0} \tag{6}$$

I_l : Intensity of diffusion – limiting current (A)
F : Faraday constant (A s m^{-1})
n : number of electrons liberated during the course of the electrochemical reaction
A_e :surface of microelectrode (m^2).
From the equation (4) we can deduce the velocity gradient:

$$S_w(\theta) = \left(1.16K_a\right)^3 . \frac{d_e}{D^2} \tag{7}$$

As the velocity gradient at the wall is related to the local shear stress (τ_w) through the relationship: $\tau_w = \eta_l . S_w$ (Newtonian liquid), η_l is the dynamic viscosity (Pa.s).
Eq. (7) may be used to compute the frictional drag force (F_f) exerted on the whole sphere:

$$F_f = 2\pi R^2 \eta_l \int_0^{\pi} S_w(\theta) \sin^2\theta \, d\theta \tag{8}$$

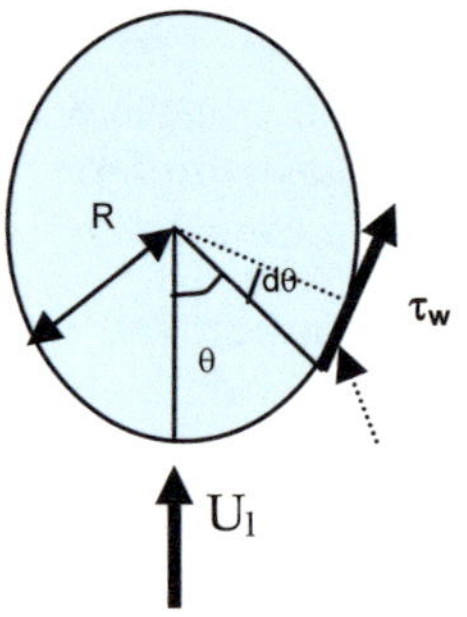

Method verification:

The relationship of Reiss and Hanratty was established for the case of a flat surface and validity for a spherical surface should therefore first be verified.

Figure 4 shows the velocity gradient at the wall for five values of the Reynolds number and with different angles between the electrode surface and the mean fluid flow direction. The curves clearly show a maximum in velocity gradients around the 45° angle which is in agreement with calculations based on numerical solution of the momentum balance equation around a sphere. The figure also shows that the velocity gradient rises with the Reynolds number at all positions on the sphere. The influence is however far more noticeable in the front than in the rear of the sphere.

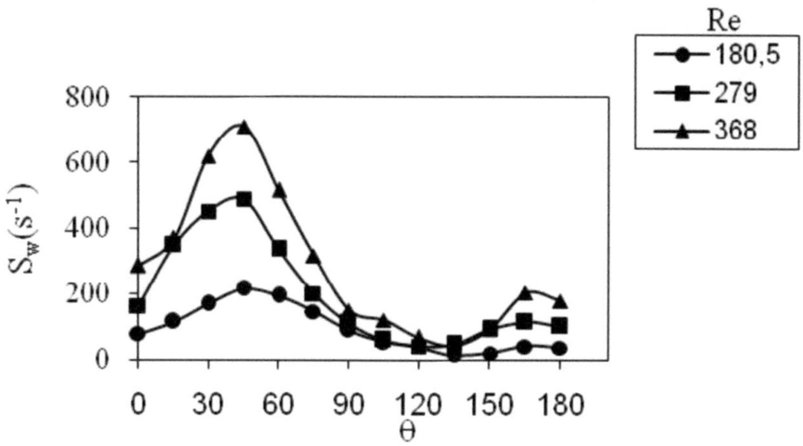

Fig. 4. Surface velocity gradient as function of position for single-phase around a sphere.

In the case of three phase fluidization, glass spheres (2mm in diameter, ρ_s= 2532 kgm^{-3}) and plastic spheres (5mm in diameter, ρ_s= 1388 $kg.m^{-3}$) are used. This choice provides very different bubbly flows due to different balances of coalescence and break-up of bubbles.

The contribution of the frictional force is more important in "coalescent" fluidized beds than in "break-up" fluidized beds. The effect of gas injection is depending on the fluidized particle effect on bubble coalescence and break-up. Correlations are developed linking frictional force to gas hold-up. The correlations recommended for frictional force in fluidized beds for both systems, (i.e., coalescence and break-up) are as follows:

- Glass spheres (2mm diameter, coalescence regime):

$F = 2.43\ Re^{0.052}\ \varepsilon_g^{0.4}$, standard deviation = 6%.

- Plastic spheres (5mm diameter, break-up regime):

$F = 0.123\ Re^{0.3}\ \varepsilon_g^{0.1}$, standard deviation = 4%.

F is a dimensionless force defined by $F = F_f/P_a$, where P_a is the effective weight of the sphere.

Re: Reynolds number defined by $d_p U_l \rho_l/\eta_l$, d_p is the particle diameter, ρ_l is the liquid density, U_l is the superficial liquid velocity, η_l is the dynamic viscosity, ε_g : gas holdup.

2.2 Bubble size measurement

The purpose is to demonstrate that the electrochemical probe can be used as a means of measuring bubble sizes. First, calibrated bubbles are used by single tubes. Then, a gas injector is used in a bubble column with homogeneous bubbling regime. In this case, the average frequency of the fluctuations of diffusion limited current detected by the probe is postulated as being equal to the bubble frequency leading to an estimation of bubble size.

2.2.1 Single orifice: bubble train

To test the possibility of measuring bubble size with the use of an electrochemical probe in a bubble column, a train of calibrated bubbles is generated in a tube in which gas is injected under closely controlled conditions. To obtain different bubble sizes, three tubes, T_1, T_2, and

T_4, with inner diameters of 1, 2, and 4 mm, respectively, are placed 5 cm below the probe. The probe is placed 30 cm above the liquid distributor and in the column center.

At low gas volumetric flow rates, the effect of the regular passage of bubbles close to the electrochemical probe on the diffusion limit current (or the velocity gradient) is studied. The stability of the bubble frequency is shown on the signal by regularly spaced peaks (fig.5). Signals similar to this are obtained for all positions of the front of the spherical probe (active surface) exposed to bubbles ($0° < \theta < 120°$). The frequency spectrum clearly shows a single peak representing the bubble frequency, f_b (fig.6).

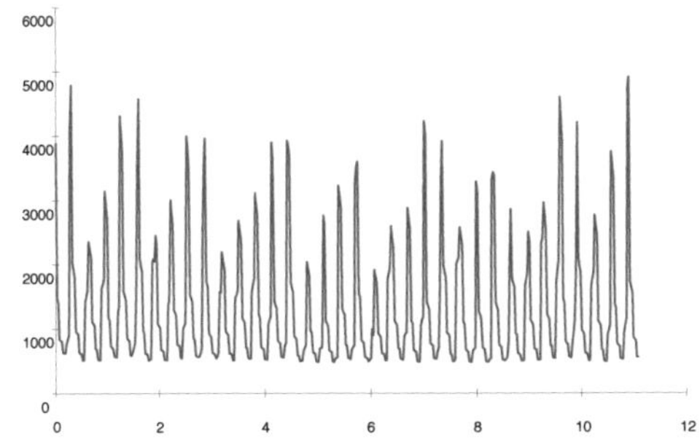

Fig. 5. History records of the velocity gradient at the wall: Effect of the regular passage of bubbles near the electrochemical probe.

Knowing the volumetric flow rate, Q_g, we can obtain the bubble volume, V_b by the relation:

$$V_b = \frac{Q_g}{f_b} \tag{9}$$

Assuming spherical bubbles, we can deduce the bubble diameter: $d_b = (6V_b/\pi)^{1/3}$.

A comparison at low gas flow rates is achieved by video recording when the bubble emission frequency and volume are obtained. Excellent agreement is observed, proving that the two methods give exactly the same frequency and the same average bubble diameter within 3%.

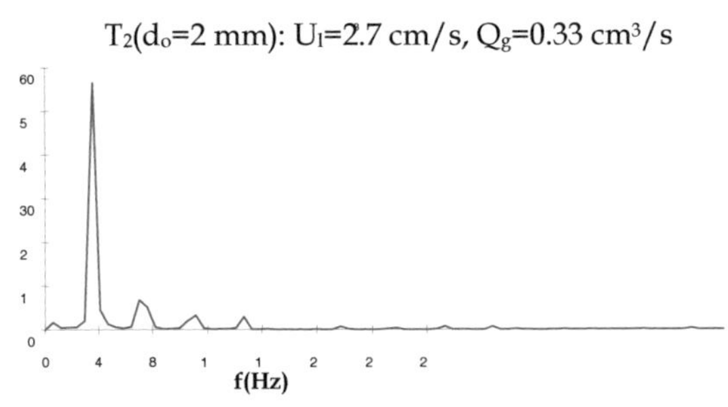

Fig. 6. Power spectral density function (P_s/s^2): Effect of the regular passage of bubbles near the electrochemical probe.

2.2.2 Gas injector

In the case of a gas injector consisting of two concentric circular tubes with regularly spaced holes, the average frequency is used (<f>) to evaluate the bubble diameter in the bubble column for the homogeneous bubbly flow. The bubble diameter is estimated by the following expression (Essadki et al., 1997):

$$d_b = 1.5\frac{U_g}{\langle f \rangle} \tag{10}$$

The possibility of measuring bubble size by electrochemical probe is possible for the homogeneous bubbling regime in any conducting medium.

3. Innovative electrochemical reactors

This part describes the electrocoagulation and electroflotation as electrochemical methods to treat waste water. The conventional reactors used and the different disposition of electrodes are pointed out. It is also explained how innovative reactors can improve the process of waste water treatment. Thus, specific energy and electrode consumptions are even smaller without the need for mechanical agitation, pumping requirements and air injection, which could not be achieved in other kinds of conventional gas-liquid contacting devices.

An application is then presented to show the efficiency of electrocoagulation/ electroflotation in removing colour from synthetic and real textile wastewater by using aluminum and iron electrodes in an external-loop airlift reactor. The defluoridation is also showed. The time, pH, conductivity and current density are the most parameters for the removal efficiency and energy consumption.

According to many authors, electrochemical technique such as electrocoagulation, electroflotation, electrodecantation have a lot of advantages comparatively to other techniques. Biological treatments are cheaper than other methods, but for example for the decolorization of dye wastewater, dye toxicity usually inhibits bacterial growth and limits therefore the efficiency of the decolorization. Physico-chemical methods include adsorption (e.g. on active carbon), coagulation-flocculation (using inorganic salts or polymers), chemical oxidation (chlorination, ozonisation, etc.) and photodegradation (UV/H_2O_2, UV/TiO_2, etc.). However, these technologies usually need additional chemicals which sometimes produce a secondary pollution and a huge volume of sludge. Water treatments based on the electrocoagulation technique have been recently proved to circumvent most of these problems, while being also economically attractive.

3.1 Electrocoagulation-electroflotation theories

This technique is based on the in situ formation of coagulant as the sacrificial anode (usually aluminum or iron) corrodes due to an applied current (figure 7). Aluminum and iron materials are usually used as anodes, the dissolution of which produces hydroxides, oxyhydroxides and polymeric hydroxides. In EC, settling is the most common option, while flotation may be achieved by H_2 (electroflotation) or assisted by air injection.

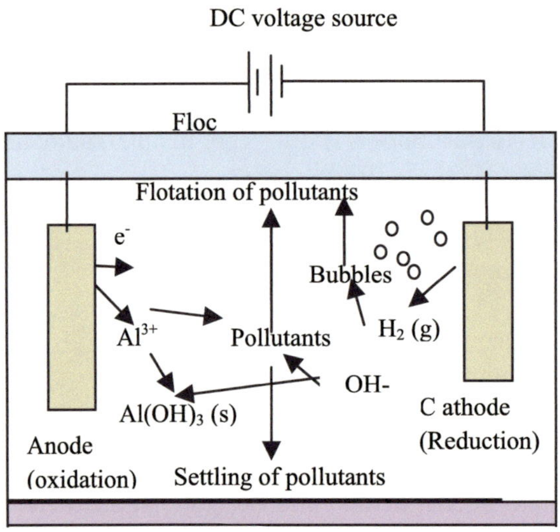

Fig. 7. Schematic diagram of two-electrode electrocoagulation (EC) cell.

Three successive stages occur during EC:

i. Formation of coagulants by electrolytic oxidation of sacrificial anode
ii. Destabilization of the pollutants, particulate suspension and breaking of emulsions
iii. Aggregation of the destabilized phases to form flocs.

The pollutants can be in form of:

- Large particles easy to separate them from water by settling.
- Colloids.
- Dissolved mineral salt and organic molecules.

It is impossible to use the decantation as a technique to eliminate the maximum of particles. This remark is especially valid for colloids. Thus, colloids are organic or mineral particles in which the size is between some nanometers and approximately 1 μ responsible for color and turbidity.

The destabilization mechanism of the pollutants can be summarized as follows:

- Compression of the diffusion double layer around the charged species by the interactions of ions generated by oxidation of the sacrificial anode.
- Charge neutralization of the ionic species present in wastewater by counter ions produced by the electrochemical dissolution of the sacrificial anode. These counter ions reduce the electrostatic inter-particle repulsion to the extent that the Van Der Walls attraction predominates, thus causing coagulation. A zero net charge results in the process.
- Floc formation; the floc formed as a result of coagulation creates a sludge blanket that entraps and bridges colloidal particles still remaining in the aqueous medium.

3.2 Reactions at electrodes and different modes of electrodes connection

In the case of aluminum electrodes, the reactions taking place at the electrodes are as follow:

At the anode, takes place oxidation:

$$Al \rightarrow Al^{3+} + 3e^-$$

For higher current density:

$$4\,OH^- \rightarrow 2\,H_2O + O_2 + 2\,e^-$$

At the cathode, takes place reduction:

$$2\,H_2O + 2\,e^- \rightarrow 2\,H_2 + 2\,OH^-$$

Although the sacrificial anodes deliver Al cation, their dissolution produces hydroxides, oxyhydroxides or polymeric hydroxides as a function of pH (Chen et al., 2004). These can adsorb or precipitate.

Al^{3+} ions may generate $Al(H_2O)_6{}^{3+}$, $Al(H_2O)_5OH^{2+}$, $Al(H_2O)_4(OH)^{2+}$. Many monomeric and polymeric species will be formed by hydrolysis such as, $Al(OH)^{2+}$, $Al(OH)^{2+}$, $Al_2(OH)_2{}^{4+}$, $Al(OH)^{4-}$, $Al_6(OH)_{15}{}^{3+}$, $Al_7(OH)_{17}{}^{4+}$, $Al_8(OH)_{20}{}^{4+}$, $Al_{13}O_4(OH)_{24}{}^{7+}$, $Al_{13}(OH)_{34}{}^{5+}$ over a wide pH range.

In the case of iron electrodes, $Fe(OH)_n$ with $n = 2$ or 3 is formed at the anode. The production of $Fe(OH)_n$ follow two mechanisms.

a. Mechanism 1:

Anode : $Fe \rightarrow Fe^{2+} + 2e^-$
$Fe^{2+} + 2OH^- \rightarrow Fe(OH)_2$

Cathode : $2\,H_2O + 2\,e^- \rightarrow 2\,H_2 + 2\,OH^-$

Overall : $Fe + 2H_2O \rightarrow Fe(OH)_2 + H_2$

b. Mechanism 2:

Anode : $4Fe \rightarrow 4Fe^{2+} + 8e^-$
$4Fe^{2+} + 10H_2O + O_2 \rightarrow 4Fe(OH)_3 + 8H^+$

Cathode: $8H^+ + 8e^- \rightarrow 4H_2$

Overall: $4Fe + 10H_2O + O_2 \rightarrow 4Fe(OH)_3 + 4H_2$

Polymeric hydroxy complexes are also generated namely : $Fe(H_2O)_6{}^{3+}$, $Fe(H_2O)_5(OH)^{2+}$, $Fe_2(H_2O)_8(OH)_{24}{}^{+}$ and $Fe_2(H_2O)_6(OH)_4{}^{4+}$, depending on the pH of aqueous medium.

Because of the workable rate of metal dissolution, a two-electrode EC is not always suitable. That's why large surface area is needed for a good performance. The electrodes configuration can be divided into three modes:

- Monopolar electrodes in parallel connections :

As observed in figure 8, the parallel arrangement consists of pairs of conductive metals plates placed between two parallel electrodes and DC power source. All cathodes are connected to each other and to negative pole of DC; in the same manner, all sacrificial anodes are connected to each other and to positive pole of DC. The electric current is divided between all the electrodes in relation to the resistance of the individual cells.

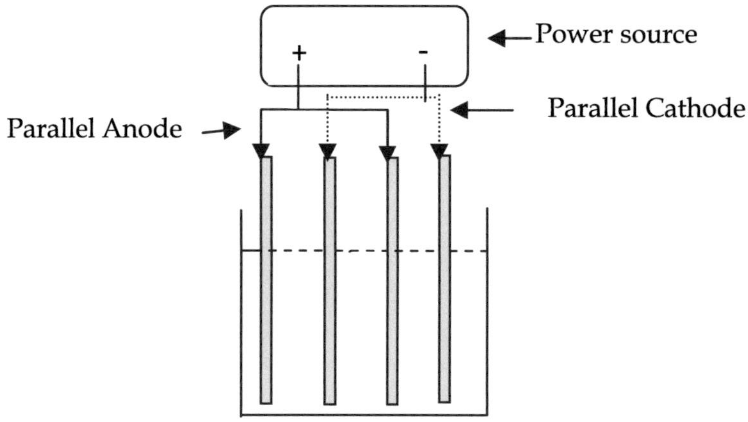

Fig. 8. Monopolar electrodes in parallel connections.

- Monopolar electrodes in series connections :

Each pair of sacrificial electrodes is internally connected with each other, and has no inter-connections with the outer electrodes (figure 9). The same current would flow through all the electrodes.

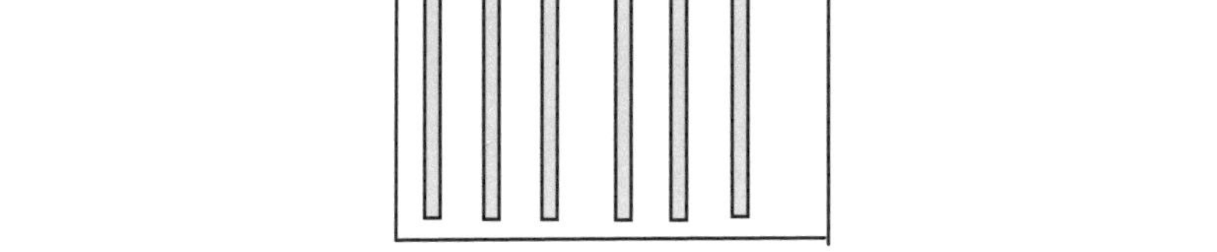

Fig. 9. Monopolar electrodes in series connections.

- Bipolar electrodes

The cells are in series. The sacrificial electrodes are placed between the two parallel electrodes without any electrical connection. Only the two monopolar electrodes are connected to the electric power source with no interconnections between the sacrificial electrodes.

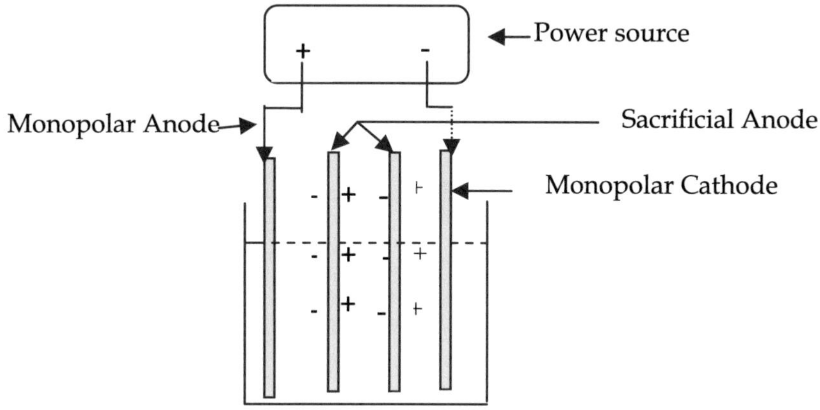

Fig. 10. Bipolar electrodes in series connections.

3.3 Parameters affecting electrocoagulation/electroflotation

Several parameters can affect the efficiency of removal by electrocoagulation/ electroflotation. The most important parameters are described bellow.

3.3.1 Current density or charge loading

The current density (j) is expected to exhibit a strong effect on EC: higher the current, shorter the treatment. This is ascribed to the fact that at high current density, the extent of anodic dissolution of aluminum (iron) increases, resulting in a greater amount of precipitate for the removal of pollutants. Moreover, bubble generation rate increases and the bubble size decreases with increasing current density. These effects are both beneficial for high pollutant removal by H_2 flotation. As a first approximation, the amount of Al released is proportional to the product $\varphi_{Al}It$. The values of the Faradic yield φ_{Al} are between 100% and 160%; they decrease with increasing time in the first minutes of the run, but also with higher current density. These trends have already been reported in the literature (Essadki et al., 2009). This mass overconsumption of aluminum electrodes may be due to the chemical hydrolysis of the cathode but it also often explained by the "corrosion pitting" phenomenon which causes holes on the electrode surface. The mechanism suggested for "corrosion pitting" involves chloride anions and can be summarized as follows:

$$2Al + 6HCl \rightarrow 2AlCl_3 + 3H_2$$

$$AlCl_3 + 3H_2O \rightarrow Al(OH)_3 + 3HCl$$

This mechanism can therefore produce more aluminum hydroxide flocs and H_2 bubbles than the equivalent current supplied should. Conversely, high current density allows the passivation of the cathode to be reduced but an increase in energy consumption that induces heating by Joule effect. As a result, too high current densities have generally to be avoided.

The specific electrode consumption per kg pollutant (μ_{Al}) is determined by the following expression:

$$\mu_{Al}(kgAl / kg \text{ pollutant}) = \frac{3600.M_{Al}.I.t.\varphi_{Al}}{3F} \cdot \frac{1}{V(C_0 Y)} \tag{11}$$

using initial pollutant concentration C_0 (kg/m^3), current intensity I (A), cell voltage U (V), electrolysis time t (h), liquid volume V (m^3), molar weight of aluminum M_{Al} = 0.02698 constant F (96,487 C/mole$^-$) and the faradic yield ϕ_{Al} of Al dissolution, Y is the removal efficiency. ϕ_{Al} is estimated as the ratio of the weight loss of the aluminum electrodes during the experiments Δm_{exp} and the amount of aluminum consumed theoretically at the anode Δm_{th}:

$$\phi_{Al} = \frac{3F}{3600.M_{Al}.I.t} . \Delta m_{\text{exp}} \tag{12}$$

This parameter depends upon the pH and the amount of other species present in solution, for example co-existing anions.

3.3.2 Conductivity

The increase of the conductivity (κ) by the addition of sodium chloride is known to reduce the cell voltage U at constant current density due to the decrease of the ohmic resistance of wastewater. Energy consumption, which is proportional to *U.I*, will therefore decrease. Chloride ions could significantly reduce the adverse effects of other anions, such as HCO_3^- and SO_4^{2-}, for instance by avoiding the precipitation of calcium carbonate in hard water that could form an insulating layer on the surface of the electrodes and increase the ohmic resistance of the electrochemical cell (Chen et al., 2004). Chloride anions can also be oxidized and give active chlorine forms, such as hypochlorite anions, that can oxidize pollutants. The main mechanism is as follows:

$$Cl_2 + 2e^- \rightarrow 2Cl^-$$

$$Cl_2 + H_2O \rightarrow Cl^- + ClO^- + 2H^+$$

However, an excessive amount of NaCl (higher than 3 g/L) induces overconsumption of the aluminum electrodes due to "corrosion pitting" described above; Al dissolution may become irregular.

3.3.3 pH effect

pH is known to play a key role on the performance of EC. An optimum has to be found for the initial pH, in order to optimize the EC process. However, the pH changed during batch EC. Its evolution depended on the initial pH. EC process exhibits some buffering capacity because of the balance between the production and the consumption of OH^- (Chen et al., 2004), which prevents high change in pH. The buffering pH seems just above 7: when the initial pH is above this value, pH decreases during EC; otherwise, the opposite behavior is observed.

The effect of pH can be explained as follows. The main reactions during EC are:

Anode: $Al0(s) \rightarrow Al^{3+} + 3e^-$

Cathode: $2H_2O + 2e^- \rightarrow H_2(g) + 2OH^-$

At low pH, such as 2–3, cationic monomeric species Al^{3+} and $Al(OH)^{2+}$ predominate. When pH is between 4–9, the Al^{3+} and OH^- ions generated by the electrodes react to form various monomeric species such as $Al(OH)^{2+}$, $Al_2(OH)_2^{2+}$, and polymeric species such as $Al_6(OH)_{15}^{3+}$, $Al_7(OH)_{17}^{4+}$, $Al_{13}(OH)_{34}^{5+}$ that finally transform into insoluble amorphous $Al(OH)_3(s)$ through complex polymerization/precipitation kinetics. When pH is higher than 10, the monomeric $Al(OH)^{4-}$ anion concentration increases at the expense of $Al(OH)_3(s)$. In addition, the cathode may be chemically attacked by OH^- ions generated together with H_2 at high pH values:

$$2Al + 6H_2O + 2OH^- \rightarrow 2Al(OH)^{4-} + 3H_2 \quad (10)$$

Two main mechanisms are generally considered: *precipitation* for pH lower than 4 and adsorption for higher pH. Adsorption may proceed on $Al(OH)_3$ or on the monomeric $Al(OH)^{4-}$

3.3.4 Power supply

From an energetic point of view, energy consumption during EC is known to vary as the product *UIt*. Energy requirements per kg of pollutant removed ($E_{\text{pollutant}}$) to achieve a certain percentage of efficiency (Y) shows a continuous increase of $E_{\text{pollutant}}$ with j.

The specific electrical energy consumption per kg pollutant removed ($E_{\text{pollutant}}$) is calculated as follows:

$$E\ (kWh\ /\ kg\ \text{pollutant}) = \frac{U.I.t.}{1000.V.(C_0 Y)} \quad (13)$$

3.3.5 Temperature

Few papers were investigated to show the effect of temperature on EC efficiency. The current efficiency of aluminum was found to be increased with temperature until about 60°C (Chen, 2004) where a maximum was found. Further increase in temperature results in a decrease of EC efficiency. The increase of temperature allows to a destruction of the aluminum oxide film on the electrode surface.

3.4 Design of electrocoagulation cell

The position of the electrodes in the reactor can be optimized as a function of hydrodynamic parameters and current density (j). Complementary rules should include the influences of electrode gap (e) and operating conditions on voltage U (and consequently on energy consumption). The measured potential is the sum of three contributions, namely the kinetic overpotential, the mass transfer overpotential and the overpotential caused by solution ohmic resistance. Kinetic and mass transfer overpotentials increase with current density, but mass transfer is mainly related to mixing conditions: if mixing is rapid enough, mass transfer overpotential should be negligible. In this case, the model described by Chen et al., (2004) is often recommended for non-passivated electrodes:

$$U = -0.76 + \frac{e}{k} j + 0.20 \ln (j) \quad (14)$$

Different typical reactors applied for electrochemical technologies are explained by Chen (2004).

3.5 Airlift reactors as innovative electrocoagulation cells

Airlift reactors constitute a particular class of bubble columns in which the difference in gas hold-up between two sections (namely, the riser and the downcomer) induces an overall liquid circulation without mechanical agitation (Chisti, 1989). They have been extensively applied in the process industry to carry out chemical and biochemical slow reactions, such as chemical oxidation using O_2, Cl_2 or aerobic fermentation, but never as EC cells, as far as we know. Airlift reactors present two main designs: external-loop and internal-loop configurations (figure 11 a -b).

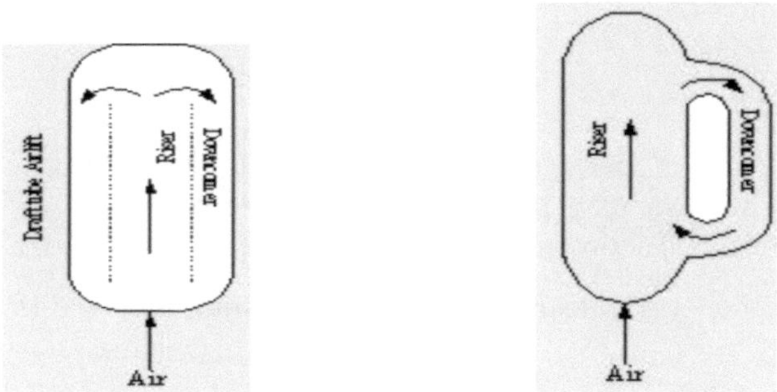

(a) (b)

Fig. 11. Airlift reactors, (a) : internal loop airlift reactor, (b) : external loop airlift reactor.

External-loop airlift reactors offer the advantage to allow various designs of the separator section, which favors gas disengagement at the top of the reactor and maximizes consequently the overall recirculation velocity at the expense of more complex reactor geometries. Their hydrodynamics has also been extensively studied in two-phase gas-liquid and three-phase gas-liquid-solid flows. In airlift reactors, the driving force of the overall liquid circulation results from the gas hold-up difference between the riser (ε_r) and the downcomer (ε_d), and also from the dispersion height. Gas hold-up is defined as the ratio of volume occupied by the gas phase over the total volume of the corresponding section. Dispersion height (h_D) corresponds to the distance from the surface in which a gas phase can be observed in the riser.

The overall liquid circulation velocity in the riser U_{Lr} can therefore be predicted from an energy balance using Equation 15 (Chisti, 1989):

$$U_{Lr} = \left[\frac{2g \cdot h_D \cdot (\varepsilon_r - \varepsilon_d)}{\frac{K_T}{(1-\varepsilon_r)^2} + \left(\frac{A_r}{A_d} \right)^2 \frac{K_B}{(1-\varepsilon_d)}} \right] \quad (15)$$

K_T coefficient taking into account the effects of pressure drop in the riser and the separator section and K_B accounts for pressure drop in the downcomer and the junction.

A_d: cross-sectional area of the downcomer (m^2) and A_r: cross-sectional area of the riser (m^2)

The superficial liquid velocity in the riser (U_{Ld}) is deduced from a mass balance on the liquid phase:

$$U_{Lr} = \frac{A_d}{A_r} U_{Ld} \tag{16}$$

In order to transform this reactor to an electrochemical one, the gas phase is not injected, but electrochemically-generated. This means that both h_D and ε_r depend on the axial position of the electrodes in the riser. At constant current density, ε_r should vary approximately as dispersion height h_D when electrode position is modified, provided ε_r <<1. One can deduce from Equations (15) and (16) that U_{Ld}~h_D.

The objective of complete flotation may be achieved only if hydrodynamic shear forces remain weak in the riser to avoid floc break-up and in the separator to limit break-up and erosion, which means low U_{Ld} values.

3.5.1 Reactor design

An external-loop airlift made of transparent plexiglas is used for this study. The reactor geometry is illustrated by figure 12. By definition, the riser is the section in which the gas phase is sparged and flows upwards. The diameters of the riser and the downcomer are respectively 94mm and 50 mm. Consequently, the riser-to-downcomer cross-sectional ratio (A_r/A_d) is about 3.5. This is a typical value when reaction takes place only in the riser section. Both are 147 cm height (H_2 +H_3) and are connected at the bottom by a junction of 50 mm diameter and at the top by a gas separator (also denoted gas disengagement section) of H_S = 20 cm height. The distance between the vertical axes of the riser and the downcomer is 675 mm, which limits the recirculation of bubbles/particles from the riser into the downcomer. At the bottom, the curvature radius of the two elbows is 12.5 cm in order to minimize friction and avoid any dead zone. The liquid volume depends on the clear liquid height (h) and can be varied between 14 L and 20 L, which corresponds to a clear liquid level between 2 cm and 14 cm in the separator section. All the experiments are conducted at room temperature (20±1 °C) and atmospheric pressure in the semi-batch mode (reactor open to the gas, closed to the liquid phase). Contrary to conventional operation in airlift reactors, no gas phase is sparged at the bottom in the riser. Only electrolytic gases induce the overall gas recirculation resulting from the density difference between the fluids in the riser and the downcomer. Two readily available aluminum flat electrodes of rectangular shape (250mm×70mm×1 mm) are used as the anode and the cathode, which corresponds to S = 175 cm^2 electrode surface area (Fig. 12). The distance between electrodes is e = 20mm, which is a typical value in EC cells. They are treated with a HCl aqueous solution for cleaning prior use to avoid passivation. The electrodes are placed in the riser, parallel to the main flow direction to minimize pressure drop in the riser and maximize the recirculation velocity. The axial position of the electrode can also be varied in the column. The distance (H_1) between the bottom of the electrodes and the bottom of the riser ranged between 7 cm and 77 cm. EC is conducted in the intensiostat mode, using a digital DC power supply (Didalab, France) and recording potential during the experiments. The width of the electrodes is maximized by taking into account riser diameter and electrode inter-distance. Current density values (j)

between 5.1 and 51 mA/cm² are investigated, which corresponds to current ($I = jS$) in the range 1.0–10 A.

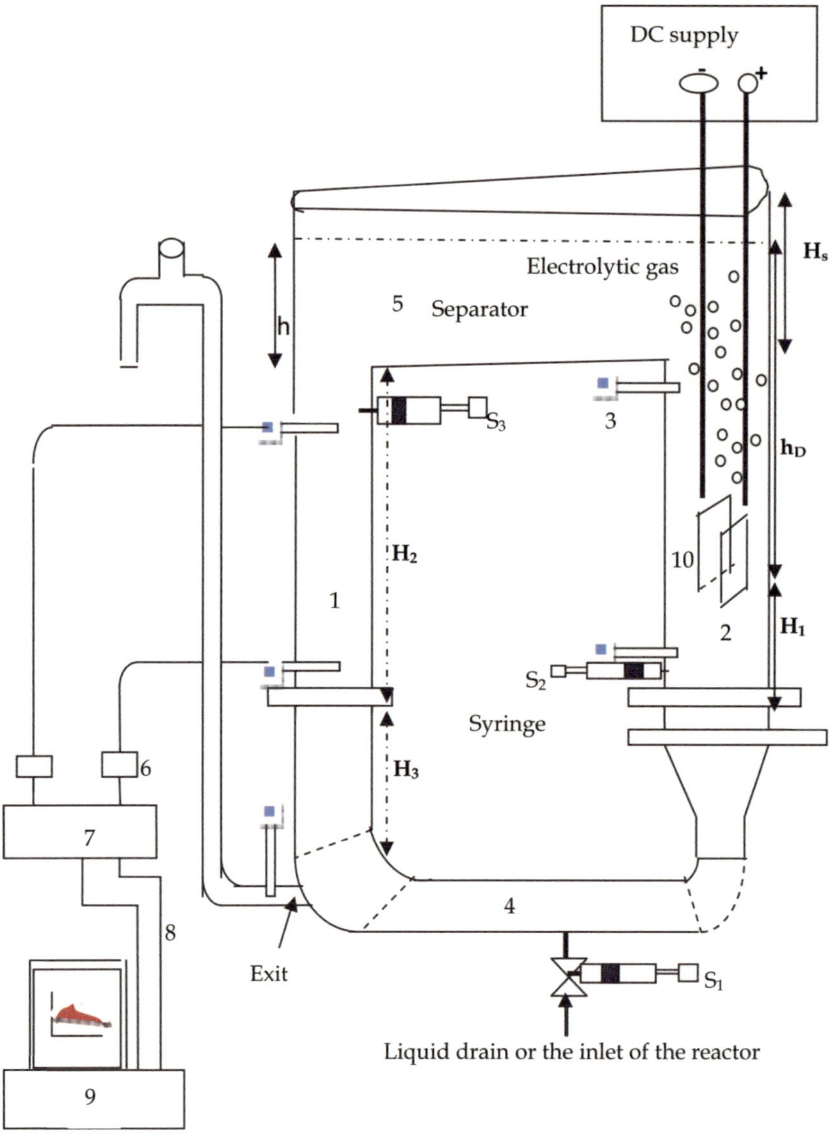

Fig. 12. External-loop airlift reactor (1: downcomer section; 2: riser section; 3: conductivity probes; 4: junction column; 5: separator 6: conductimeter; 7: analog output/input terminal panel (acquisition system); 8: 50-way ribbon cable kit; 9: data acquisition system; 10: electrodes).

3.5.2 Chemicals and methods

The average liquid velocity in the downcomer (U_{Ld}) is measured using the conductivity tracer technique. Two conductivity probes placed in the downcomer section were used to record the tracer concentration resulting from the injection of 5mL of a saturated NaCl solution at the top of the downcomer using a data acquisition system based on a PC computer equipped with UEI-815 A/D converter. The distance between the probes is 90 cm. Liquid velocity is estimated using the ratio of the mean transit time between the tracer peaks detected successively by the two electrodes and the distance between the probes. The superficial liquid velocity in the riser (U_{Lr}) is deduced from a mass balance on the liquid phase: $U_{Lr} = A_d / A_r U_{Ld}$.

An example of experimental data provided by the conductivity tracer technique is reported in figure 13.

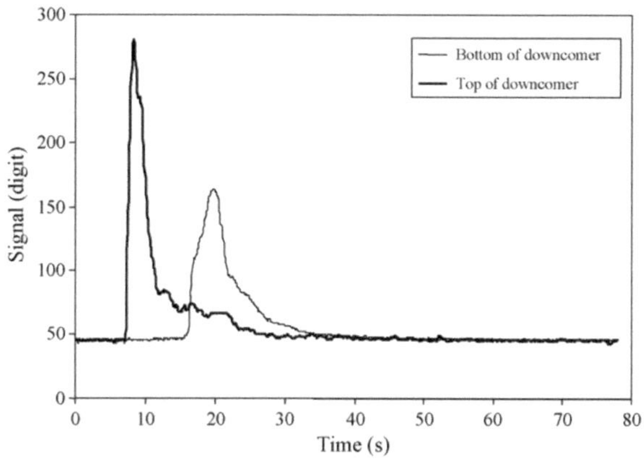

Fig. 13. Example of experimental data from conductivity tracer experiments in the downcomer section when current density j=21.4 mA/cm^2.

Experiments are carried out using a red dye solution consisting of a mixture of 2-naphthoic acid and 2-naphtol with a total concentration C_0 = 20 mg/L. Synthetic solutions were prepared by dissolving the dye in tap water. Solution conductivity and pH are measured using a CD810 conductimeter (Radiometer Analytical, France) and a Profil Line pH197i pHmeter (WTW, Germany).

Dye concentration is estimated from its absorbance characteristics in the UV-Vis range at maximum wavelength A_{450} (λ_{max}=450 nm) using a UV-Vis spectrophotometer (Pye Unicam, SP8-400, UK).

The objective of complete flotation may be achieved only if hydrodynamic shear forces remain weak in the riser to avoid floc break-up and in the separator to limit break-up and erosion, which means low U_{Ld} values.

At constant electrode geometry, the solution consists of an adequate selection of U_{Ld} resulting from a compromise between mixing and floc stability. This can be obtained first by optimizing the axial position of the electrodes using both the conductivity tracer technique for U_{Ld} estimation and turbidity measurements to estimate the amount of dispersed Al particles in the downcomer. Experimental results show that no liquid overall circulation can be detected when the electrodes were placed in the upper part of riser, for H_1 approximately higher than 60 cm. For 7 cm <H_1 <60 cm, U_{Ld} decreases when H_1 increases. Overall liquid velocities in the downcomer for two axial positions are reported in figure 14 at various current densities.

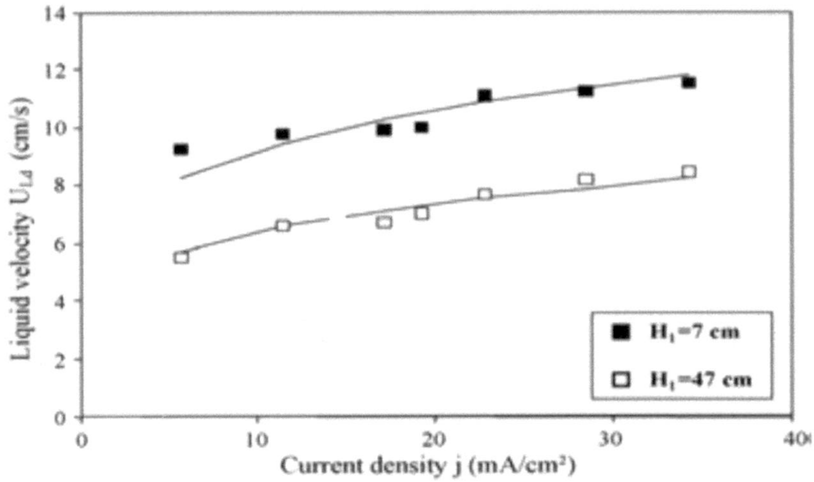

Fig. 14. Influence of the axial position of the electrodes (H_1) and current density (j) on the overall liquid recirculation U_{Ld} (h=14 cm; initial pH: 8.3; initial conductivity: κ=2.4 mS/cm).

Two-parameter model is used to fit the data. The results are given in equation (17):

$$U_{Ld} = 5.8 \cdot \left(\frac{h_D}{h_{D\max}} \right) \cdot j^{0.20} \text{ (cm/s)} \tag{17}$$

in which h_{Dmax} is the maximum dispersion height corresponding to H_1=7cm. This equation confirms the key role of the axial position of the electrodes on reactor hydrodynamics and mixing properties, and the weaker influence of current.

The corresponding turbidity data based on A_{450} without filtration is reported in figure 15 for an electrolysis time of 10 min and two axial positions of the electrodes. In all cases, flocs occupy nearly 1 cm thickness at the free surface of the disengagement section and no settling is reported in the junction and in the separator. Figure 15 shows however that turbidity rose when H_1 =7cm for current density higher than 15 mA/cm^2. This corresponds to U_{Ld} values in the range 9–10 cm/s in figure 15. Conversely, complete flotation is always observed for H_1 = 47 cm, as A_{450} remains low, about 0.06. As a result, electrode position must be chosen in order to maintain U_{Ld} always lower than 9 cm/s, regardless of current density in the range of j studied (Fig. 15). Such a condition is achieved for H_1 = 47 cm, as shown in figure 15, which corresponds nearly to mid-height in the riser. However, H_1 = 47 cm is only a coarse approximation of the optimum electrode height, but the simple way to optimize mixing conditions does not consist in adjusting precisely H_1 because it is easier from a practical point of view to adjust the clear liquid height h. Indeed, h affects simultaneously h_D and h_{Dmax} in Eq. (17), but its range (2 cm < h < 14 cm in this work) is usually far smaller than H_1 (between 7 and 77 cm).

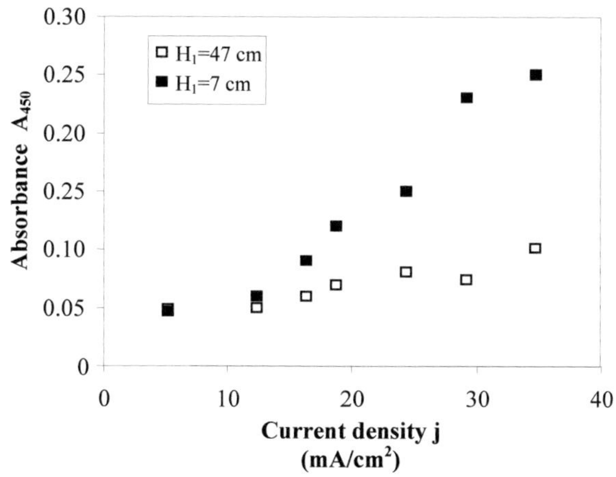

Fig. 15. Evolution of absorbance A_{450} of non-filtered samples after 10 min operation with 1 cm of floc thickness already formed as a function of superficial liquid velocity U_{Ld} in the downcomer (initial pH: 8.3, conductivity: κ=2.4 mS/cm).

3.5.3 Some results

The efficiency of electrocoagulation/electroflotation in removing colour from synthetic textile wastewater by using aluminum and iron electrodes in an external-loop airlift reactor is presented. Disperse, reactive and the mixture are used to determine the optimized parameters.

The real textile wastewater is then used using the optimized parameters. Three effluents were also used: disperse, reactive and the mixture. Energy is determined for each kind of dye.

3.5.3.1 Synthetic dye

For this synthetic dye, chemical oxygen demand (COD) is measured using the standard closed reflux colorimetric method. Initial COD was about 2500 mg O_2/L. Initial pH is varied between 5 and 10 using minute addition of 0.1 M H_2SO_4 or NaOH solutions. The conductivity κ (i.e. the ionic strength) of dye solutions is adjusted by sodium chloride addition in the range 1.0–29 mS/cm, which covers the range usually explored in the literature.

COD, color removal and turbidity efficiencies (Y_{COD}, Y_{COL} and Y_{ABS}) are expressed as percentage and defined as:

$$Y_{COD} = \frac{COD(t=0) - COD(t)}{COD(t=0)} \tag{18a}$$

$$Y_{COL} = \frac{A_{450}(t=0) - A_{450}(t)}{A_{450}(t=0)} \tag{18b}$$

Y_{COL} and Y_{ABS} are obtained using the same equation (Eq.18b), but Y_{COL} was based on absorbance measurements at 450 nm after filtration, while Y_{ABS} is measured without filtration/decantation. Y_{ABS} is used to analyze qualitatively the evolution of turbidity over time. This parameter shows whether the flocs flotate or are destroyed and driven by the liquid flow. The influence of the initial pH on COD and turbidity removals is illustrated in Figure 16 at constant current density and initial conductivity. An optimum is found for the initial pH, which is between 7.0 and 8.0, although it differs slightly between COD and color removal yields.

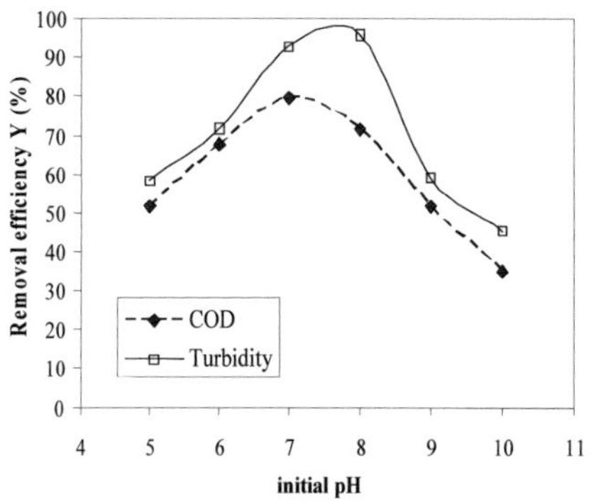

Fig. 16. Influence of the initial pH on COD removal and decolorization after 8 minutes operation (conductivity: κ=2.4 mS/cm, current density: j=28.5 mA/cm^2).

The influence of conductivity is illustrated in Figure 17 shows an increase of Y_{COD} and Y_{COL} with κ for the red dye between 2 and 28 mS/cm. Y_{COL} enhancement becomes however slight when κ is higher than 15 mS/cm.

The decrease in specific energy consumption E_{dye} due to the increase of conductivity is illustrated by Figure 18. This figure indicates that E_{dye} can be divided roughly by a factor 13 when conductivity is multiplied by a factor 12.

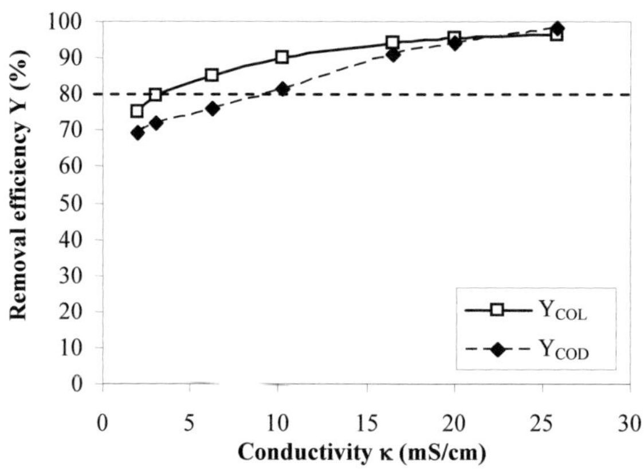

Fig. 17. Influence of conductivity κ on COD and color removal efficiencies (initial pH: 8.3; current density: j=28.6 mS/cm).

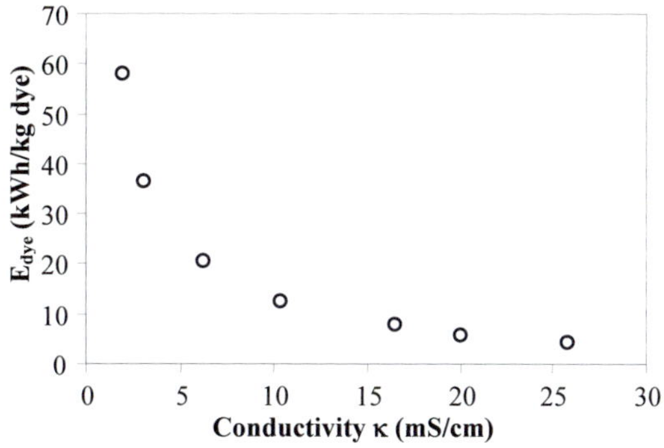

Fig. 18. Influence of conductivity κ on energy consumption E_{dye} at constant current density j and operation time (initial pH: 8.3).

3.5.3.2 Real textile dye

The real textile wastewater is then used. Three effluents are also used: disperse, reactive and the mixture. The color efficiency is between 70 and 90% and COD efficiency reached 78%. The specific electrical energy consumption per kg dye removed (E_{dye}) in optimal conditions for real effluent is calculated. 170 kWh/kg dye is required for a reactive dye, 120 kWh/kg dye for disperse and 50 kWh/kg dye for the mixture.

For disperse dye, the removal efficiency is better using aluminium electrodes, whereas, the iron electrodes show more efficiency for removing colour for reactive dye and mixed synthetic dye.

Figure 19 shows a photo representing a real effluent before EC treatment and 10 minutes after EC treatment.

Before EC treatment 10 minutes after EC treatment.

Fig. 19. Photo showing decolourization by EC in external loop airlift reactor of real textile dye.

3.5.3.3 Defluoridation of drinking water

An excess amount of fluoride anions in drinking water has been known to cause adverse effects on human health. To prevent these harmful consequences, especially problems resulting from fluorosis, the World Health Organization (WHO) fixed the maximum acceptable concentration of fluoride anions in drinking water to 1.5 mg/L (Essadki et al., 2009).

The defluoridation of drinking water by EC is studied in the same reactor. Current density values j between 2.8 and 17 mA/cm^2 are investigated, which corresponded to current ($I=j\cdot S$) in the range 0.5-3 A. Samples are filtered and the concentrations of the remaining fluoride content are determined in the solution by means of a combined selective fluoride electrode ISEC301F and a PhM240 ion-meter (Radiometer Analytical, France), using the addition of a TISAB II buffer solution to prevent interference from other ions.

Experiments are carried using an initial fluoride concentration $[F^-]_0$ between 10-20 mg/L by adding sodium fluoride NaF (Carlo Erba Réactifs, France). The efficiency of fluoride removal can be calculated as follows:

$$Y(\%) = 100\,\mathrm{x}\,\frac{\left[F^-\right]_0 \left[F^-\right]}{\left[F^-\right]_0} \tag{19}$$

$[F^-]_0$: the initial fluoride concentration .

$[F^-]$: the remaining concentration of fluoride .

Figure 20 shows the effect of the current intensity on the evolution of the fluoride concentration. For I = 0.5A corresponding to a current density of 2.85mA/cm^2, fluoride concentration reaches only 4.5 mg/L for an electrolysis time of 30 min. Conversely, for exceeding 1A, i.e. for a current density higher than 5.7mA/cm^2, one converges towards a concentration of 1mg/L and more rapidly as current density is increased. This confirms that defluoridation can be achieved at low current density. The relatively low efficiency observed at 0.5A can be attributed to the weak charge loading produced in this case, 0.47 F/m^3. As expected, the efficiency of EC depends on the amount of coagulant produced in situ.

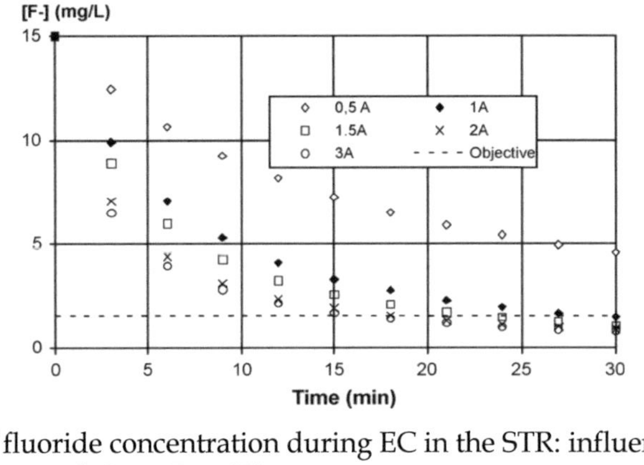

Fig. 20. Evolution of fluoride concentration during EC in the STR: influence of current intensity (initial pH 7.4 and C_0 = 15mg/L).

3.5.4 External loop airlift reactor as a continuous reactor

The reactor is operated in a continuous flow of liquid. The inlet volumetric liquid flow-rate Q_L varied between 0.1 and 2 L/min.

A study of the residence time distribution (RTD) analysis of liquid phase has been performed. The liquid RTD is determined by means of the tracer response technique. An approximated δ-Dirac pulse of tracer solution (NaCl) is injected into the reactors at a certain time (t = 0) and the outlet signal is detected by conductivity probe and recorded by the acquisition system. The tracer is injected as quickly as possible to obtain as closely as practical a δ-pulse of tracer at the inlet.

The syringe (S_1) was introduced in the drain that is open to the flow and representing the inlet of the whole reactor. The probe conductivity is placed at the exit position of the reactor (Fig.12). Examples of RTD measurements are shown in in figure 21. Two kinds of signal are observed in this figure. One showed two peaks for the case of Q_L = 0.36 L/min, I = 6 A, H_1 = 7cm and the other showed one peak for the case of Q_L = 0.73 L/min, I = 1 A, H_1 = 47 cm.

The main flow (Q_L) is divided into two flows: one exit directly the rector by crossing the junction and the other crosses the riser, the separator zone and the downcomer to exit. The percentage of flow that quit the reactor without reacting increased when the main flow increased and the current intensity decreased. The experiments confirm also that the liquid crosses the reactor without achieving loops in the case of the continuous flow.

So to support the reaction during the electrocoagulation, it is necessary to amplify the current intensity and to decrease the inlet flow.

Interesting results are also obtained:

- The superficial liquid velocity (U_{Ld}) at the downcomer, decreases when the volume inlet flow increases ($0 < Q_L < 2$ L/min).

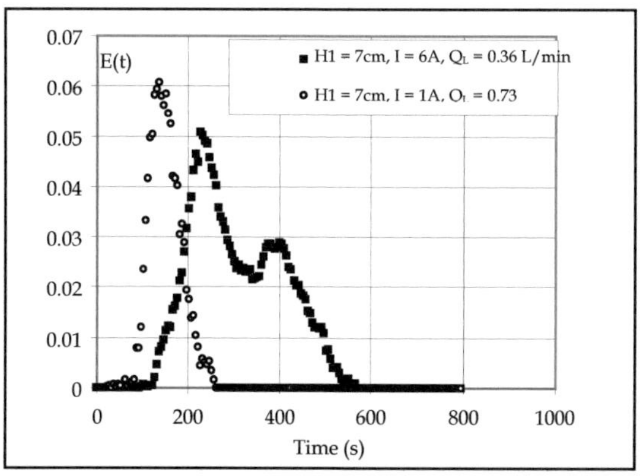

Fig. 21. E-curve as a global RTD in External-loop airlift reactor for (Q_L = 0.36 L/min, I = 6 A, H1 = 7cm: 2 peaks) and (Q_L = 0.73 L/min , H1 = 7 cm, I = 1 A, 47 cm: 1 peak).

4. Conclusion

External-loop airlift reactors have been shown to be versatile tools to carry out EC with complete flotation, using only electrochemically generated H_2 bubbles to achieve an overall liquid circulation and good mixing conditions. Consequently, the use of mechanical agitation, pumping or compressed air was not necessary. This could not be achieved in other kinds of conventional gas-liquid contacting devices than airlift reactors. External-loop devices are particularly adapted because they offer specific designs for the disengagement section that allow large distance between riser and downcomer. This improves flotation by minimizing the recirculation of aluminum or iron particles in the downcomer. These results were obtained by the adequate selection of the axial position of the electrodes (H_1) and the liquid height in the separator section (h) in order to avoid floc break-up in the riser and floc erosion at the free surface. A limiting value of the liquid velocity in the downcomer was defined, while U_{Ld} was correlated to dispersion height h_D and current density j (Equation 17). These can be used at constant j and A_r/A_d ratio for scale-up purpose.

To increase the efficiency of EC in a continuous reactor, the mean residence time should be increased. The experiments showed that this effect is reached in the case of a relatively high value of current density and weak value of the inlet flow-rate. This study highlighted the hydrodynamic aspect of the flow in the external airlift reactor functioning as a batch and continuous reactor. The design of this kind of reactor should be improved to allow the reactant to follow the compartment in which the reaction takes place (riser).

5. References

Chen G., (2004), Sep. Purif. Technol. 38, 11–41.

Chisti Y., (1989), *Airlift Bioreactors*, Elsevier, London.

Essadki H., Nikov I., Delmas H., (1997), *Electrochemical Probe for Bubble Size Prediction in a Bubble Column*, Experimental Thermal and Fluid Science, 243-250.

Essadki A.H., Gourich B., Vial Ch., Delmas H., Bennajah M., (2009), *Defluoridation of drinking water by electrocoagulation/electroflotation in a stirred reactor with a comparative performance to an external-loop airlift reactor,* Journal of Hazardous Materials 168, 1325-133.

Levart E., and Schumann D., (1974), *Analyse du Transport Transitoire sur un Disque Tournant en Régime Hydrodynamique Laminaire et Permanent*, Int. J. Heat Mass Transfer 17, 555–566.

L.P.Reiss and T.I.Hanratty , *An experimental study of the unsteady nature of the viscous sublayer*, AIChE J, 9, (1963), 154-160.

4

Electrochemical Cells with the Liquid Electrolyte in the Study of Semiconductor, Metallic and Oxide Systems

Valery Vassiliev[1] and Weiping Gong[2]
[1]*Chemistry Department, Lomonosov University, Moscow,*
[2]*Institute of Huizhou,*
[1]*Russia*
[2]*China*

1. Introduction

The first publications devoted to a study of the thermodynamic properties of metallic alloys, using electrochemical cells (EMF method) was known since 1936 year (Strikler & Seltz, 1936). This was the groundwork for all the next studies.

Co-workers from Moscow State University (Geyderih et al., 1969) have considered some questions about this experimental method.

A new attempt to generalize the electrochemical methods on thermodynamic studies of metallic systems was made again in the book (Moratchevsky, 1987). The general aspects of the thermodynamics of nonstoichiometric compounds were presented there and the methods for experimental studies of the thermodynamic properties of molten metal and salt systems were discussed.

The different types of electrochemical cells with solid and liquid electrolytes and dynamic EMF methods were examined in the recent book (Moratchevsky et al., 2003). A separate chapter of this book deals with methods of treatment and presentation of experimental data. In recent decades the important step of qualitative development of EMF method had been made and it was not considered in this book.

In the present chapter we focus on those experimental techniques that help to increase significantly the experimental result precision.

The knowledge of thermodynamic properties and phase diagrams of binary, ternary and multi component systems is necessary for solving materials science problems and for designing new products and technologies fitted to actual needs. A rational study of equilibria among phases and of the given system thermodynamic properties not only leads to the discovery of unknown phases but also to the determination of phase thermodynamic stability, to homogeneity domain boundaries, and finally to the elaboration of analytical description of the system by using thermodynamic models which are based on the dependence of phase Gibbs energies on such parameters as temperature, concentration and pressure.

Experimental studies are the primary information sources for thermodynamic properties and phase diagrams of all systems. The method of electromotive force (EMF) is one of the most important methods of the physicochemical analysis. One peculiarity of the EMF is its proportionality to the chemical potential:

$\Delta\mu_i = n \cdot F \cdot E$ of one of the system components,

where n is a charge of the ion responsible for the potential,

F=96485.34 C/mol is the constant of Faraday,

E is electromotive force.

Improving the accuracy and reproducibility of measurements leads to the increase of the quality and quantity of information about the system. Values of $\Delta\mu_i$ (T, x_i) versus temperature (T) and atomic fraction (x_i) obtained with uncertainties of ±500 J/mol (especially in a narrow temperature range) give only rough estimates of partial entropy and enthalpy of the components.

An accuracy improvement in determining of the chemical potential $(\Delta\mu_i(T, x_i)$ versus temperature (T) and atomic fraction (x_i) from ±10 to 50 J/mol not only leads to the various thermodynamic properties of the system (partial entropies ($\Delta_f\bar{S}_i$) and enthalpies ($\Delta_f\bar{H}_i$) of components, phase enthalpies of transformation ($\Delta_f\bar{H}_{tr}$), partial enthalpies at infinite dilution ($\Delta_f\bar{H}_i^{\infty}$), thermal capacities (C_p), but also gives a possibility to study the phase diagram in detail (liquidus and solidus, miscibility gaps, invariant points, stoichiometry deviations, ordering, etc...)

2. Principals of the method

We perform measurements of the EMF or chemical potential across the electrochemical cells with liquid electrolyte, such as:

$$(-)\ A \mid A^{(n+)} \text{ in the electrolyte} \mid A_xB_{(1-x)}\ (+) \qquad (I)$$

x represents a molar fraction of component A in liquid or solid alloy $A_xB_{(1-x)}$. The component A (usually a pure metal) is the negative electrode, the alloy $A_xB_{(1-x)}$ where the component B is more noble than A, is the positive electrode.

The chemical potential of pure metal A $\mu_A^{(A)}$, is always higher than its chemical potential in the alloy $A_xB_{(1-x)}$ or ($\mu_A^{(A)}$)> $\mu_A^{(AxB(1-x))}$)

$$\Delta\mu_A = RT\, ln(a_A^{(AxB(1-x))} / a_A^{(A)}) = RT\, ln\, a''/a' \qquad (1)$$

If $a_A^{(A)} = 1$, we can simplify the equation (1) and we have the equation (2).

$$\Delta\mu_A = RT\, ln(a_A^{(AxB(1-x))}) \qquad (2)$$

$\mu_A^{(A)}$ is the change of the chemical potential of component A in its transition from a pure metal A into an alloy $A_xB_{(1-x)}$ in reference conditions. The measurement of EMF as a function of temperature leads to partial thermodynamic functions.

$$\Delta\mu_A = \Delta\bar{H}_A - T\cdot\Delta\bar{S}_A \quad (3)$$

$$\Delta\mu_A = -n\cdot F\cdot E \quad (4)$$

$$\Delta\bar{H}_A = -n\cdot F\cdot a \quad (5)$$

$$\Delta\bar{S}_A = n\cdot F\cdot b \quad (6)$$

$$\Delta\bar{S}_A = -(\partial\Delta\mu_A / \partial T)_p = nF(\partial E / \partial T)_P \quad (7)$$

where b is $tg(\alpha)$ (see Fig.1)

$$\Delta\bar{H}_A = \Delta\mu_A + T\Delta\bar{S}_A = nF\left[T(\partial E / \partial T)_p - E\right] \quad (8)$$

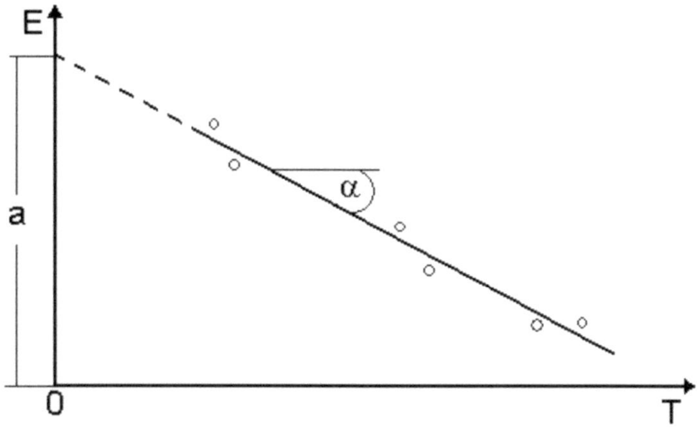

Fig. 1. Graphical relation of measured values E(T) with partials thermodynamic functions a = - $\Delta\bar{H}_A$ and b = $\Delta\bar{S}_A$.

The integral thermodynamic functions can be calculated with help of Gibbs-Duhem or Gibbs-Duhem- Margules equations:

Gibbs-Duhem equation for the two-component system AB is:

$$\Delta\Phi = x_1\Delta\bar{\Phi}_1 + (1-x_1)\Delta\bar{\Phi}_2 \quad (9)$$

The Margules equation is generalized by Gibbs-Duhem equation:

$$\Delta\Phi = (1-x_i)\int_0^{x_1/1-x_1} \Delta\bar{\Phi}_A d\left(\frac{x_1}{1-x_1}\right) \quad (10)$$

The partial and integral thermodynamic values are presented by the terms $\Delta\bar{\Phi}_i$ and $\Delta\Phi$, respectively.

If we can ignore the homogeneity regions of the intermediate phases of certain binary phase diagram, we can calculate the integral properties of these phases by combining equations.

For example, the system lutetium-indium in the region 0-50 at % Lu has two intermediate phases of 1:2.5 and 1:1. Calculation of the thermodynamic integral properties is easily carried out, if we have in our possession, the partial thermodynamic properties of these phases:

Lu+ 2.5In → $LuIn_{2.5}$ $\Delta_f\bar{\Phi}'_{Lu}$ is partial thermodynamic value of formation $LuIn_{2.5}$ for one mole of lutetium

Lu +0.667 LuIn 2.5 → $\Delta_f\bar{\Phi}''_{Lu}$ is partial thermodynamic value of formation LuIn for one mole of lutetium.

Combining these equations permits to determine the integral properties of LuIn phase.

$$0.667 \mid \text{Lu} + 2.5\ \text{In} \rightarrow \text{LiIn2.5}\ \Delta_f\bar{\Phi}'_{Lu} \equiv \Delta\Phi(LuIn_{2.5})$$

$$\mid \text{Lu}+0.667\text{In} \rightarrow 1.667\ \text{LuIn}\ \Delta_f\bar{\Phi}''_{Lu}$$

$$1.667\text{Lu}+ 1.667\text{In} \rightarrow 1.667\ \text{LuIn}\ \Delta_f\bar{\Phi}'_{Lu} + \Delta_f\bar{\Phi}''_{Lu}$$

The integral functions of Lu_2In_5 and LuIn phases are equal for one mole-atom:

$$\Delta_f\Phi'(Lu_2In_5) = 2\Delta\bar{\Phi}'_{Lu} / 7$$

$$\Delta_f\Phi''(LuIn) = (0.667\Delta_f\bar{\Phi}'_{Lu} + \Delta_f\bar{\Phi}''_{Lu}) / 1.667 \cdot 2$$

3. Main experimental steps

Here are the main experimental steps of the EMF method:

- synthesis of alloys and preparation of the electrodes,
- dehydration of salt mixture and preparation of the electrolyte,
- different types of electrochemical cell and its assembly.

3.1 Synthesis of alloys and preparation of the electrodes

The alloy preparation techniques are different and depend on the work objectives. A study of the systems in the liquid state does not require such special treatment as annealing, while a study of alloys in solid state requires annealing for several days. And then it is necessary to avoid working with alloy ingots. As a rule we use the pellets fabricated from powedered alloys. For this reason the dismountable mold is best suited. For this reason it is necessary to use the dismountable mold (Fig.2). The internal party of this mold consists of dismountable block cylinder from high strength tempered steel of four sections (Fig.3). To fabricate a pellet we introduce the tungsten wire (diameter 0.5 mm) throughout special groove of support. Then we untroduce the powedered alloys and press the pellet by punch, using the hydrolic press. The contact between the pellet and the tungsten wire must be well secured.

If the alloy is sufficiently plastic, its can be drilled with a hole a little larger than the diameter the conductor wire and compress with the vise. Sometimes, the reference electrodes or measuring one are extremely fragile, and it is impossible to ensure good contact between the wire and the sample. In this case, we prepare the mechanic mixture of powder studied alloys with powder or filling of a more plastic inert metal and then we press the pellets. This added inert plastic metal serves as a matrix of the studied material. For this reason tantalum fillings are mixed in proportion 1:1. We used this procedure in forming the reference manganese electrodes (Vassiliev et al., 1993). It is better to work with the pellets (Vassiliev et al., 1968). Sometimes using the samples in the form of ingot leads to distortion of the measurement results of the EMF, especially if these measurements are made lower of solidus temperature (Terpilowski et al., 1965). If the study of alloys is carried out over a wide temperature range, from liquid homogeneous state to mixed solid-liquid and then to solid states, the EMF measurements are reliable if they are carried out at the complete solide state (of the studied phases) but the temperature should not be more than 100-150 K below the solidus line.

Fig. 2.The four-part dismountable mold is shown in this picture: 1-punch, 2-four-section dismountable block cylinder from high strength tempered steel, 3 - constricting clamp with twotightening bolts, 4 – support with groove for tungsten wire.

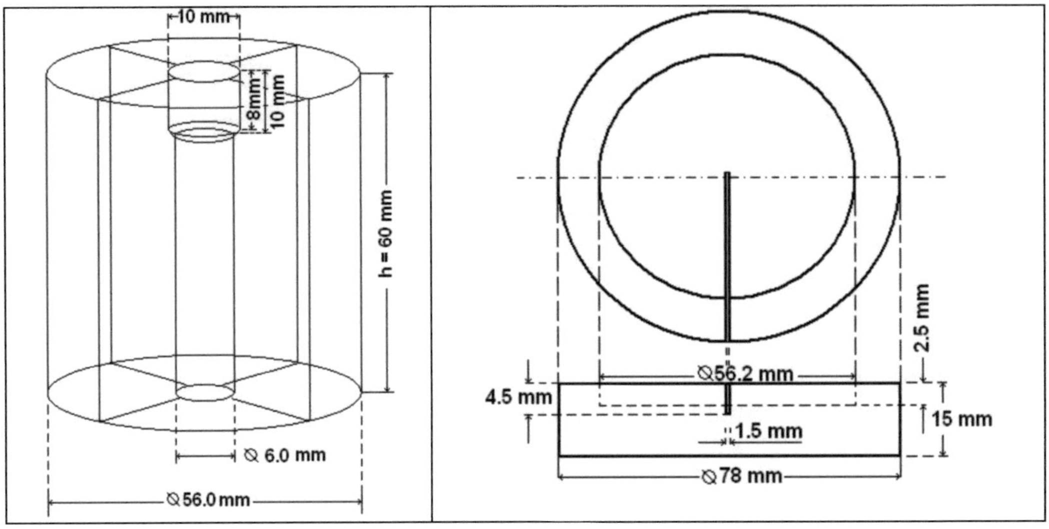

Fig. 3. Internal four-section dismountable block cylinder of molder (2) and support with groove (4).

3.2 Salt mixture dehydratation and preparation of the electrolyte

It is necessary to pay special attention to the preparation of salts of the electrolyte. They must be dried very carefully. It applies especially to Li, Ca, Zn and Al halides. These salts are extremely hygroscopic, and their melting without special dehydration leads to the formation oxyhalogenides which presence must be avoided. The used electrolyte in the liquid state must be completely transparent and shows no disorder or heterogeneity. The ingots of electrolyte can be stored in sealed Pyrex ampoules. Upon introduction of the electrolyte in a cell, contact with air should be minimal (no more than 10 seconds). Dehydration salts (eg LiCl + RbCl) must pass under pumping with a slowly increasing heating for 5 days to prevent formation of hydroxides. Then the dehydrated salt mixture is transferred to a silica beaker preheated to 500°C in an electrical furnace. To remove oxychlorides, the molten salts are treated with dry hydrochloride gas (HCl). Hydrogen chloride can easily be synthesized by reacting potassium chloride (KCl) or sodium chloride (NaCl) by reaction of concentrated sulfuric acid (H_2SO_4) (Fig.4):

$$KCl + H_2SO_4 \rightarrow KHSO_4 + HCl\uparrow$$

$$KHSO_4 + H_2SO_4 \rightarrow K_2SO_4 + HCl\uparrow \text{ (with gentle heating).}$$

Hydrogen chloride must be dried using zeolites loaded into a U-tube. The gas is bubbled through the melt until there are no suspended particles (about 1h).The melt prepared in this way is poured into Pyrex ampoules with a neck, which are sealed then. The electrolyte can be stored indefinitely in sealed Pyrex ampoules and may be used as required. In practice, it is possible to use the different eutectic mixtures of halides (See Table 1).

The study of some systems such as chalcogenides of zinc, cadmium, mercury, thallium, bismuth, etc. (binary or multicomponent) can be achieved in "low temperature" cell. In this case the calcium chloride ($CaCl_2$) is dissolved in glycerol between 40 and 180°C. Calcium

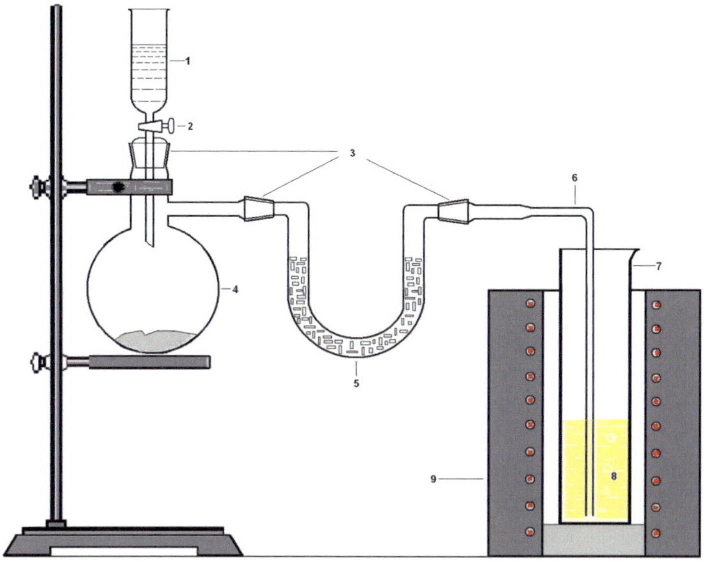

Fig. 4. Device for obteing of gazeous hydrogen cloride: 1- Container of concentrated sulfuric acide, 2- Tap, 3- Glass grandings, 4- Vial for potassium cloride, 5- U-shaped tube for zeolites, 6- Quartz tube, 7- Quartz beaker with a spout, 8- Molten electrolyte, 9- Furnace.

Mixtures of salts	**T_m°C**
55.5 NaI - 44.5 KI	585
38.0 NaCl - 62.0 $CaCl_2$	500
32.9 LiCl - 34.8 NaCl - 32.3 KCl	357
46.0 LiCl - 54.0 KCl	352
48.LiBr – 52.0 KBr	348
29.75 LiCl – 64.77 KCl - 5.48 $CaCl_2$	320
30.3 LiCl – 69.7 RbCl	312
28.97 LiCl – 4.42 NaCl – 66.61 CsCl	299
45.0 LiBr – 55.0 RbBr	270
52.7 LiI – 47.3 KI	260
54.0 CH_3COOH – 46.0 CH_3COONa	233
70.0 $ZnCl_2$ – 18 KCl – 12 NaCl	206
69.5 $AlCl_3$ – 30.5 NaCl	152
Solution of $CaCl_2$ in the glycerole	*40 - 180*

Table 1. Eutectic mixtures of halides (weight percent) used for preparation of electrolyte.

chloride (dried under vacuum at temperature of 200°C) is used as a water absorbent for glycerol and also for creating an ionic conductivity of the electrolyte with chloride (ACl_x) (Vassiliev et al., 1968).

3.3 Different types of electrochemical cells and their assembly

There are many examples of construction of electrochemical cells, proposed in the literature (Morachevski et al. 2003).

We have tested the different types of alloys with three types of cells from 'Pyrex' glass, which had been made by ourselves. One of three cell works well up to 200°C (Vassiliev et al., 1968, 1971) using the salts solution of glicerine as electrolyte. We can also use the same types of cells at the high temperature to 1000°C (Vassiliev et al., 1998a) if we use the refractory material as quartz or alumina glasses. The other two types of cell are operated at high temperatures to softening up 'Pyrex' glass (800-900 K) (Vassiliev et al., 1980, 1993, 1995, 1998b, 2001). When we experiment with a liquid electrolyte, we can use the various electrochemical cells. One of them represents a double H-shaped vessel is suspended on a central tube of 6-8 mm in diameter, fitted with hooks, and which also is served as a cover for the thermocouple. This construction is incorporated inside a protective glass cylinder which is equipped with ground-in cap and two vacuum valves on the sides. This valves permit to control the vacuum and the pression of inert gas inside of the cell. The tungsten current leads with electrodes are soldered in inlet tubes 8 mm in diameter. Height and diameter of the protective cylinder depend on the internal diameter and depth of a using electric furnace. Described construction of the cell is convenient for working with glycerine electrolyte (Pyrex material) and for salt melts (quartz material) (See Fig.5).

Figure 6 shows a scheme of the isothermal Pyrex cell. The lower part of the cell (below the dashed line) is 54 - 60 mm in diameter and about 90 mm in height. The tungsten current leads and the electrodes attached to them are soldered in inlet tubes 8 mm in diameter. The bottom of the cell has cruciblelike holes, which are enable to study both solid and liquid alloys, with no risk of accidental mixing. A calibrated Pt/Pt–10% Rh thermocouple is introduced into the casing, which is soldered in the centre of the cell at the level of the electrodes. Such cells can operate indefinitely between the solidification temperature of the eutectic melt and the onset of softening of Pyrex glass (about 900 K). The offtake of the cell is about 400 mm in length and 25 mm in diameter is fitted with a ground-glass joint. It is served as a container for the electrolyte. The time needed to withdraw the ingot from the storage ampoule, to introduce into the container, and to connect it to the vacuum system does not exceed ten seconds. At the first we pump the cell (10^{-3} to 10^{-4} Pa) for a day, then we flush with purified argon, and then the ingot is melted under dynamic vacuum using a portable gas torch. The melt drains down into the lower part of the cell, which is introduced into a microfurnace heated from 50 to 100°C above the melting point of the eutectic mixture. Next, the cell is sealed off at the neck under vacuum and transferred to a preheated working furnace. The electrochemical cell in running order is presented in Fig.7.

The third type of cell (Fig. 8) is a modification of the previous. The bottom of that cell is the same as the previous one. The difference concerns a technic of the electrolyte charging into the cell. For certain systems, such as chacogenides, it is necessary to avoid the vacuum heating of the bottom part of cell that causes an evaporation of volatile metals such as Hg (Vassiliev et al., 1990) or chalcogens (Vassiliev et al., 1980). To put the electrolyte in the vessel we should proceed in this way. An ingot of electrolyte, sealed in a Pyrex ampoule, and a massive porcelain mortar were warmed previously in an oven to 200°C. The ingot was removed gently from the ampoule with a special knife and a Pyrex glass stick as follows. One end of the stick was heated up to temperature of softening with help of the torch. After we have applied this part of stick on the stripe traced with a knife on the ampoule. The ampoule was broken easily and the ingot became free. Then the ingot was grounded into small pieces in a mortar and as soon as possible the pieces were loaded in a small offtake flask and it was connected with cell and vacuum.This flask was heated gradually in a microfurnace under vacuum (0.1 Pa) up to 200 ° C for 24 hours. Then the cell was rinsed by purified argon and the electrolyte was poured out at the bottom by turn of the flask. Vacuum-sealed cells can be preserved for a long time and can be used reasonably. This cells permit to obtain the reproducible results of measurements during many months.

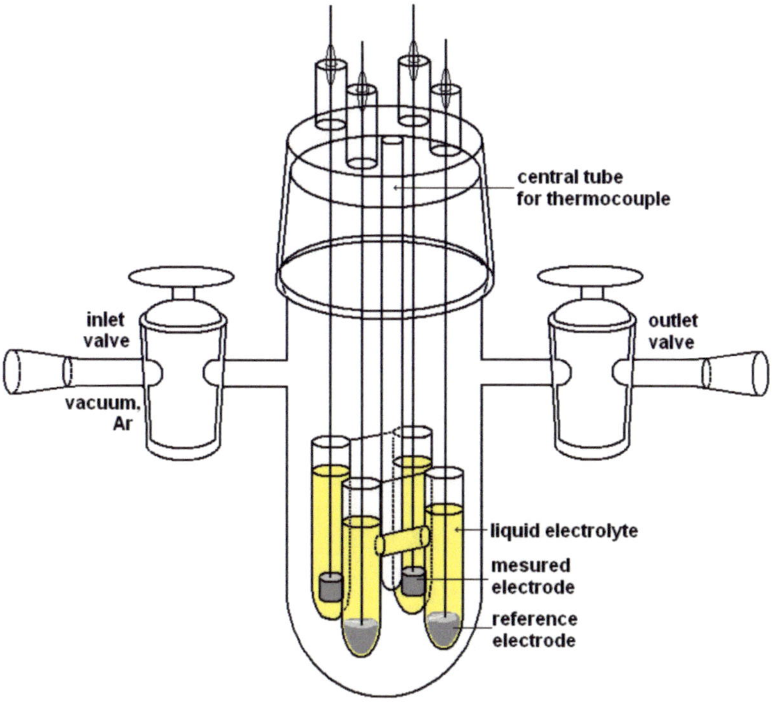

Fig. 5. Cell with control of vacuum and inert gas pression inside.

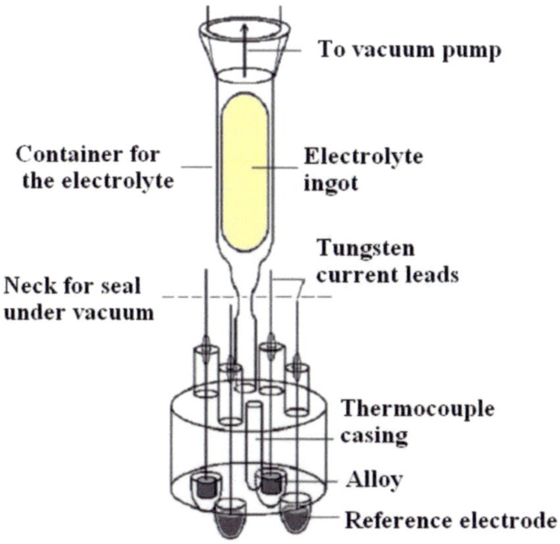

Fig. 6.Vacuum isothermic cell (firste variant).

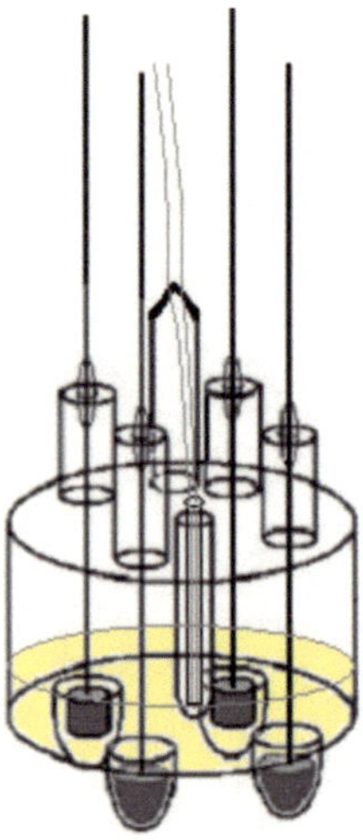

Fig. 7. Electrochemical cell in running order.

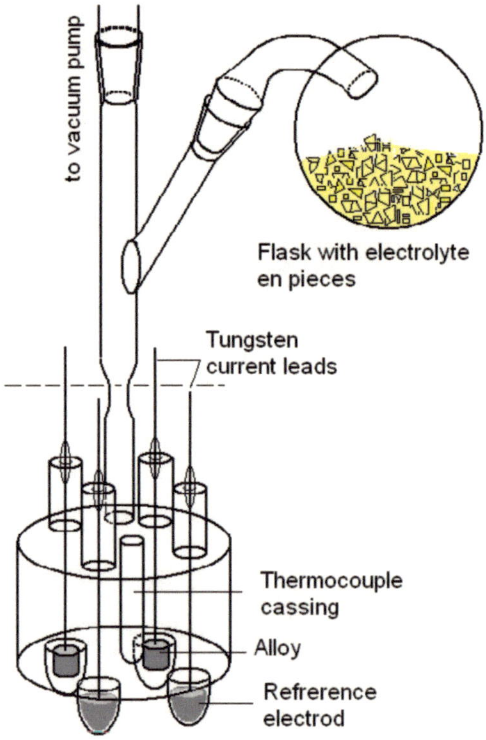

Fig. 8.Vacuum isothermic cell (second variant).

3.4 Cell with diaphragm

In some cases the decrease in activity of metal A does not lead to expected results. This concernes to the metals such as titanium, zirconium, hafnium, uranium, and beryllium ...

The ions of these metals have simultaneously two different charges in the electrolyte. Then we have an average charge of the ion-forming potential:

$$n_A = (mC'_m + nC'_n)/(C'_m+C'_n), \tag{11}$$

where C'_m and C'_n are the concentrations of ions A with charges m and n in equilibrium with the electrode $A_xB_{(1-x)}$. However we can prevent this transfer by separating the electrodes by a diaphragm. Fig. 9 gives an idea of this arrangement. The reference electrodes and electrode comprising of the alloy are studied in the same vessel. These electrodes are separated from each other by the tubes with capillaries that are closed with asbestos plug. In this case using asbestos must be very pure (Shourov, 1974,1984).

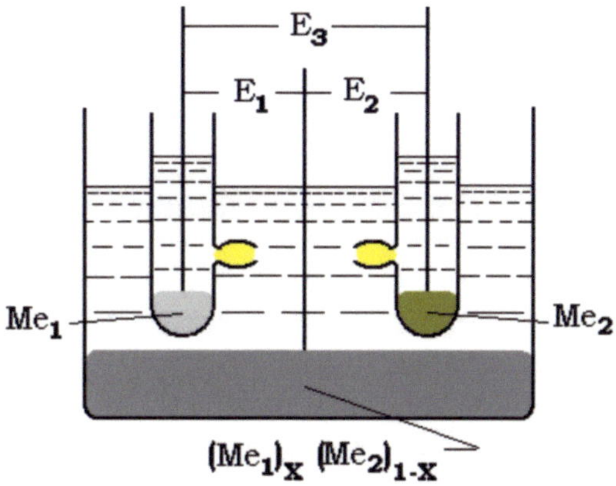

Fig. 9. Scheme of electrochemical cell with diaphragms.

$$\longleftarrow\text{-----------------------------}E_3\text{-----------------------------}\longrightarrow$$

$$(-)Me_1 \Bigg\vert \begin{matrix}\left[Me_1^{m+}\right]^0\\ \left[Me_1^{n+}\right]^0\\ +solvent\end{matrix} \Bigg\Vert \begin{matrix}\left[Me_1^{m+}\right]'\left[Me_2^{z+}\right]\\ \left[Me_1^{n+}\right]'\left[Me_2^{p+}\right]\\ +solvent\end{matrix} \Bigg\vert Me_1Me_2 \Bigg\vert \begin{matrix}\left[Me_1^{m+}\right]'\left[Me_2^{z+}\right]\\ \left[Me_1^{n+}\right]'\left[Me_2^{p+}\right]\\ +solvent\end{matrix} \Bigg\Vert \begin{matrix}\left[Me_2^{z+}\right]^0\\ \left[Me_2^{p+}\right]^0\\ +solvent\end{matrix} \Bigg\vert Me_2(+)$$

$$\longleftarrow\text{-----------}E_1\text{-----------}\longrightarrow \qquad \longleftarrow\text{-----------}E_2\text{-----------}\longrightarrow$$

m+ }
} charges of superior ion
z+ }

charges of ions m, z > n, p

n+ }
} charges of lower ion
p+ }

The solvent is mixture of molten salts ACl+BCl.

M1 is more electronegative than Me2 in a series of electrode potentials.

$$\left[Me_1^{m+}\right]^0, \left[Me_1^{n+}\right]^0, \left[Me_2^{z+}\right]^0, \left[Me_2^{p+}\right]^0$$

- equilibrium concentration in the ion mole fraction

$$E_1 = \varphi_{all.} - \varphi_{Me1}\,;\ E_2 = \varphi_{Me2} - \varphi_{all.}$$

One compare $E_1 + E_2$ with E_3 ($E_1 + E_2 = E_3$) to verify the accuracy of the electromotive forces of the chain.

4. Required apparatus and materials

In this paragraph we cite the main required apparatus and materials:

Simple vacuum post is presented in Fig.10.

- Electric furnaces with precise control of temperature. To speed up the work it is desirable to use several furnaces.
- Microfurnaces. Microfurnace presents a porcelain tube with an inner diameter of 65-70 mm and a length of 100-120 mm. This microfurnace can be easily made in the laboratory.
- Laboratory transformers.
- Primary and secondary vacuum pumps or turbomolecular one.
- Gauge (Vacuummeter)
- Vacuum baker.
- Digital millivoltmeter at high impedance.
- Hand or automatic commutation.
- Computer in case of automatic measurement system.
- Manuel or automatic press.
- Thermostat for cold ends of the thermocouples or Dewor bottle.
- Analytic balance.
- Vacuum post with multiple outputs.
- Vacuum comb for making of alloys.
- Hard steel mold of the four sections.
- Pyrex cells. In perspective, using the intermediate glass between Pyrex and quartz is possible for applying the electrochemical cells up to 1200-1300K. The coefficient of expansion of this type of glass must be as closer to the tungsten.
- Electrode leak detector
- Pt/Pt–10% Rh thermocouples.
- Oxygen and propane cylinders.
- Portable gas torch.
- Tungsten wire 0.5 mm. Simple vacuum post is presented in Fig.10.

Trap with liquid nitrogen (Fig.11) serves to catch aggressive substances vapors and it can be used as the sorption pump. In the second case it is recommended to use the zeolytes annealed at 300°C.

Sorption column serves to purificate the argon against the traces of oxigen and moisture. Sorbtion is performed by the activated copper deposited on silicagel at 170°C (Brauer, 1954).

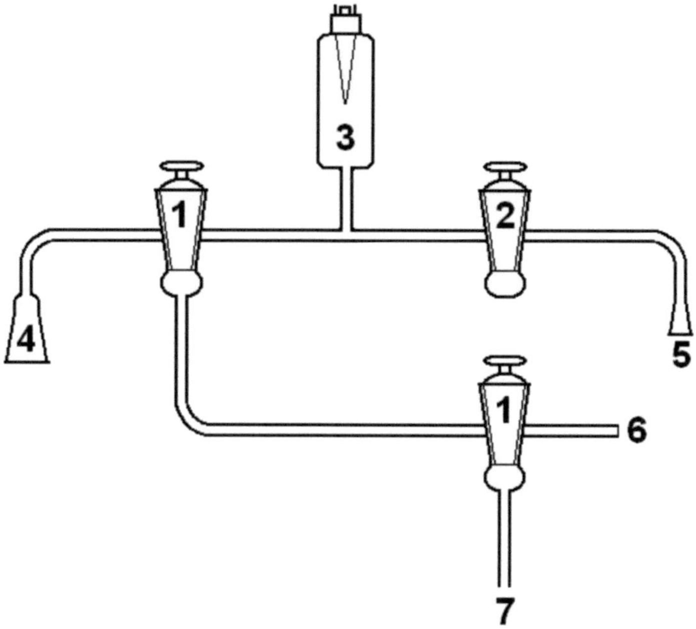

Fig. 10. Simple vacuum post: 1- three-way valve (T=valve), 2- two way valve, 3- gauge, 4- cell sleeve 5- ampule sleeve, 6- to air, 7- to pump.

Vacuum comb permits to accelerate the preparation of alloys (Fig.12). The block-scheme of the EMF measuring is presented in the Fig.13. The configuration of the vacuum post and related details may be changed by the experimenter (Fig.10).

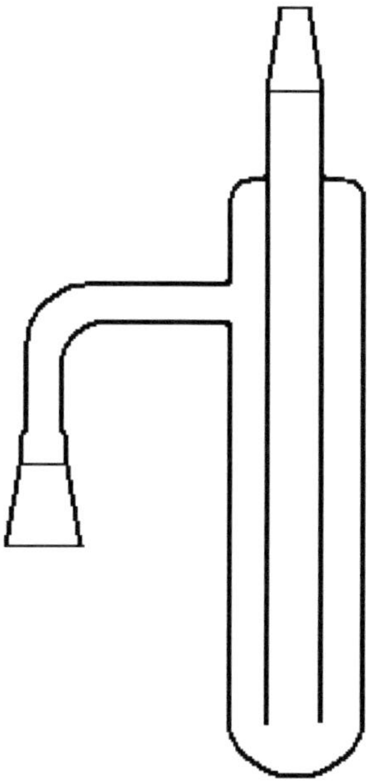

Fig. 11. Liquid-nitrogen trap.

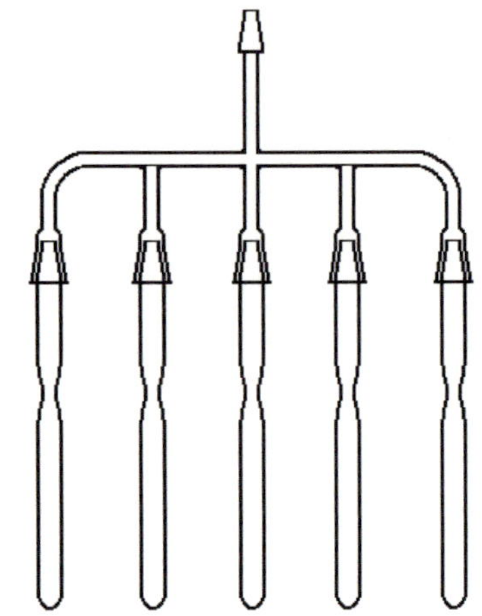

Fig. 12. Vacuum comb for alloy preparation.

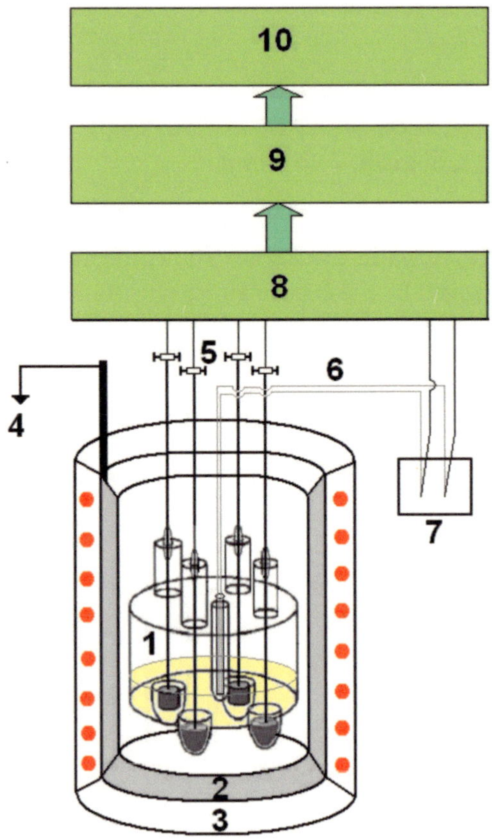

Fig. 13. Block-scheme of the measuring assembly: 1- cell, 2- stainless steel protector nozzle, 3- furnace, 4- ground, 5 – clamp of current leafs, 6-thermocouple, 7- thermostat for cold ends of the thermocouples, 8- hand or automatic commutation, 9- digital millivoltmeter at high impedance, 10- computer.

5. Methodological problems most frequently encountered in EMF method

On the difficult path of the experimental study we overcame the various methodological problems that had significantly and successfully improved the method of the EMF. Let us consider these main methodological problems of the EMF method, having a significant influence on the potentiometric measurements:

1. Exchange reaction between electrodes caused by a small difference of electrode potentials components of the alloy in the given electrolyte (Hladik, 1972).
2. Spontaneous reactions due to a large chemical potential difference between measured electrodes and reference.
3. Presence of multiple oxidation states of ions involved in redox reaction generating the EMF of the cell.
4. Liquation (or phase separation) of liquid alloys.
5. Evaporation of element and the temperature of the experiment.

5.1 Choice of different types of halides as the electrolyte

It is possible to use different types of halides as the electrolyte: chlorides, bromides and iodides. All these salts are hygroscopic and their treatment by the corresponding hydrogen halide (HCl, HBr, HI) is necessary to avoid water marks. If one type of halide unsuitable due to the exchange reaction, it can be replaced by another.

For the study of the Ni-Cu (Geyderih et al. 1969), measurements of EMF were made from two reference electrodes

$$(-)\ Ni(sol)\ |\ KCl + LiCl + NiCl_2\ |\ Ni_xCu_{(1-x)}\ (sol)\ (+)$$

$$(-)\ Cu(sol)\ |\ KCl + LiCl + CuCl\ |\ Cu_xNi_{(1-x)}\ (sol)\ (+)$$

In both cases there is a reduction of the second metal; the electrodes of pure metals are covered with crystals of the second metal. So, there are two oxidoreduction reactions:

$$NiCl_2 + 2\ Cu\ (\text{from alloy}) \rightarrow Ni + 2CuCl_2$$

$$2\ CuCl + Ni\ (\text{from alloy}) \rightarrow 2\ Cu + NiCl_2$$

$$(1/nB)\ B\ (\text{in } A_xB_{(1-x)}) + (1/nA)\ A^{n+} \rightarrow\ (1/nA)\ A + (1/nB)B^{n+} \qquad \text{(II)}$$

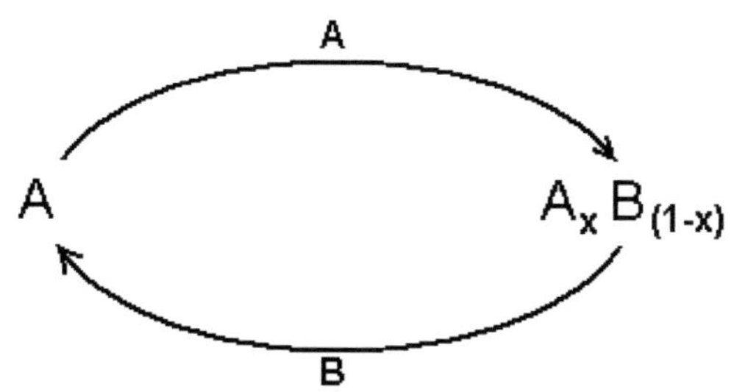

The replacement of the chloride on iodide electrolyte permits to determine the thermodynamics of the Cu-Ni system.

(-) Cu(sol.) | KI + NaI + CuI | Cu_xNiI_{-x} (sol.) (+) This experiment have made possible to formulate an empiric rule:

By working with electrochemical cells without separation of electrodes, enthalpy of formation *($\Delta_f H$)* salt of the metal B must not exceed 75% of enthalpy of formation of metal salt A (Geyderih et al. 1969).

As the equilibrium constant of the reaction Kp (II) has a finite value, there will be always an exchange reaction, even partial, between the salt ions A^{n+} and alloy $A_xB_{(1-x)}$. The authors have attempted to assess the relative error in determining the activity of the less noble metal by the EMF method based on constant exchange reaction (II) (Wagner& Werner, 1963). Simplified equation, admitting a relative error, expressed by the expression:

$$® = (n_B/n_A)\,(Y°/Y_A)\,(n_A/n_B)\,(1/x)\,(I + (n_A/n_B))\,(D°/D)^{1/2}.\,(V'_m/V''_m)\exp(-(n_B E°F)/RT) \quad (12)$$

where Y_A is the coefficient activity of component A, and Y° is the molar salt fraction of component A in the melt.

D ° and D are the diffusion coefficients of salt BX_n in the electrolyte and the component A in the alloy respectively,

E ° is difference of the reference potentials of the components,

V'_m and V''_m molar volumes of alloy and electrolyte.

In the formula (12) we can see:

1. that the error is maximum for the minimum concentration of component A,
2. that the higher is potential difference of the electrodes the smaller is the relative error,
3. that increasing of the salt concentration AX_n causes the increase of error ®, nevertheless, too small concentration may give a relative error greater than the reaction (II). The optimal concentration of salt in the molten electrolyte AX_n, according to Wagner, is 1 - 3%

In concordance to our experience, the concentration of AX_n salt in liquid electrolytes must not exceed 0.1%. In some cases, this concentration can be reduced if the salt AX_n is slightly soluble in the electrolyte.

If there is a problem of chemical stability or hygroscopicity AX_n of certain salts (e.g. as the indium chloride or zinc chloride) we can do forming potential without salt. Synthesis of indium monochloride (InCl) is carried out inside the cell by the interaction of hydrogen chloride absorbed by the electrolyte, with metallic indium:

$$2In + 2HCl\ (gas) = 2InCl + H_2.$$

Contact of indium monochloride (InCl) with moist air provokes the formation of indium ions with different valence states: $InCl+O_2+H_2O \rightarrow In(OH)Cl_2 + In(OH)_2Cl + In(OH)Cl + InOCl + \ldots$,

that leads to the exchange reaction between the electrodes of electrochemical cell.

5.2 Exchange reactions

The interaction between the electrodes via the electrolyte is one of the most important problems. In the electrolyte, the A and B elements are characterized by different potentials against a reference electrode (Hladik, 1972). We must choose an electrolyte which causes a potential difference of A and B as large as possible. The more the difference is between of the electrode potentials, the smaller is the exchange reaction between the component B and the A^{n+} ions in the melted electrolyte. Information about a possible exchange reaction between electrodes can be found from the electrode potentials for different halide melts (Hladik, 1972). This set of the chemical potentials characterizes activity of metals against each other for the system in study for giving electrolyte. The more metal is electronegative the more it is chemically active. In particular, in the set of the electrochemical potentials each metal replaces in an electrolyte all metals with lower potential. And, in turn, it will be replaced in the same electrolyte by metals with greater potential. So, the metal with the most negative potential, in the given electrolyte, replaces all those with more positive potentials. If we study the binary system Zn-Sb or In-Sb by the potentiometric method we do not see any problem of exchange reactions. So we can note that if the difference of the electrochemical potential reaches 0.4 V, the exchange reactions do not exist. And opposite, this problem appears for the systems Zn-In and In-Sn if the difference of the electrochemical potential is 0.19 V.

If we study the cells of the type:

$$(-)\ Zn \mid Zn^{2+} \text{ in an electrolyte} \mid Zn_xIn_{(1-x)}\ (+)$$

$$(-)\ In \mid In^{+} \text{ in an electrolyte} \mid In_xSn_{(1-x)}\ (+)$$

the exchange reactions take place easily when the concentration of the second element has reached 90%. The continuous drop of the EMF of the cells with alloys $x \leq 0.1$ is observed if the duration of the experiment is over several weeks (Vassiliev et al., 1998b; Mozer, 1972). The rate of the exchange reaction increases with increasing temperature especially in liquid systems. The speed of the exchange reaction depends on:

1. the difference of electrochemical potentials of the elements,
2. the difference in alloy composition,
3. the presence of metal ions of different charges ($A^{(n+)}$ et $A^{(m+)}$; m+ > n+) in the electrolyte,
4. the temperature in the cell.

This set of electrochemical potentials characterizes the chemical activity of metals against each other under consideration system and a given electrolyte. The more metal is electronegative the more this metal is chemically active. Especially, in the set of the electrochemical potentials each metal replaces in the electrolytes of all metals with inferior potential. In turn, it was replaced in the same electrolyte by metals with superior potential. If the data for the electrode potentials are incomplete, it is possible to judge about the relative chemical activity of two elements by comparing the Gibbs energies or enthalpy of formation of salts AX_n and BX_n (X being the corresponding a salt anion). See Table 4 bottom. We will consider some binary systems based on elements of Table 2 and 3.

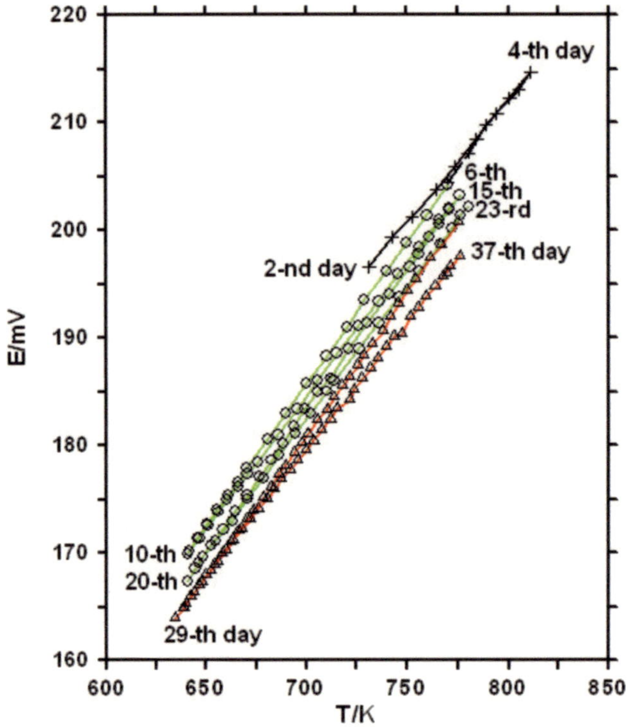

Fig. 14. Effect of exchange reaction between pure indium and its alloy $In_{0/05}Sn_{0.95}$ on the EMF mesures as a function of temperature and time experience in the cell (-) In | In^+ in electrolyte | In_xSn_{1-x} (+) (- ○ -) - used points (-+- and -Δ-) - unused points.

Electrochemical system	φ / V			
	450°C	700°C	800°C	1000°C
Li^+ | Li	- 3.684	- 3.514	- 3.457	- 3.352
Na^+ | Na	- 3.566	- 3.332	- 3.250	- 3.019
La^{3+} | La	- 3 .241	- 3.016	- 2.997	- 2.876
Ce^{3+} | Ce	- 3.193	- 3.014	- 2.945	- 2821
Nd^{3+} | Nd	- 3.103	-	- 2.856	- 2.736
Gd^{3+} | Gd	- 3.013	-	- 2.807	- 2.709
Mg^{2+} | Mg	- 2.720	- 2.536	- 2.460	- 2.346
Sc^{3+} | Sc	- 2.621	- 2.455	- 2.375	- 2.264
U^{3+} | U	- 2.530	- 2.350	- 2.280	- 2.162
Be^{2+} | Be	- 2.167	-	-	-
Al^{3+} | Al	- 2.018	-	-	-
Mn^{2+} | Mn	- 1.999	- 1.854	- 1.807	- 1.725
V^{2+} | V	- 1.794	- 1.794	- 1.636	- 1.441
Zn^{2+} | Zn	- 1.629	- 1.512	- 1.476	
Tl^+ | Tl	- 1.629	- 1.512	- 1.473	
Cd^{2+} | Cd	- 1.442	-1.262	- 1.193	- 1.002
In^+ | In	- 1.43	-	-	-
Cr^{3+} | Cr	- 1374	-	- 1.113	- 1.006
Sn^{2+} | Sn	- 1.34	- 1.264	- 1.259	

Electrochemical system	**φ/V**			
	450°C	700°C	800°C	1000°C
Pb^{2+} | Pb	- 1.30	- 1.163	- 1.112	- 1.039
Fe^{2+} | Fe	- 1.297	-	- 1.118	- 1.050
Co^{2+} | Co	- 1.171	- 1.028	- 0.977	- 0.900
Ni^{2+} | Ni	- 1.104	- 0.939	- 0.875	- 0.763
Sb^{3+} | Sb	- 1.019 (300°C)	-	-	-
Cu^{+} | Cu	- 1.035	- 0.987	- 0.970	- 0.943
Ag^{+} | Ag	- 0.911	- 0.848	- 0.826	- 0.784
Bi^{3+} | Bi	- 0.844	- 0.817	-	-
Pd^{2+} | Pd	- 0.487	- 0.487	- 0.340	- 0.285
Pt^{2+} | Pt	- 0.299	- 0.180	-	-
Au^{+} | Au	+ 0.223	-	-	-

Table 2. Calculated electrochemical potential (φ Cl_2/Cl^-) in individual molten chlorides, calculated from thermodynamic data.

Electrochemical system	**φ / V for the anions in Volt**			
	F^-	**Cl^-**	**Br^-**	**I^-**
Ba^{2+} | Ba	- 1.31	- 2.59	- 2.62	
K^{+} | K	- 0.62	-2.50	- 2.53	- 1.98
Sr^{2+} | Sr	- 1.16	- 2.51	- 2.41	- 1.94
Li^{+} | Li	- 0.08	- 2.39	- 2.40	- 1.95
Na^{+} | Na	- 1.84	- 2.36	- 2.35	- 1.81
Ca^{2+} | Ca	- 1.13	- 2.35	- 2.25	- 1.53
Mg^{2+} | Mg	- 0.28	- 1.58	- 1.58	- 1.01
Mn^{2+} | Mn	- 0.17	- 0.85	- 0.83	- 0.44
Zn^{2+} | Zn	- 1. 09	- 0.40	- 0.50	- 0.27
Cd^{2+} | Cd	- 0.91	- 0.25	- 0.46	- 0.19
Tl^{+} | Tl		- 0.44	- 0.29	- 0.41
Sn^{2+} | Sn	-	**- 1.270**	- 0.981	**- 0.462**
Pb^{2+} | Pb	-	**- 1.215**	- 0.976	**-0.620**
Cu^{+} | Cu	+ 1.37	+ 0.29	- 0.06	- 0.17
Co^{2+} | Co	+ 0.52	+ 0.06	- 0.05	+ 0.43
$2H^{+}$ | H_2	0	0	0	0
Bi^{3+} | Bi	- 0.14	+ 0.39	+ 1.19	+ 0.33

Table 3. Electrochemical potential of metals in molten liquid halides at the temperature 700°C compared with reference electrode φ H_2.

Elecrtocemical system	**φ/V**	
	T=300°C	T=500C°
Zn^{2+} | Zn	1.706	1.603
In^{+} | In	1.520	1.414
Sn^{2+} | Sn	1.428	1.320
Pb^{2+} | Pb	1.420	1.271
Sb^{3+} | Sb	1.019	-

Table 4. Comparison of certain electrochemical potentials of metals in molten liquid chloride.

So, if we study the electrochemical cell of the type:

$$(-)\ Zn \mid Zn^{2+}\ \text{in electrolyte} \mid Zn_xIn_{(1-x)}\ (+)$$

$$(-)\ In \mid In^{+}\ \text{in electrolyte} \mid In_xSn_{(1-x)}\ (+)\ ,$$

the exchange reactions occur readily in these systems if the concentration of the second element has reached 90% (Fig.14). We observed a continuous fall of the EMF for alloys with $x \leq 0.1$ when the duration of the experiment was a few weeks (Vassiliev et al., 1998b; Mozer, 1972). The rate of the exchange reaction was augmented with increasing temperature, especially in the case of liquid systems.

5.3 Influence of a third component on the exchange reaction

Exchange reactions occur not only between a pure A metal and alloy $A_xB_{(1-x)}$ but also between alloys of different compositions, if the activities a_A of them are very different. This phenomenon is very pronounced in the liquid ternary system In-Sn-Sb (Vassiliev, 1998). Table 4 shows that the attraction between atoms In and Sb is greater than between In and Sn or between Sn and Sb. Accordingly, the tin atoms of the ternary alloy are very free, and exchange reactions between electrodes of different compositions of alloys, with indium, occur easily.

5.4 Selection of different types of halides such as electrolytes

It is possible to use different halides such as electrolytes: chlorides, bromides and iodides. All these salts are hygroscopic and their treatment by the corresponding hydrogen halide (HCl, HBr, HI) is needed to avoid water marks. If some halogenade is not appropriate due to an exchange reaction it is replaced by another.

The choice of the electrolyte is determined by its melting temperature and by the need to minimize the exchange reaction. The study of Sn-Pb system is not possible in molten chlorides (the difference of potential electrodes is 0. 05 V at 500°C ; Table 4).The substitution of chloride by iodide significantly increases this difference to 0.168 V (see Tabl.3) and decreases the exchange reactions:

$$SnI_2 + Pb \rightarrow PbI_2 + Sn$$

$$PbI_2 + Sn \rightarrow SnI_2 + Pb,$$

Although it is impossible to eliminate their influence completely, especially for tin-rich alloys, the electrochemical chain:

$$(-)\ Pb \mid KI + LiI + PbI_2 \mid Pb_xSn_{1-x}\ (+)$$

can be studied.

5.5 Effect of exchange reaction on the EMF measurements in ternary system In-Sn-Sb

Let us consider the example of spontaneous exchange reaction in the cell that contains the series of four alloys with low indium content, number 1-4, (x_{In} from 0.05 to 0.11), and one indium-rich alloy № 5 (x_{In}=0.5) at presence of two electrodes of pure indium. (See Table 5)

№	x_{In}	x_{Sn}	x_{Sb}
1	0.0500	0.4751	0.4749
2	0.0503	0.4497	0.5000
3	0.1002	0.3996	0.5003
4	0.1112	0.4444	0.4444
5	0.5000	0.2499	0.2501

Table 5. Composition of In-Sn-Sb alloys used for detection of kinetic of spontaneous exchange reaction in the electrochemical cell.

We can state that the measured values E(T, x_{In}) for alloys with number 1-4, and slightly for number 5, are exposed to such reactions. So, we used only the first points of the measurements E(T, x_{In}), which were less susceptible to this influence. Exchange reaction is more pronounced for alloy 1 and 2. Dynamics of a regular drift of EMF values for alloys № 2 and 5 versus time and temperature were shown in Fig. 15 and Fig. 16. Experimental points are divided into two series. The gap of EMF values between two series is connected with the study of other phases at lower temperatures are not indicated in Fig. 15 and 16. We took in consideration only the black dots. The different stages of the experiment are marked in time. Fig. 15 and 16 show that the exchange reaction depends on the time and temperature. Two main types of exchange reactions (*a* and *b*) take place in cell:

(-) In| In^{2+} in electrolyte| In-Sn-Sb (+)

a) In (pure) $^{In\rightarrow}{}_{\leftarrow Sn}$ InSnSb (№ 1-4) (13)

$a_{In}=1$ $\qquad$ $a'_{In}<1$

b) InSnSb (№5) $^{In\rightarrow}{}_{\leftarrow Sn}$ InSnSb (№ 1-4) (14)

a''_{In} $\quad > \quad$ a_{In}'

Reactions *a* and *b* lead to a decrease of the EMF values for alloys (№№ 1-4) and the reaction of *b* increases the EMF values of the alloy number 5 in relation to the reference electrode made of pure indium. The rate of exchange reaction prevails over the reaction (13) and (14). Kinetics of exchange reactions is shown in Fig. 17 and Fig.18 in accordance with 4 passes at the same temperature 755K versus the time. We did not observe exchange reactions for alloys with x_{In}> 0.1, although the duration of the experiment exceeded more than two months, the maximum temperature reached 822K.

5.6 Issues are related to the valence of A^{n+} ion

To obtain good experimental results, it is necessary to know rigorously the ion charge that is responsible for the EMF in cell of type (I). However, the difference in activities of pure metal A and alloy $A_xB_{(1-x)}$ can lead to different charges of the A^{n+} ion in the vicinity of the electrodes A and $A_xB_{(1-x)}$. In this case, even in open circuit, a spontaneous transfer of component A to alloys $A_xB_{(1-x)}$ is possible and a constant drift of the EMF occurs over time.

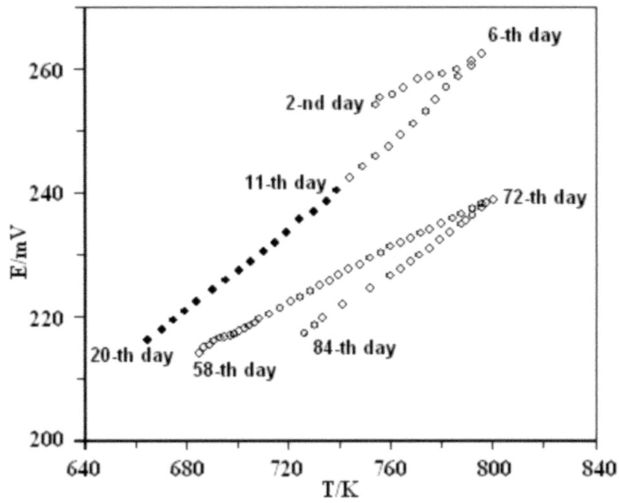

Fig. 15. EMF of alloy number 2 versus temperature and time.

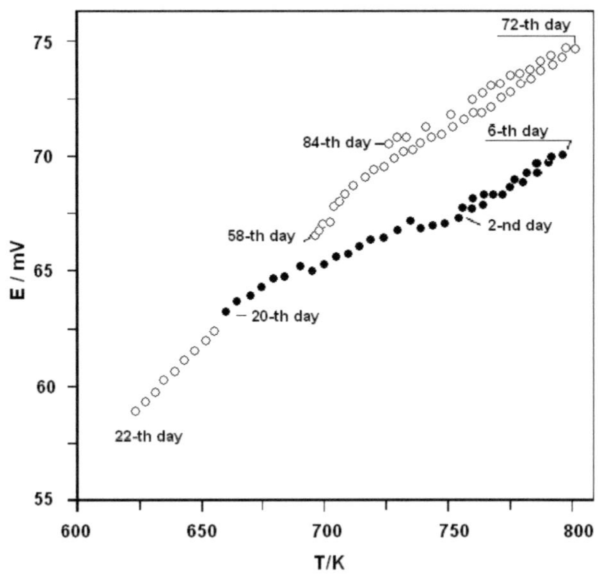

Fig. 16. EMF of alloy number 5 versus temperature and time.

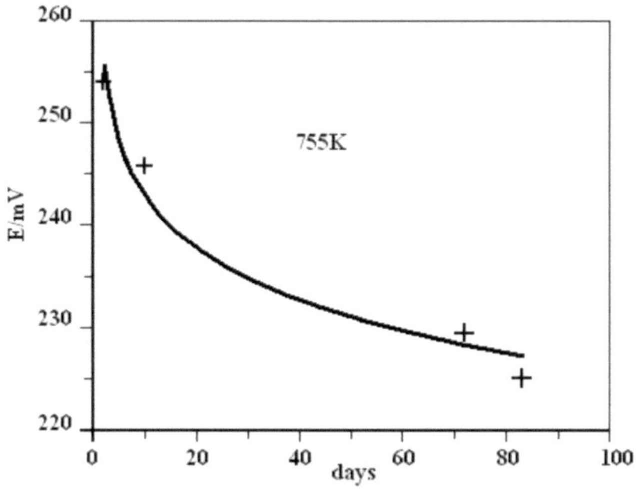

Fig. 17. Kinetics of the exchange reaction of the alloy number 2 at the T=755K.

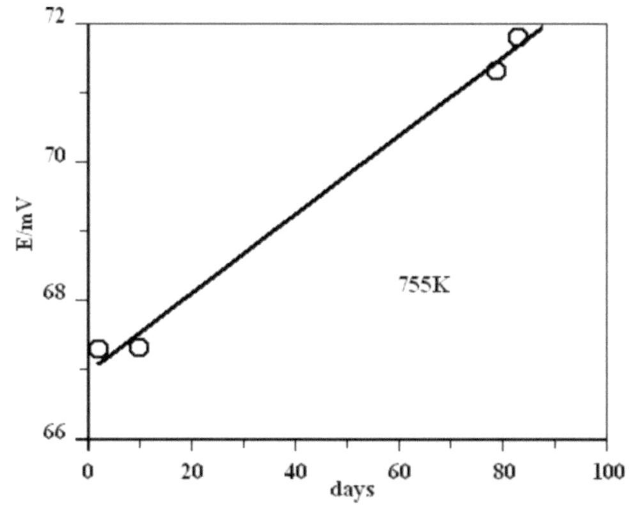

Fig. 18. Kinetics of the exchange reaction of the alloy number 5 at T=755K.

This relates to the ions of metal that may have the different charges of ions (m> n) in different parts of cell in electrolyte:

On the cathode: $(m-n)\ A + nA^{m+} \rightarrow mA^{n+}$

On the anode: $mA^{n+} \rightarrow (m-n)\ A + nA^{m+}$

So, in the case of open circuit, there is a transfer of component A from pure metal to its alloy $A_xB_{(1-x)}$. Near the cathode, the fraction of charged ions (n +) will be greater and near anode around of the alloy $A_xB_{(1-x)}$ will be smaller. In another words, there is dissolution of pure metal and deposit on the surface of alloy $A_xB_{(1-x)}$. The transfer is even faster than the temperature T and the concentration of A^{m+} ions is higher and the distance between electrodes is smaller. A metal is transferred through an electrolyte by ions of lower charge.

If an activity of component A in the cell of type (I) is reduced, it is possible to determine the thermodynamic properties even in cases of spontaneous transfer of the component A. In this case, it is necessary to take as reference electrode some electrode binary alloy $A_xC_{(1-x)}$, which the thermodynamic properties are well known, and it can substitute the pure metal A.

Then we have

$$\Delta\mu_A = RT\cdot \ln (a''/a') \tag{15}$$

In this case the activity (a') is less than 1. The difference of the chemical potential of component A between electrodes, and hence, the electromotive force will be smaller.

This problem is encountered when the system Ln -Te (Ln = lanthanide) was studied with the cell:

$$(-)\ Ln \mid Ln^{(3+)} \text{in electrolyte} \mid Ln_xTe_{(1-x)}\ (+)$$

$$\Delta\mu_A = RT\cdot Ln\left(a_A^{A_xB_{(1-x)}} \,/\, a_A^{A_xC_{(1-x)}} \right)$$

Alloys, rich in metal B (in this case by tellurium) generate an electromotive force E near 1.5 V. The substitution of the lanthanide by one of its alloys of Ln-In system (Vassiliev et al., 2009), allows to make the measurements with the following electrochemical cell:

$$a_A^{A_xC_{(1-x)}} < a_A^{(A)} = 1$$

$$(-)\ LnIn_{2,5}+In \mid La^{(3+)} \text{ in electrolyte } \mid Ln_xTe_{(1-x)}\ (+) \tag{III}$$

The electromotive force of the cell is as twice as smaller and spontaneous transfer becomes negligible.

5.7 Problem of liquation (or phase separation) of liquid alloys

If the metal liquid system does not contain any components that provoke the exchange reactions in an electrochemical cell, the EMF method is well suited for such a system. The problem of liquation (or phase separation) is less serious, but the big difference of the specific density of alloying effects on the rate of establishment of thermodynamic equilibrium and distorts the potentiometric measurements.

Element	Specific density (g/cm³)	Element	Specific density (g/cm³)
Al	2.7	Cu	8.94
Sn	5.85	Bi	9.80
Zn	7.13	Ag	10.50
In	7.31	Pb	11.34

Table 6. Specific density (g/cm³) of some metals.

The more the difference of the specific density of elements and their chemical interaction are weaker, the liquation influence is greater. As a consequence, the EMF measurements do not correspond of thermodynamic equilibrium, because the surface of alloy is depleted by the elements of the smaller density. Especially it should be taken into account that the metallic systems with a miscibility gap. Table 6 shows the specific density of some metals. The liquation problem was noted by us in the investigation of systems based on ternary zinc-lead alloys (Zn-Pb-Sn, Zn-Pb-In) (David at al., 2004) and In-Sn-Ag and In-Bi-Ag (Vassiliev et al., 1998c). We observed hysteresis loops during heating and cooling of the cells in a wide temperature range for a number of alloys. In the case of the hysteresis loop, it is necessary to choose measurement results only when the temperature is lowered, starting from the homogeneous liquid region. The isothermal cell is also well serves for the thermodynamic study of metals with high vapor pressure (mercury, zinc, etc.) (Vassiliev et al., 1990; David at al., 2004). We did not observe any evaporation of zinc, even though long-term measurements (more than two months and the temperature to 780 K) were taken.

In the systems based on zinc, the latter serves as the reference electrode and the part of both the internal calibration of the thermocouple in the cell, because the measured values E (T) give a clear kink at the melting point of zinc. The kinetic curve of solidification or melting of the metal also exhibits a characteristic jump EMF at the melting point. Such curve is easily obtained by continuous measurement of the EMF of the cell at the phase transition sol $\leftrightarrows$ liq. The zinc chloride must not be added to electrolyte previously. We found that the ions, forming the potential, appeared in a few hours inside the cell after the experiment began. We used the metals of 99.999% and chlorides of lithium and potassium 99.99% purity. The type of electrochemical cell for EMF measuring has been used:

$$(-)\ Zn \mid Zn^{2+} \text{ in } (LiCl + KCl) \text{ liquid eutectic} \mid Pb_xSn_yZn_Z\ (+) \qquad (IV)$$

Temperature range of research is limited one hand by the crystallization of the electrolyte, and on the other hand by the softening Pyrex glass. Control of the state reference electrodes of pure zinc in the course of the experiment was carried out by measuring the difference of EMF between such electrodes. If the cell functions normally, this difference is about 5 μV. Temperature correction is performed by the melting point of zinc, located in the cell.

5.8 How to study the systems with closely spaced electrode potentials. The EMF measurements of ternary Pb-Sn-Sb system

EMF measurements of the ternary system Pb-Sn-Sb (Vassiliev et al., 1995) were carried out for five alloys along the isopleth (Pb-$Sn_{0.5}Sb_{0.5}$) with x_{Pb} = 0.15, 0.20, 0.25, 0.30, 0.333 at temperatures 690 - 820K. Alloy $Sn_{0.5}Sb_{0.5}$ has been used as a reference electrode. Scheme for EMF measuring and the definition of excess functions of mixing in the ternary system Pb-Sn-Sb are shown in Fig. 19 and 20. To measure the EMF we can not use the alloys of the ternary Pb-Sn-Sb system as reference electrodes from pure metals. The using of tin or lead in a cell with undivided space is impossible, because their electrode potentials are almost equal and the exchange reactions take place. Cell with a diaphragm (Shourov, 1974, 1984) allows solving this problem, but the experiment is very laborious. In order to avoid the exchange reaction, we choose an alloy of tin and antimony with a component ratio 1:1 ($Sn_{0.5}Sb_{0.5}$) as the reference electrode, which chemical potential we studied in this paper using an isothermal cell without diaphragm. Activity of tin a_{Sn} in the $Sn_{0.5}Sb_{0.5}$ alloy at 900K is equal 0.41, that is much less than its activity in pure tin (a_{Sn}=1). So, the activity of lead becomes comparable with the activity of tin when we study the ternary alloys $(Sn_{0.5}Sb_{0.5})_{1-x}Pb_x$. Thus, we have practically eliminated the exchange reaction between the electrodes with different contents of tin and lead. The choice of electrode $Sn_{0.5}Sb_{0.5}$ allowed us to obtain stable and reproducible results E (T) (Fig.21).

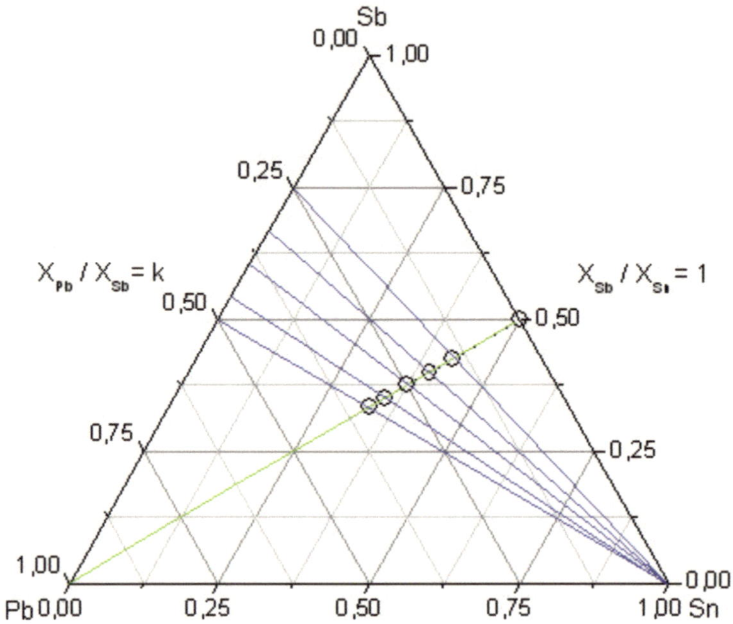

Fig. 19. Measurements of EMF (μ^E_{Sn})in the ternary Pb-Sn-Sb system.

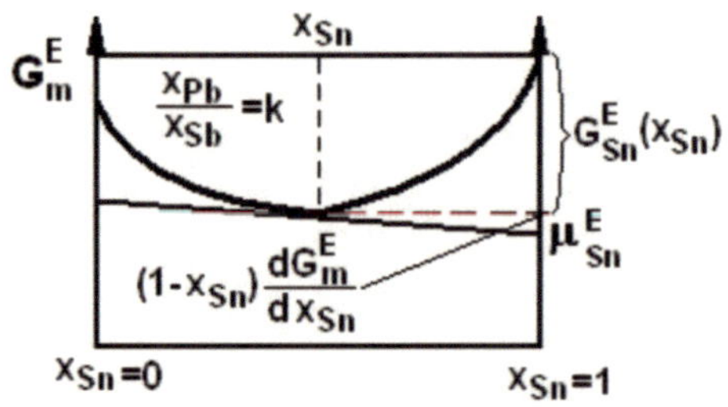

Fig. 20. Scheme for determining the excess functions of mixing of the ternary Pb-Sn-Sb system.

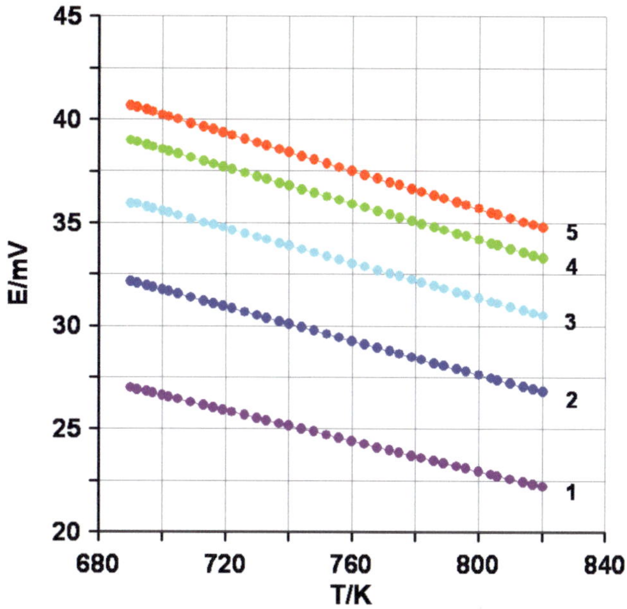

Fig. 21. Dependence E (T) cell and compositions of alloys (-)$Sn_{0.5}Sb_{0.5}$ | $KCl+LiCl+Sn^{2+}$ | $(Sn_{0.5}Sb_{0.5})_{1-x}Pb_x$(+) 1-$(Sn_{0.5}Sb_{0.5})_{0.666}Pb_{0.333}$, 2- $(Sn_{0.5}Sb_{0.5})_{0.70}Pb_{0.30}$, 3- $(Sn_{0.5}Sb_{0.5})_{0.75}Pb_{0.25}$, 4- $(Sn_{0.5}Sb_{0.5})_{0.80}Pb_{0.20}$, 5- $(Sn_{0.5}Sb_{0.5})_{0.85}Pb_{0.15}$

There were obtained 39 experimental points E(T) for each ternary alloy with a maximum error ± 0.1 mV. The coefficients of linear equations $E_1(T)$ of ternary alloys (1-5) was obtained with respect to the reference electrode $Sn_{0.5}Sb_{0.5}$. The correction was performed using the ratio (16) to find the chemical potential of ternary alloys relatively to pure tin:

$$\mu_{Sn} = E_1(T)+E_2(T) \quad (16)$$

Where the dependence $E_2(T)$ (E_2/mV= 10.52+26.5·10^{-3}·T) was obtained between the liquid alloy $Sn_{0.5}Sb_{0.5}$ and the reference electrode from pure tin.

5.9 Pb-Pd system

Finally, let us consider an example of the Pb-Pd phase diagram that has several intermediate phases Fig.22 (Vassiliev et al., 1998a). We have established a narrow region of homogeneity of the phases Pb_2Pd and Pb_9Pd_{13}, and the deviation from stoichiometry of PbPd phase. We have determined also the coordinates of the gamma phase, and refined the melting point of Pb_2Pd phase and the coordinates of the eutectic point.

General trend of experimental EMF results versus temperature and composition for the Pb-Pd system is shown in the Fig.23. The points at the right of curves L2-L6 correspond to the homogeneous liquid alloys with x_{Pd} from 0.10 to 0.60). The curves L2-L6 correspond to the heterogeneous alloys (liquid and solid states). The lines at the left of the curves L_2-L_6 correspond to the heterogeneous solid alloys from S_2 to S_8 (see, please, Fig.23). Monophasic supercooled liquid alloys are on the left of lines L_2-L_4 with x_{Pd} from 0.25 to 0.40. Points of intersections of the lines S_5 and S_8, S_2 and S_7 , S_4 and S_6 correspond to the phase transformations in solid state. Line S'1 corresponds to solid solution on the base of Pb_2Pd phase. Points of intersections of the lines S_i with curves L_i correspond to the eutectic or peritectic transformations. Points of intersections of the E(T) for liquid homogeneous region with curves L_2-L_6 correspond to the points of liquidus. Invariants points (T_e = 721K and T_m (Pb_2Pd) = 725K) were shown in the Fig.22 and 23.

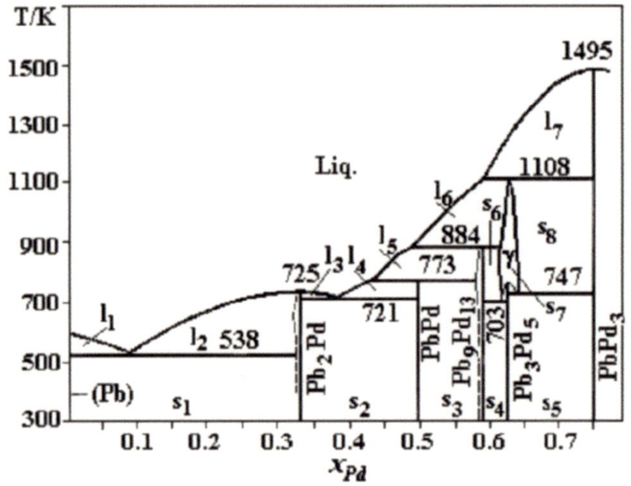

Fig. 22. Equilibrium phase diagram of the palladium-lead with our corrections.

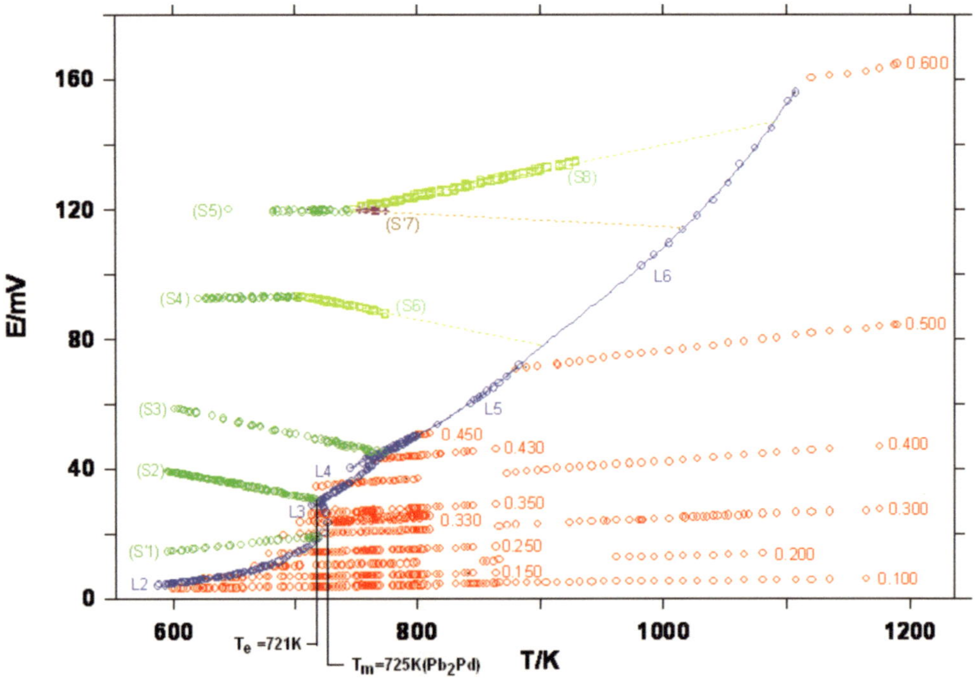

Fig. 23. General trend of experimental EMF results versus temperature and composition for the Pb-Pd system.

We applied successfully the described technique for different binary, ternary and quaternary systems:

Tl-S (Vassiliev et al., 1971,1973a, 1973b, 2008), Tl-Se (Vassiliev et al., 1967,1969,1971), Tl-Te (Vassiliev et al., 1968); Cd-Te (Vassiliev et al., 1990); **Pb-Pd** (Vassiliev et al., 1998a); In-P (Vassiliev & Gachon, 2006); In-As (Abbasov et al., 1964); In-Sb (Vassiliev, 2004); In-Sn (Vassiliev et al., 1998b); Ga-As (Abbasov et al., 1964); Ga-Sb (Abbasov et al., 1964); Sn-Sb (Vassiliev 1997, 2005); Bi-Se (Vassiliev et al., 1968); Ga-Te (Abbasov et al., 1964); Mn-Te (Vassiliev et al., 1993); Rare Earth Metals (REM) with In (Vassiliev et al., 2009), Sb (Gorjacheva et al.,1981), Pb (Vassiliev et al., 1993), Te (Vassiliev et al., 1980); Cd-Hg-Te (Vassiliev et al., 1990,2004); In-Sn-Sb (Vassiliev et al., 2001); **Pb-Sn-Sb** (Vassiliev et al., 1995); Pb-Sn-Zn (David at al., 2004) In-Bi-Ag (Vassiliev et al., 1998c); In-Ni-Sb Vassiliev et al., 2003); Pb-Cu-Zn; Pb-In-Zn; Pb-In-Zn-Sn (Hertz at al., 1998). As well as, this technic can be applied for the oxide systems also.

6. Conclusion

The accuracy of the proposed experimental technique does not yield the best calorimetric measurements and completeness of the information obtained by the EMF exceeds the calorimetric methods in some case. So, the proposed variant of the EMF can be called as universal and self-sufficient, and the cost of equipment used in the EMF method is much lower than the calorimetric one. EMF remains one of the most important methods in metallurgical thermodynamics.The method of electromotive force (EMF) with proposed electrocemical cells is a powerful tool to study the thermodynamic properties of metallic, semiconductor and oxyde systems. It permits to identify a set of values of the chemical potentials of one of the component of the different phases. The final result of this work consist of the thermodynamic optimization of the studied system.

7. Acknowledgements

This work was financially supported by the Russian Foundation for Basic Research, project **11-08-01154** as well as the Talents Project of Guangdong Province, Guangdong, P.R. China. We would like to thank Dr. S. Kulinich for helpful discussion.

8. References

Abbasov A.S. ,Nikoliskaja A.V., Gerassimov Ja.I., Vassiliev V.P. (1964) The thermodynamic properties of gallium antimonide investigated by the electromotive force method. Dokl. AN SSSR, Vol.156, No.6, pp.1399-1402. (in Russian)

Abbasov A.S., Nikoliskaja A.V., Gerassimov Ja.I, Vassiliev V.P. (1964) The thermodynamic properties of indium arsenide investigated by the electromotive force method. Dokl.AN SSSR, Vol.156, No.1, p.118-121. (in Russian)

Abbasov A.S., Nikoliskaja A.V., Gerassimov Ja.I., Vassiliev V.P. (1964) The thermodynamic properties of gallium tellurides investigated by the electromotive force method. Dokl. AN SSSR, Vol.156, No.5, pp.1140-1143. (in Russian)

Abbasov A.S.,Mamedov K.N.,Nikoliskaja A.V.,Gerassimov Ja.I., Vassiliev V.P. (1966) The thermodynamic properties of gallium arsenide investigated by means of electromotive forces. Dokl.AN SSSR, Vol.170, No.5, pp.1110-1113. (in Russian)

Brauer G. (1954) Handbuch der Präparathieven Anorganischen Chemie. Ferdinand Enke Chemie. Stuttgart.

Chourov N.I. (1984) Institut of Electochemistry, Science Center of Ural "Modification of electromotive forces method by the determination of thermodynamic properties in case of exchange reacton". Deposided document VINITI № 2422-84. (in Russian).

David N., Vassiliev V., Hertz J., Fiorani J.M., Vilasi M. (2004) Measurements EMF and thermodynamic description of Pb-Sn-Zn liquid phase, Z. Metallkunde, Vol.95B, pp.1-8.

Geyderih V.A., Nikolskaya A.V., Vasilyeva I.A., (1969) in book *Coedineniya peremennogo sostava»* pod red. Ormonta B.F., Izdatelstvo Himiya.-Leningrad, glava 4, 210-261. *Nonstoicheometric Compounds*, Edited by B.F. Ormont, Publishing House Chemistry, Leningrad, chapter 4, pp.210-261. (in Russian)

Goryacheva V.I., Gerassimov Ja.I., Vassiliev V.P. (1981) Thermodynamic study of monoantimonides of holmium and erbium by EMF method. Vol.55, No.4, pp.1080-1083. (in Russian)

Hertz J., Naguet Ch., Bourkba A., Fiorani J.M., Vassiliev V., Sauvanaud J. (1998) A strategy to establish an industrial thermodynamic data bank. The ternary (Pb, Sn, Zn) liquid phase in the zinc purification process. Extension to quaternary system (M, Pb, Sn, Zn). Thermochim. Acta, Vol.314, No.1-2, pp.55-68.

Hladik J. (1972) Physics of Electrolytes. Thermodynamics and Electrode Processing in Solid State Electrolytes, Vol.2, Academic Press, London.

Morachevsky A.G. (1987) Thermodynamics of Molten Metal and Salt Systems. Publisher "Metallurgy" Moscow.1987. 240 C (in Russian)

Morachevsky A.G., Voronin G.F., Geyderih V.A., Kutsenok I.B. (2003) "Electrochemical methods in the thermodynamics of metallic systems", Edition "Akademrkniga", 334 p. (in Russian)

Mozer Z. (1972) Thermodynamic studies of liquid Zn-In solution. Metal. Transaction, Vol.2, pp.2175-2183.

Shourov N.I. (1974). Ph. Diss. Thesis in Physical Sciences "Potentiometric method for the study of thermodynamic properties of alloys whose components are electrochemically very close to each other. Sverdlovsk.

Strikler H.S., Seltz H.A. (1936) Thermodynamic Study of the Lead-Bismuth System,Vol.58, pp.2084-2090.

Terpilowski J., Zaleska E., Gawel W. (1965) Characterystyka termodynamiczna ukladu stalego tal-tellur Roczniki Chemii. Vol.39, pp. 1367-1375.

Vasil'ev V. P. (2004) Thermodynamic properties of alloys and phase equilibria in the In-Sb system. Inorganic Materials Vol.40, No.5, pp.524-529.

Vasil'ev V.P. and Gachon J. C. (2006) Thermodynamic properties of InP. Inorganic Materials. Vol.42, No.11, pp.1171-1175.

Vasil'ev V.P.(2005) A complex study of the phase diagram of Sn-Sb system Rus.J Phys.Chem. Vol.79, No.1, pp.26-35.

Vasilliev V. P., Mamontov M. N., Bykov M. A. (1990). Thermodynamic properties and stability of solid solutions CdTe-HgTe-Te system, Vestnik Moskovskogo Universiteta, serie2, Chemistry, Vol. 31, No.3, pp. 211-218. (in Russian)

Vassiliev V., Borzone G., Gambino M., Bros J.P. (2003) Thermodynamic properties of ternary system InSb-NiSb-Sb in the temperature range 640-860K. Intermetallics, Vol.11, pp.1211-1215.

Vassiliev V.,Lelaurain M.,Hertz J. (1997) A new proposal for binary (Sn, Sb) phase diagram and its thermodynamic properties based on a new e.m.f. study. J. Alloys Comp., Vol.247, p.223-233.

Vassiliev V., Alaoui-Elbelgeti M., Zrineh A., Gambino M., Bros J.P. (1998) Thermodynamic of Ag-Bi-In system with $0<x_{Ag}< 0.5$) J. Alloys Comp., Vol.265, pp.160-170.

Vassiliev V., Azzaoui M., Hertz J. (1995) EMF study of ternary (Pb, Sn, Sb) liuquid phase Z. Metallkunde, Vol. 86, pp.545-551.

Vassiliev V., Bykov M., Gambino M., Bros J.P. (1993). Thermodynamic properties of the intermetallic compounds MnTe and $MnTe_2$. Z. Metallkunde, Vol. 84, pp. 461-468.

Vassiliev V., Feutelais Y., Sghaier M., Legendre B. (1998). Liquid State Electrochemical Study of the System Indium-Tin Thermochim. Acta, Vol.315, pp.129-134.

Vassiliev V., Feutelais Y., Sghaier M. , Legendre B.,(2001) Thermodynamic Investigation in In-Sb, Sn-Sb and In-Sn-Sb liquid systems J. Alloys Comp., Vol. 314, pp.197-205.

Vassiliev V., Gambino M., Bros J.P., Borzone G., Cacciamani G., Ferro R. (1993) Thermodynamic investigation and optimisation of the Y-Pb alloys system. J. Phase Equilibria. Vol.14, No.2, pp.142-149.

Vassiliev V., Voronin G.F., Borzone G., Mathon M., Gambino M., Bros J.P. (1998) Thermodynamics of the Pb-Pd system. J. Alloys Comp., Vol. 269, Vol.123-132.

Vassiliev V.P., Goryatcheva V.I., Gerassimov Ja.I. (1980) The phase equilibria and thermodynamic properties of solid alloys of erbium with tellurium.Vestnik Moskovskogo Universiteta, serie 2, chimie, 21, No 4, pp. 339-345. (in Russian)

Vassiliev V.P., Goryacheva V.I., Gerassimov Ja.I., Lazareva T.S. (1980) A study of phase equilibrium and thermodynamic properties of solid alloys of erbium with tellurium. Vest. Mosk. Univ. Ser.2. Chim., Vol.21, No.4, pp.339-345. (in Russian)

Vassiliev V.P., Minaev V.S. (2008) Tl-S phase diagram, structure and thermodynamic properties. J. Optoelectronics Advanced Materials. Vol.10. No.6. pp.1299-1305

Vassiliev V.P., Nikoliaskaja A.V., Chernyshov V.Y., Gerassimov Ja.I. (1973) Thermodynamic properties of thallium sulphides. Izv.AN SSSR, Neorg. Material., Vol.9, No.6, pp.900-904. (in Russian)

Vassiliev V.P., Nikoliskaja A.V., Bachinskaja A.G., Gerassimov Ja.I. (1967). The thermodynamic properties of thallium monoselenide. Dokl.AN SSSR, Vol.176, No.6, pp.1335-1338. (in Russian)

Vassiliev V.P., Nikoliskaja A.V., Gerassimov Ja.I. (1969) The thermodynamic properties of the lowest thallium selenide (Tl_2Se), Dokl. AN SSSR. Vol.188, No 6, pp. 1318-1321 (in Russian)

Vassiliev V.P., Nikoliskaja A.V., Gerassimov Ja.I. (1971) Thermodynamic investigation of thallium-selenium system by the electromotive force method. J. Phys. Chem., (russ), Vol.45, No.8, pp.2061-2064. (in Russian)

Vassiliev V.P., Nikoliskaja A.V., Gerassimov Ja.I. (1971), Thermodynamic characteristics of higher thallium sulfides, and certain correlation for IIIB subgroups of monochalcogenides Docl.AN SSSR. Vol.199, Vol.5, pp.1094-1098. (in Russian)

Vassiliev V.P., Nikoliskaja A.V., Gerassimov Ja.I. (1973) Phase equilibriums of thallium-sulphur system in the solid state. Izv.AN SSSR, Neorg. Material., Vol.9, No.4, pp.553-557. (in Russian)

Vassiliev V.P., Nikoliskaja A.V., Gerassimov Ja.I., Kuznestov A.F. (1968), Thermodynamic study of thallium tellurides by electromotive forces Izv. AN SSSR, Neorg. Material. Vol.4, No 7, pp. 1040-1046. (in Russian)

Vassiliev V.P., Pentine I.V., Voronine G.F. (2004) Conditions de stabilité thermodynamique de la solution solide du système CdTe-HgTe. J.Phys. IV, France. Vol.113,. pp. 97-100.

Vassiliev V.P., Somov A.P., Nikoliskaja A.V., Gerassimov Ja.I. (1968) Investigation of the thermodynamic properties of bismuth selenide by the electromotive force method. J. Phys.Chem., Vol.42, No.3, pp.675-678. (in russian)

Vassiliev V.P., Taldrik A.F., Legendre B., Thermodynamics analysis of the rare earth solid metals and their alloys $REIn_3$. XXXV JEEP (Journées d'Etude des Equilibres entre Phases), 1-3 avril 2009, Annecy. France, pp.103-107.

Wagner C., Werner A. (1963) The role of displacement reaction in the determination of activities in alloys with the aid of galvanic cells J.Electrochem.Soc. Vol.110, No 4, pp.326-332.

Part 2

Recent Developments for Applications of Electrochemical Cells

5

Cold Plasma – A Promising Tool for the Development of Electrochemical Cells

Jacek Tyczkowski
Technical University of Lodz
Poland

1. Introduction

The growing concern over the environment on a global basis is driving extensive research into new materials for next generation of clean, efficient, alternative energy systems, to a large extent based on various types of electrochemical cells. An especially useful method that has paved the way to novel materials with unique properties, which cannot be fabricated by other methods, is the cold plasma technology. Both integrally new materials in the form of thin films with unusual molecular structure and conventional materials modified by cold plasma, which reveal surprising surface properties, can be prepared in this way. Some of these materials turned out to be very interesting for the innovative electrochemical cells.

In this Chapter, a brief review of investigations performed on materials produced by the cold plasma technology and their testing in electrochemical cells, such as fuel cells, rechargeable cells and related devices as solar cells and water splitting systems, is presented. Some attention is paid to research conducted in our laboratory.[1] Recent results of our studies on plasma deposited thin films of nanocatalysts based on metal oxides, such as cobalt and cupper oxides, which have been tested as electrodes for fuel cells, are discussed in more detail. Our efforts directed towards the development of new materials for solar cells and water splitting systems are also mentioned. However, before turning to the details of these issues, we should first familiarize ourselves, even cursory, with the cold plasma technology.

2. What is the cold plasma?

As it is known, in the Universe, matter exists in four different states from the molecular interrelations point of view, namely: solid, liquid, gas and plasma. Simply speaking, plasma is a kind of ionized gas, into which sufficient energy is provided to free electrons from atoms or molecules and to allow both species, ions and electrons, to coexist. Generally, the plasma state can be divided into two main types (Fig. 1): low-temperature plasma - that is the state in which only a part of gas molecules is ionized and the gas is a mixture of electrons, ions, free radicals, excided and neutral molecules - and high-temperature plasma, in which all atoms are fully ionized. The latter type of plasma can be found, for example, in

---

[1] Division of Molecular Engineering (DME), Faculty of Process and Environmental Engineering, Technical University of Lodz, Wolczanska 213, 90-924 Lodz, Poland, E-mail: jatyczko@wipos.p.lodz.pl

the Sun or in laboratories involved in nuclear fusion research, but this type of plasma is rather not interesting as a technology for the preparation of new materials.

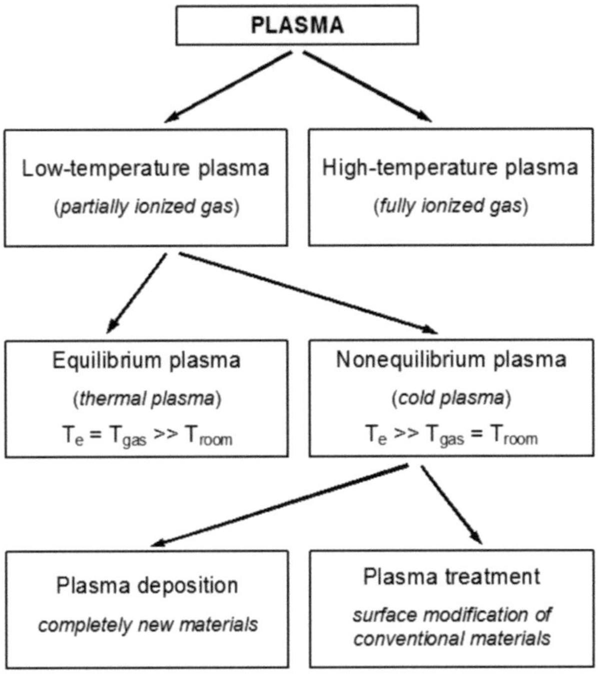

Fig. 1. Classification of the plasma types.

The low-temperature plasma can be divided, in turn, into two further types (Fig. 1): equilibrium and non-equilibrium plasmas. In the equilibrium plasma, often called the thermal plasma, electrons and the rest of plasma species have nearly the same temperature ($T_e \approx T_{gas}$), much higher than the room temperature. Such plasma is generated, for example, in plasma jets and torches. On the other hand, the non-equilibrium plasma, called sometimes the cold plasma, is characterized by the lack of thermal equilibrium between electrons and the rest of plasma species. In this case the electron temperature is in the range of 10^4–10^5 K, whereas the rest of the species are at temperature close to the room temperature. Under such conditions, chemical processes (e.g. chemical synthesis of new materials) can be performed at the room temperature using energetic electrons to cleavage covalent bonds in the gas molecules. By contrast, very high temperature of all the species in the thermal plasma considerably limits its application for the chemical syntheses and the surface modifications of thermal-degradable materials.

As one can see, among the various types of plasmas, the cold plasma is especially recognized as a promising tool on the road towards the search for new materials. The creation of such materials by the cold plasma technology can be carried out in two ways (Fig. 1). The first one is the deposition of completely new materials in the form of thin films, which is mainly accomplished by plasma polymerization processes (sometimes not quite correctly called plasma-enhanced chemical vapor deposition (PECVD)), and also, but relatively more rarely, by reactive sputtering processes. Thin-film materials with unusual chemical constitution, molecular construction and nanostructure can be obtained in this way (Gordillo-Vázquez et al., 2007; Konuma, 1992).

The second way consists in the modification of conventional materials, performed by their cold plasma treatment. Generally, such a treatment triggers three basic processes occurring mainly on the surface. It can create new functional groups by implantation of atoms present in the plasma, it can generate free radicals that then react with atmospheric oxygen and water molecules giving additional functional groups or can be used in grafting processes, and finally it can modify the microporous structure by the etching and degradation effects (Inagaki, 1996).

The cold plasma is most often generated in laboratories and industry by an electric glow discharge under low pressure using various frequencies of the applied electric field: audio frequencies (AF, mainly in the range of 10–50 kHz), radio frequencies (RF, mainly 13.56 MHz), and microwave frequencies (MW, mainly 2.45 GHz). Sometimes, a direct current (DC) discharge is also used. An example of typical parallel plate plasma reactor, one of those being used in our laboratory for deposition of thin films, is sketched in Fig. 2.

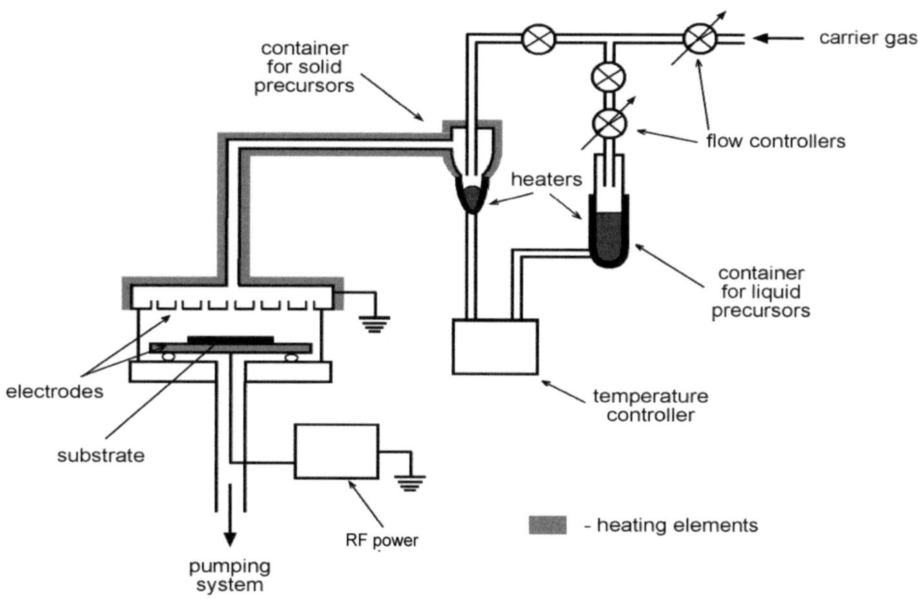

Fig. 2. A sketch of a typical parallel plate plasma reactor.

At first, the reactor chamber is evacuated down to 10^{-1}–10^{-3} Pa. Then, a precursor of plasma deposition is introduced to the chamber in the form of gas or vapor under controlled flow. Organic and inorganic gases, sublimating solids and evaporating liquids can be used as precursors. They are supplied as pure compounds or as mixtures with an inert carrier gas (e.g. argon). The carrier gas enables to generate plasma in the presence of compounds with very low vapor pressure. It is also possible to perform the plasma deposition process using a mixture of two or more precursors. Suitable selection of these compounds and their concentrations in the reactor chamber make it possible to control the molecular structure of deposited films, and consequently – properties of the films. In the reactor chamber with reactive gases, the glow discharge is generated between two internal metal electrodes by means of an appropriate generator. As a result of chemical processes proceeding in the plasma, a thin film of a new material, commonly referred to as "plasma polymer", is deposited on the electrode surfaces as well as on any substrate being in the plasma region. The same reactor can be also used for the modification of conventional materials. In this procedure, however, only "non-polymerizable" gases (e.g. Ar, O_2, N_2, NH_3) are utilized as precursors of plasma processes.

Another important plasma reactor employed in our study for the preparation of very interesting new materials is the so-called "three-electrode AF reactor". A schematic view of this reactor is shown in Fig. 3. A small electrode, on which films are deposited, is placed horizontally between two main perpendicular electrodes maintaining a glow discharge (10–50 kHz). The small electrode is coupled with the powered main electrode by a variable capacitor. The coupling capacitance controls the sheath voltage of the small electrode ($V_{(-)}$) and, in consequence, the impact energy of ions bombarding the growing film, independently of the plasma chemistry processes proceeding in the gas phase. It is especially striking that in some cases a very small variation of the bombarding ion energy in a defined range of its values is sufficient to create a drastic change in the electronic structure of deposited films (Tyczkowski, 1999). Thin films of this type will be discussed later in this Chapter (*Sec. 4.1.*).

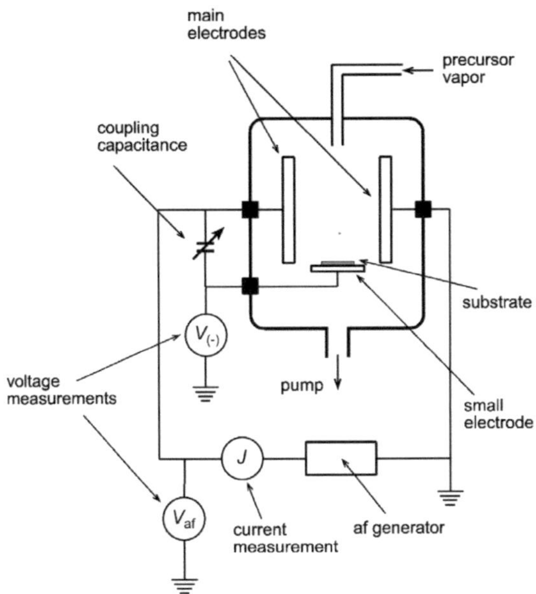

Fig. 3. A schematic diagram of the three-electrode AF reactor.

A useful variant of the fabrication of new materials by the cold plasma technique is the reactive sputtering. A schematic diagram of a typical set-up for sputtering deposition is sketched in Fig. 4. Positive ions that are produced in RF plasma generated in an inert gas, for example Ar, bombard the target surface (supplied with the negative self-bias) and cause the sputtering of its material. The sputtered material condenses on the substrate that is located out of the plasma region. If we use some reactive gas (e.g. O_2, N_2, CH_4), the target material takes part in chemical processes after sputtering and finally a new converted material is deposited. This is the so-called reactive plasma sputtering.

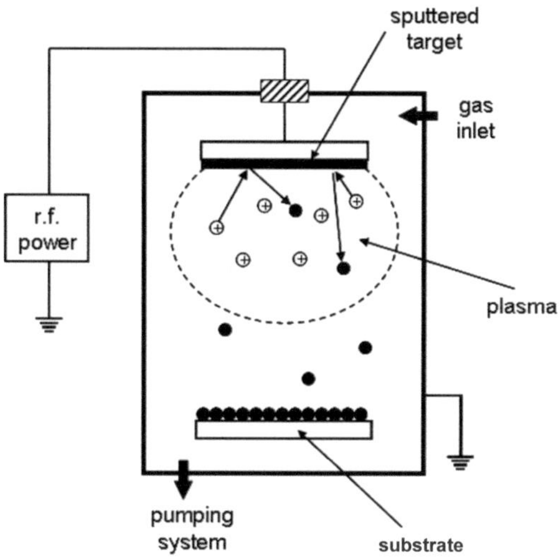

Fig. 4. A schematic diagram of a typical set-up for sputtering deposition.

For the sake of formality, it should be added that recently more and more attention has been focused on the cold plasma processes performed under atmospheric pressure conditions (Belmonte et al., 2011). However, plasma processes carried out under low pressure are, so far, still dominating in the cold plasma technology.

3. Plasma deposition of new materials

Since the first literature reports describing - nearly 140 years ago - the formation of solid products during electrical discharge in a tube filled with acetylene, many researches working in the field of plasma chemistry have observed the presence of high molecular weight materials as reaction by-products. These products were usually considered disadvantageous - they were deposited on the reactor walls and, due to their good adhesion to glass and insolubility in organic solvents, were not easily removable. They began to stimulate a scientific interest as late as in the sixties of the 20th century only, after Goodman (Goodman, 1960) had reported a successful application of plasma polymerized styrene films as an insulating layer in nuclear batteries. A vast amount of literature concerning plasma deposited films (plasma polymers), their properties, structure and mechanism of formation as well as potential application in various technologies has been published since then. The application ability of plasma polymers originates from both their often unique properties and relative simplicity of their production (Biederman, 2004; and references therein).

At the beginning, the interest was mainly limited to classical monomers, i.e. substances known to be able to polymerize in the conventional way (e.g. ethylene, styrene, butadiene). Hence, at that time the term "plasma polymer" was introduced as the result of the supposed analogy to conventional polymerized materials. In soon turned out, however, that many other low molecular weight organic, organometallic and inorganic compounds undergo plasma polymerization as well. Plasma polymers and conventional polymers, even though they should be produced from the same precursor (monomer), have practically nothing in common. The fundamental difference is that mer units cannot be defined in the case of plasma polymers. A large variety of chemical species created in the plasma, statistical combination them into high molecular structures and generally a high degree of their crosslinking cause that the structure of such a material is very often much closer to that of covalent glasses than that of conventional polymers.

Frequently plasma polymers are not classified in respect of the type of monomer but from a point of view of their chemical composition and morphology. For example, amorphous (a-) covalent material obtained by plasma polymerization of silane (SiH_4), which is composed of silicon and hydrogen, can be termed as a-Si:H. In turn, amorphous plasma polymer deposited from acrylonitrile (C_3H_3N) can be called as plasma-polymerized (pp-) acrylonitrile or a-C_XN_Y:H. It is usually met, but it is not a rule, that if the plasma polymer structure is close to a covalent glass structure, the latter notation is used. If plasma polymer reveals nano- or microcrystalline structures, prefixes nc- or μc- are put in the place of a-.

The structure and properties of plasma polymers are closely connected with a thin-film form, in which they are produced. In general, the thickness of the films is between a few nanometers and a few micrometers. Appropriate choice of precursors and plasma process parameters allow for the preparation of such thin films with a huge variety of structure and properties. Hence, there is a wide and diverse range of their current and anticipated applications, such as electronic and photoelectronic materials, insulating coatings, catalytic films, semi-permeable and electrolyte membranes, protecting layers, and many others. Some of these uses are also related to the electrochemical systems.

4. Optoelectronically active materials

Optical and electrical properties of plasma deposited films, sometimes unique indeed, as well as the easy of their deposition, at low temperature and low cost, on inexpensive substrates of almost any size and shape, render these materials very attractive for optoelectronic applications. The possibility to tailor optical parameters, such as refractive index and extinction coefficient, and what is particularly important - the ability to adjust parameters of the electronic structure, such as transport gap, optical gap, density of localized states, etc., recommend these plasma films as active photoelectric elements, e.g. for solar cells and water splitting cells.

4.1 Thin-film solar cells

Among the plasma deposited amorphous semiconductors, which have so far been studied in greatest detail and also appear to be of greatest applied interest, are hydrogenated amorphous silicon films (a-Si:H). Initial studies of this material were done in the sixties of the 20th century, when the RF glow discharge deposition of a-Si:H from silane was demonstrated. This was followed by the very important works that reported on successful *n*- and *p*-doping a-Si:H. It was really surprising, because previous attempts to dope thermally evaporated a-Si had failed. As it turned out, hydrogen was responsible for this effect. In 1975 and 1976, research confirmed that plasma-deposited films from silane contained hydrogen. Hydrogen serves primarily to passivate the dangling-bond defects and thus to decrease the density of localized states in the mid-gap.[2] Thermally evaporated a-Si has about 10^{26} $eV^{-1}m^{-3}$ states in the mid-gap, whereas the density of states for typical a-Si:H films is about 10^{21} $eV^{-1}m^{-3}$, i.e. five orders of magnitude lower. This just explains why one can control the electronic properties of a-Si:H by doping with donor and acceptor centers, contrary to a-Si (LeComber & Spear, 1979).

The most frequent method used to prepare doped films is plasma copolymerization from a mixture of the film precursor and the dopant agent. Diborane (B_2H_6) and phosphine (PH_3) are often used as sources of acceptor centers (boron atoms) and donor centers (phosphorus atoms), respectively. Recently, liquid compounds instead of these gases, such as triethylboron ($B(C_2H_5)_3$) and trimethylphosphine ($P(CH_3)_3$) have become more and more popular dopant agents. They are less toxic, more stable and their low vapor pressure offers facilities for precise controlling of the doping process. In turn, as a-Si:H film precursor, one can use not only SiH_4, but also, for example, a mixture of $SiCl_4$ and H_2, disilane (Si_2H_6), trisilane (Si_3H_8), cyclohexasilane (Si_6H_{12}), etc. (Pokhodnya et al., 2009; Searle, 1998; Tyczkowski, 2004). So, as one can see, the possibilities of designing and controlling the molecular and electronic structure of plasma deposited films are indeed enormous.

As an example of designing the electronic structure of a-Si:H films, the electrical conductivity of these films doped with acceptors (boron) and donors (phosphorus) is shown in Fig. 5. The room-temperature conductivity σ of the films is plotted against the ratio of the number of dopant agent molecules to the number of silane molecules in the gaseous mixture. In the center of the graph, the conductivity around 10^{-6} S/m is representative of undoped a-Si:H films, which typically are n-type material. Thus, even a small quantity of P atoms (donors) increases σ rapidly. In the case of B atoms (acceptors), however, we see that initially σ decreases to about 10^{-10} S/m. This is connected with the transition from n-type to p-type material (LeComber & Spear, 1979; Tyczkowski, 2004).

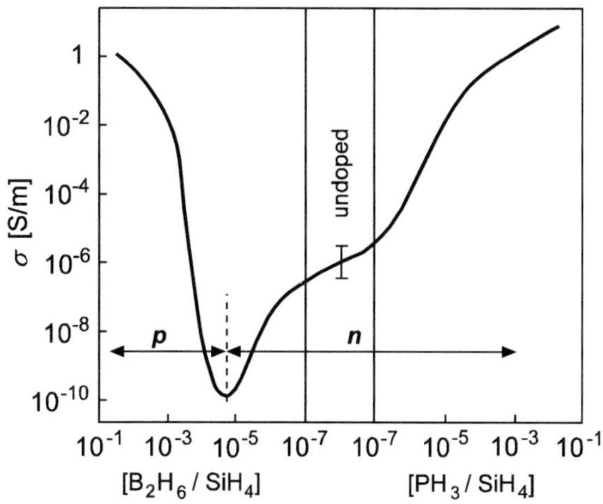

Fig. 5. Room temperature conductivity σ of n- and p-type a-Si:H, plotted as a function of the gaseous precursor ratio (Tyczkowski, 2004).

[2] For more detailed description of the electronic structure models for plasma deposited amorphous films, see, for example, (Tyczkowski, 2004).

Thin-film solar cells are without doubt one of the most spectacular application (except thin-film transistors) of a-Si:H films obtained by plasma polymerization processes. Since the demonstration of the first a-Si:H photovoltaic devices at RCA laboratories in 1976 (Carlson & Wronski, 1976) there has been remarkable progress in the development of a-Si:H solar cells, spurred, first of all, by the wide demand for a low cost, clean and safe energy (Mueller, 2009).

The most extensively studied a-Si:H solar cells, due to their highest conversion efficiencies, are those fabricated in the form of *p-i-n* devices. Typically, the *p*-a-Si:H film is less than 10 nm, the undoped *i*-a-Si:H is between 200 and 700 nm and *n*-a-Si:H layer is approx. 30–50 nm. The layers are deposited on each other in successive plasma reactor chambers connected by vacuum locks. A metallic electrode is used as a substrate. An opposite optically transparent and electrically conducting electrode (e.g. ITO film) is deposited from the top. Today, the commercial large-area solar cell modules based on a-Si:H are fabricated with a stabilized conversion efficiency (the ratio of the maximum power output to the solar energy input) in the 4–6 % range (Green, 2007).

To improve the efficiency and stability of a-Si:H solar cells a lot of various experimental investigations coupled with theoretical device modeling and design analysis have been carried out for the past two decades. Among the most important ideas are multiple-junction solar cell structures fabricated from different amorphous semiconducting films, in addition to a-Si:H, such as a-Si_XC_Y:H, a-Si_XGe_Y:H, and other related films. These films, classified as silicon-based alloys, can be plasma deposited from single precursors, e.g. tetramethylsilane ($Si(CH_3)_4$) or mixtures of precursors such as SiH_4 and CH_4, SiH_4 and GeH_4, etc. The films can be also doped with donor or acceptor atoms introduced to their structure during the plasma deposition process by addition of an appropriate dopant agent. An example of such a multiple-junction solar cell could be a triple-junction system composed of three *p-i-n* structures plasma-deposited on top of one another, prepared consecutively with the following amorphous semiconductors: a-Si:H, a-$Si_{X1}Ge_{Y1}$:H and a-$Si_{X2}Ge_{Y2}$:H. The system gives the stable efficiency of 10.4 % for a 900 cm^2 module (Green et al., 2011).

Apart from the thin films of amorphous semiconductors, nanocrystalline form of these materials has also attracted much attention due to its higher efficiency and stability. While an amorphous semiconductor, for example, a-Si:H is a single phase material, its nanocrystalline form (nc-Si:H, also very often, but less correctly, called μc-Si:H) can be described as a bi-phasic material consisting of a dispersion of silicon nanocrystals embedded in silicon or other silicon-based hydrogenated amorphous matrices, whose volume fraction could be varied by selecting the proper plasma deposition conditions. The nc-Si:H films have been recently demonstrated to be an interesting alternative to a-Si:H films (Conibeer et al., 2006).

Although the efficiency of a-(or nc-)Si:H based solar cells is considerably lower than, for example, that for Si crystalline cells (25.0±0.5 %) or GaAs crystalline cells (26.1±0.8 %) (Green et al., 2011), much lower cost and easy of production are the major arguments in favor of the plasma-deposited solar cells. It should be noted that the last word has not been said yet in this regard and further significant progress is expected, provided that the substantial evolution in the field of new materials will be achieved.

Among the promising materials in this respect are thin films (e.g. a-Ge_XC_Y:H, a-Si_XC_Y:H, a-Sn_XC_Y:H, a-Pb_XC_Y:H), which can exist in two forms with totally different electronic structures, namely, the semiconducting (a-S) and insulating (a-I) form. In general, a typical amorphous semiconductor is characterized by localized states that form only short tiles in the mid-gap above and below the valence and conduction bands of extended states whereas in amorphous insulators all states are localized. In consequence, drastic differences in electrical and optical properties of a-S and a-I films are observed. It turned out that a particularly useful system for the production of these two qualitatively different materials is the three-electrode reactor presented in *Sec. 2*. Both a-S and a-I can be fabricated in this reactor from a single precursor (organometallic or organosemimetallic compound, e.g. tetramethylsilane, tetramethyltin, etc.) in the same deposition process, only by changing the impact energy of ions bombarding the growing film. The ion energy is controlled by the sheath voltage of the small electrode ($V_{(-)}$), which in turn is governed by the coupling capacitance (Fig. 3). It is especially striking that very small variation of the sheath voltage in a defined range of its values is sufficient to create the step change in the electronic structure of deposited film. This "transition" between two forms of the film (a-I and a-S) has been called the a-I–a-S transition. As an example, changes in electrical conductivity σ of selected films deposited in the three-electrode reactor are shown in Fig. 6 (Tyczkowski, 2004, 2006).

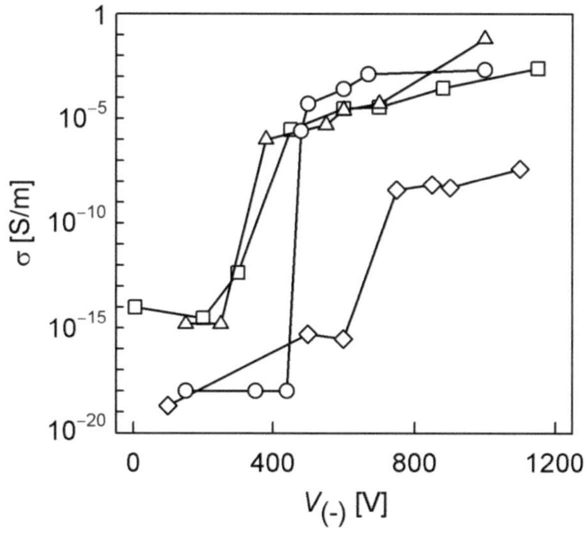

Fig. 6. Electrical conductivity σ of films deposited in the three-electrode AF reactor from organic derivatives of the carbon family as a function of $V_{(-)}$: (◇) a-Si_XC_Y:H, (○) a-Ge_XC_Y:H, (△) a-Sn_XC_Y:H, (□) a-Pb_XC_Y:H (Tyczkowski, 2006).

The existence of two qualitatively different, from the electronic structure point of view, types of plasma deposited films offers new possibilities for material technology consisting in preparation of a novel class nanocomposites formed from insulating and semiconducting fractions (in the form of clusters or layers) deposited in the same plasma process from the same single-source precursor. Without a doubt, such nanocomposites are not only interesting for solar cells, but also for other electrochemical systems. An attempt to use a-Ge_XC_Y:H films for water splitting are being undertaken now in our laboratory.

4.2 Water splitting systems

The most desirable method for production of hydrogen, which represents a sustainable fuel of the future, is photoelectrochemical (PEC) splitting of water by visible light. Theoretically, the PEC production of hydrogen has the capacity to provide global energy security at potentially low cost (James et al., 2009).

The most critical issue in PEC hydrogen generation is the development of a high-performance photoelectrode that exhibits high efficiency in the conversion of solar energy into chemical energy, resistance to corrosion in aqueous environment, and low processing costs. Metal oxides are most promising in this regard (Walter et al., 2010). After four decades of intensive research, however, no material has been found to simultaneously satisfy all the criteria required for widespread PEC application. No wonder that a broad search for new materials for photoelectrodes is still ongoing. Cold plasma technology is also involved in this activity (Randeniya et al., 2007; Slavcheva et al., 2007; Walsh et al., 2009; Zhu, F. et al., 2009).

A typical simple PEC cell with schematic representation of charge transfer is shown in Fig. 7. The cell is constructed from a semiconducting photoanode and metal cathode. The basic reaction steps involved in the PEC process in the cell are as follows (Nowotny et al., 2006): *1*. Photoionization over the band gap of the semiconductor:

$$h\nu \rightarrow e^- + h^+ , \quad (1)$$

where h is the Planck constant and ν is the light frequency; *2*. Charge separation:

$$e^- + h^+ \rightarrow e^-_{\text{bulk}} + h^+_{\text{surface}} ; \quad (2)$$

3. Reaction between water molecules and holes at the surface of the photoanode:

$$H_2O + 2h^+ \rightarrow 2H^+ + \tfrac{1}{2}\, O_2; \quad (3)$$

4. Transport of hydrogen ions from the photoanode to the cathode through the liquid electrolyte; *5*. Transport of electrons to the cathode thorough the external circuit; and *6*. Reaction between electrons and hydrogen ions at the cathode:

$$2e^- + 2H^+ \rightarrow H_2 . \quad (4)$$

The first PEC cell for water splitting, with a rutile TiO_2 photoanode and Pt counter cathode, was reported in 1972 (Fujishima & Honda, 1972). Following this discovery, intensive studies aiming at increasing the energy conversion efficiency of solar energy into chemical energy have been carried out, mainly on the analogous PEC cells, using TiO_2 as the photoanode. Then other oxides, e.g. Fe_2O_3 and WO_3, have been also tested. Despite the good catalytic activity of oxides such as TiO_2, they are generally limited by too large band gaps (approx. 3 eV), which fail to absorb a significant fraction of visible light, resulting in poor solar to hydrogen conversion efficiencies under terrestrial conditions. This value should be reduced to 1.7–2.0 eV. Besides, there is a range of other problems, like incorrect alignment of band edges with respect to the water redox potentials, energy losses due to charge recombination, low density of surface active sites reacting with water molecules, low corrosion resistance, etc. Thus, the majority of PEC oxide research has focused on trying to solve these problems by the modification of known photoactive oxides (through their doping or alloying), and very recently also by creation of new materials. Invaluable in this respect seems to be the cold plasma technology, which allows to design the structure of fabricated materials in a very wide range (Walsh et al., 2009).

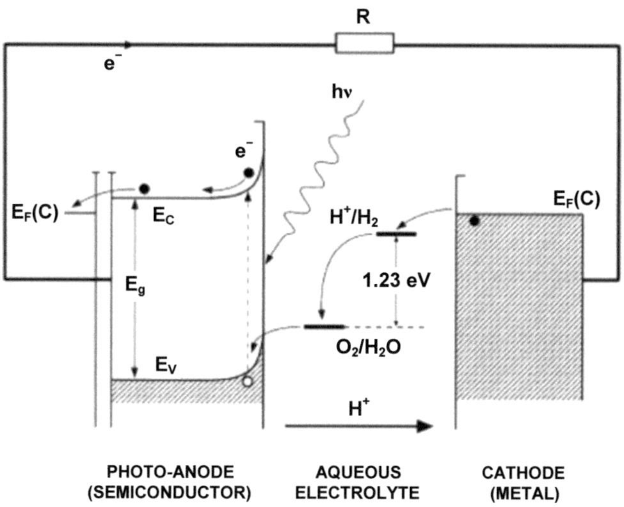

Fig. 7. Schematic representation of charge transfer within a photoelectrochemical cell involving a semiconducting photoanode and metal cathode (Nowotny et al., 2006).

The most attractive oxide to date, namely TiO_2, has already been repeatedly produced by the cold plasma technology. In most reported works, either $TiCl_4$ or Ti alkoxides (mainly titanium tetraisopropoxide, $Ti(OC_3H_7)_4$) are used as the Ti-containing precursors of the plasma polymerization (PECVD) process, resulting in amorphous or crystalline films, with the nonstoichiometric (TiO_X) or stoichiometric (TiO_2) structure. For all of these films, their physicochemical properties are strongly dependent on the film structure, which can be effectively controlled by the deposition conditions (Battiston et al., 2000; Borrás et al., 2009; daCruz et al., 2000; Maeda & Watanabe, 2005; Nakamura et al., 2001).

TiO_2 films are also obtained in a wide range by the reactive sputtering, usually using pure titanium as a target, and O_2 as reactive gas. Similarly, as in the PECVD technique, also in this case, the sputtering process conditions control the structure of the deposited films, which in turn affects to a large extend the optical and photoelectrochemical properties of the films (Brudnik et al., 2007; Dang et al., 2011; Huang et al., 2011).

A particularly useful feature of the cold plasma technology is the possibility of co-deposition either by copolymerization of a mixture of precursors or by co-sputtering using more than one target or a mixture of several reactive gases. In this way we can get doped films as well as films with alloy-type structures. Numerous studies have been already done on the introduction into the TiO_2 structure other atoms (e.g. C, N, S). For example, PECVD with the DC discharge carried out using mixtures of $Ti(OC_3H_7)_4$ and nitrogen led to a Ti(OCN) film structure (Randeniya et al., 2007; Wierzchoń et al., 1993). Instead of nitrogen, ammonia can be introduced to plasma reactors (Weber et al., 1995). Precursors containing nitrogen in their chemical structure, e.g. tetrakis(dimethylamido) or (diethylamido)-titanium ($Ti(N(CH_3)_2)_4$ or

$Ti(N(C_2H_5)_2)_4$) were also utilized (Raaijmakers, 1994). By properly adjusting the composition of the reaction mixture and the conditions of the plasma, a film structure similar to stoichiometric TiN can be obtained (Weber et al., 1995). In turn, the films composed mainly of titanium and carbon (TiC_X) were fabricated leading PECVD process in a mixture of $TiCl_4$ and hydrocarbons (Täschner et al., 1991). TiO_2 films with N and C atoms were also obtained by reactive sputtering of titanium in an appropriate gas mixture. This method also proved to be useful for the production of nanocomposite thin films for photoanodes, e.g. Au:TiO_2 films sputtered (using an RF discharge) from Ti and Au targets in O_2 as the reactive gas (Naseri et al., 2011).

The cold plasma deposition method has been used to produce, in addition to films based on TiO_2, other films that constitute an interesting material for the photoelectrodes. For example, iridium oxide (IrO_2) (Slavcheva et al., 2007), tantalum nitride (Ta_3N_5) (Yokoyama et al., 2011), ruthenium sulfide (RuS_2) (Licht et al., 2002), and tungsten trioxide (WO_3) (Garg et al., 2005) films were prepared by reactive sputtering. WO_3 was also deposited by PECVD technique, feeding the RF plasma reactor with a gas mixture of tungsten hexafluoride (WF_6) and oxygen (Garg et al., 2005).

Recently, a proposal to employ a-Si_XC_Y:H films as photoelectrodes for PEC cells has been presented (Zhu, F. et al., 2009). The films were fabricated by PECVD using a SiH_4, H_2 and CH_4 gas mixture. It was found that the a-Si_XC_Y:H photoelectrode behaves as a photocathode, where the photo-generated electrons are injected into the electrolyte and reduce H^+ ions for hydrogen evolution. The use of this photoelectrode led to a solar-to-hydrogen conversion efficiency higher than 10%. It should be noted that a-Si_XC_Y:H films deposited by PECVD technique have been extensively investigated for a long time. A lot of gas mixtures (e.g. SiH_4 and hydrocarbons) or single compounds (e.g. tetramethylsilane ($Si(CH_3)_4$)) are used as precursors of the deposition process. In some cases, dopant agents are also added. In fact, the films can be accurately produced according to the designed electronic structure and photoelectronic properties, which can change over a very wide range. For instance, their electrical conductivity changes from 10^{-18} to 0.1 S/m and the optical gap shifts between 1.8 and 3.2 eV (Tyczkowski, 2004).

In the past five years, cobalt has emerged as the most versatile non-noble metal for the development of synthetic H_2- and O_2-evolving catalysts. Among the various structures containing cobalt atoms, cobalt oxides appear to be particularly promising materials. The possibility of using such oxides to catalyze water oxidation in neutral aqueous solutions has recently experienced a burst of interest (Artero et al., 2011). There have been many reports in the literature concerning the use of cobalt oxides, mainly Co_3O_4, as electrode coatings that catalyze water oxidation. Many different methods have been applied to prepare these coatings, among others, also the cold plasma deposition technique - both PECVD and reactive sputtering. In PECVD, cobalt oxide films were obtained from volatile precursors such as bis(acetylacetonate)cobalt(II) (Fujii et al., 1995), bis(2,2,6,6-tetramethylheptan-3,5-dionato)cobalt(II) (Barreca et al., 2011), bis(cyklopentadienyl)cobalt(II) (Donders et al., 2011) or cyclopentadienyl(dicarbonyl)cobalt(I) (Tyczkowski et al., 2007). Other volatile cobalt complexes (e.g. amidinates and cyclodexrtins) are now also proposed as the precursors (Li et al., 2008; Papadopoulos et al., 2010). The sputtering process, in turn, was conducted in the presence of pure Co or Co_3O_4 as targets and plasma generated in a gas mixture containing O_2 (Ingler Jr et al., 2006; Schumacher et al., 1990). The reactive sputtering technique was also used to produce ternary cobalt spinel oxides of the type CoX_2O_4 (X = Al, Ga, In), taking the targets from Co_3O_4 and Al, Ga_2O_3 or In_2O_3. Preliminary research showed that although these materials combine excellent stability in solution and good visible light absorption properties, their performance as photoelectrochemical catalysts for water splitting is limited by the poor electrical transport properties. It is hoped, however, that the broad capabilities of plasma technology will help overcome this problem (Walsh et al., 2009).

Particularly significant is the finding that nanoclusters of Co_3O_4 are much more efficient in the PEC process than larger objects (e.g. micrometer-sized particles) of this oxide. Thus, it is very important to develop methods for producing cobalt oxides films containing Co_3O_4 nanoclusters. Recently, mesoporous silica has been used as a scaffold for growing Co_3O_4 nanocrystals within its naturally parallel nanoscale channels via a wet impregnation technique. It has been found that rod-shaped crystals measuring 8 nm in diameter and 50 nm in length are interconnected by short bridges to form bundled clusters. The bundles are shaped like a sphere with a diameter of 35 nm (Jiao & Frei, 2009). This report aroused great scientific interest. However, it should be noted that films composed of nanoclusters can be easily deposited on a flat surface without any special mesoporous structure, only involving the plasma deposition technique for this purpose. Using this method, nanocrystalline films of cobalt oxides have been already obtained. For example, small particles of CoO_X in the range of 2-10 nm in diameter were deposited in this way on TiO_2 support (Dittmar et al., 2004). If CoO_X films were fabricated on a substrate at elevated temperature (150–400°C), then columnar grains with average diameter size at the film surface of 35-60 nm were formed (Fujii et al., 1995). Research conducted recently in our laboratory has led to the cobalt oxide films containing 4–8 nm sized Co_3O_4 crystals, whose size can be controlled by the plasma deposition process (Tyczkowski, 2011; Tyczkowski et al., 2007). These films will be discussed in more detail in *Sec. 5.1*. It should be noted, however, that we are also now starting work on their application to the water splitting.

Although research on water splitting, conducted using the cold plasma technology to produce thin-film coatings on photoelectrodes, is only beginning, the obtained results give cause for great hope. Plasma deposited films by both PECVD and reactive sputtering can reveal very high incident photon conversion efficiency (Randeniya et al., 2007). These films also appear to be better as photoelectrodes than the corresponding materials produced by other method (Naseri et al., 2011).

The PEC process can be realized not only in cells with two electrodes separated from each other, which was discussed above, but also when the electrodes are in direct close contact. Much attention in this regard has been paid for systems where both electrodes are located within a single grain. Such a structure may be considered as a microsized PEC cell. The best analogue of the PEC cell shown in Fig. 7 is a microsized cell formed of a small semiconductor grain (e.g. TiO_2) and noble metal islets (e.g. Pt) deposited on its surface. Then, the surface of the semiconductor grain and the metal act as photoelectrode and counter electrode, respectively. More sophisticated systems with a bifunctional catalyst have been also proposed. In this case, anodic and cathodic photoactive islets (several nm in size) are deposited onto the same semiconducting nanoparticle (tens of nanometers in size) (James et al., 2009). These microsized PEC cells produce, however, a mixture of oxygen and hydrogen. To receive these gases separately, reactors composed of two chambers connected by a diffusion bridge are used. In one chamber there is a suspension of nanoparticles only with an anodic catalyst while in the second one - only with a cathodic catalyst (James et al., 2009).

In the field of microsized PEC cells, the plasma techniques have also proved to be very useful, both in the synthesis of semiconducting particles and in the deposition of active catalysts of them. For the production of powders, in a very wide range of grain sizes (from single nanometers to tens of micrometers) mainly the thermal plasma (see: Fig. 1) is utilized (Ctibor & Hrabovský, 2010; Karthikeyan et al., 1997). The PECVD technique is, however, also engaged for this purpose. By this technique, TiO_2 nanocrystalline powder was prepared, using an AF glow discharge (40 kHz) and titanium tetraisopropoxide ($Ti(OC_3H_7)_4$) with oxygen as a reactive mixture. The obtained nanocrystalline particles, with mean size of about 25–55 nm, revealed good photocatalytic activity (Ayllón et al., 1999). Another example is the synthesis of carbon-supported ultrafine metal particles by MW plasma from metal carbonyls (e.g. $Fe(CO)_5$, $Co_2(CO)_8$, $Mo(CO)_6$) as precursors (Brenner et al., 1997). In turn, noncrystalline organosilicon powder was produced by plasma polymerization of tetramethylsilane ($Si(CH_3)_4$). The ratio of elements (Si/C) as well as the chemical structure of the grains was highly dependent on the plasma process conditions (Fonseca et al., 1993).

As already mentioned earlier, an important feature of the PECVD method is the possibility of copolymerization. This route was used to prepare TiO_2 nanoparticles doped with Sn^{4+} ions. The plasma process was performed in a mixture of $TiCl_4$ and $SnCl_4$ with an appropriate molar ratio. It was found that photocatalytic activity of TiO_2–Sn^{4+} nanoparticles was much higher than those of the pure TiO_2 (Cao et al., 2004). Wide possibilities of the plasma technique also allow to produce nanoparticles of doped semiconducting organic polymers, e.g. polypyrrole plasma doped with iodine (Cruz et al., 2010). This is only a matter of time when such organic nanoparticles with anchored a molecular water oxidation catalyst will be produced by plasma polymerization.

The deposition of catalytically active coatings onto the surface of already prepared particles has also been performed. In this case, however, to achieve efficient coating of the nanoparticles, a special construction of the plasma reactor chamber is needed, for example, with 360° continuous rotation. In such a reactor, TiO_2 nanoparticles were coated with thin film produced by plasma polymerization of tetramethyltin ($Sn(CH_3)_4$). Subsequently, the coated particles were heated in air to remove the carbonaceous material while, simultaneously, oxidizing tin atoms to tin oxide. To obtain partially fluorinated tin oxide, hexafluoropropylene oxide (C_3F_6O) was added to $Sn(CH_3)_4$. It should be noted that significantly increased photocatalytic activity of TiO_2 nanoparticles was achieved using the PECVD approach (Cho, J. et al., 2006).

Finally, one more type of water splitting cells should be mentioned, namely integrated photovoltaic-electrolysis (PV-PEC) cells. In this type of devices, the photovoltaic cell and the electrolyser are combined into a single system, in which the light-harvesting solar cell is one of the electrodes. Very often, thin-film solar cells fabricated by the cold plasma deposition method are employed in the PV-PEC devices (Kelly & Gibson, 2006). A diagram of such a system with a simply a-Si:H solar cell is shown in Fig. 8. There is no doubt that the role played by the cold plasma deposition technique in the creation of such systems is unquestionable (see: *Sec. 4.1.*).

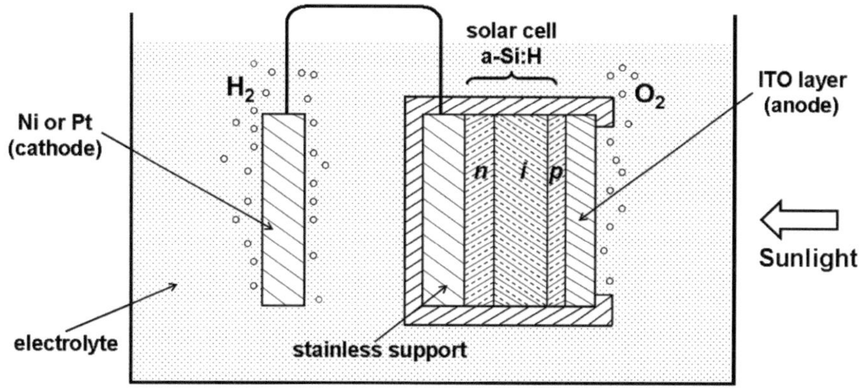

Fig. 8. Schematic representation of a photovoltaic–electrolysis system based on a simply a-Si:H solar cell (Kelly & Gibson, 2006).

5. New types of electrodes for electrochemical cells

The potential applications of plasma polymers as materials for electrochemistry are primarily associated with the possibility of designing their electronic structure and thereby with the designing of their electrical properties. Plasma deposition from diverse precursors and their mixtures, performed under various conditions of the process, leads to a huge variety of thin films characterized by a broad spectrum of electrical properties. In Fig. 9 a diagram of the conductivity typical for different plasma polymer types is presented. As one can see, the whole range from 10^{-18} S/m to 10^{6} S/m is covered by plasma deposited films. Insulating, semiconducting (of different types of conductivity) and metallic films can be obtained in this way (Tyczkowski, 2004). Some of these films can reveal a significant activity in electrochemical processes. Photocatalytic activity of such films was already discussed in the previous section (*Sec. 4.2.*).

Taking into account the electrical properties mentioned above as well as another important feature of plasma polymers, namely the membrane nature, which is very often revealed by these films, a new fascinating electrocatalyst structure has been proposed. An effective electrocatalyst must satisfy many requirements, such as high activity, high electrical conductivity, and long-term stability, which may be in conflict with each other. One possible way to solve these conflicts is the use of composite materials, where the matrix and the dispersed phase are independently selected. A successful approach is that of associating a highly conducting (though catalytically inert) matrix with an active (though less conducting) dispersed phase. When the matrix is permeable, one has a three-dimensional (3D) catalyst: all catalytic particles are active, irrespective of their position in the composite. Fig. 10 schematically represents the operation of such a system.

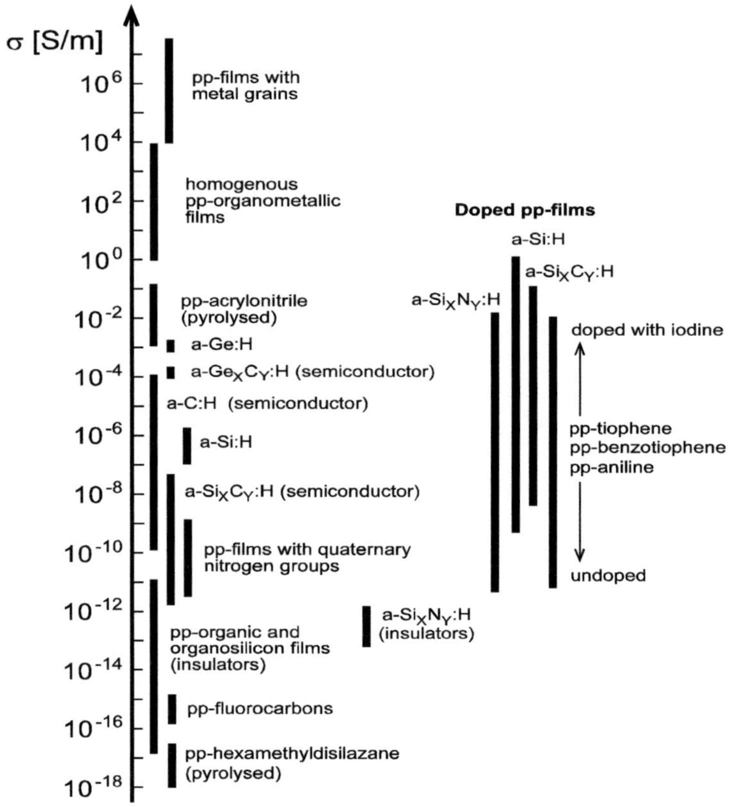

Fig. 9. Room-temperature conductivity σ of the main groups of plasma polymers (Tyczkowski, 2004).

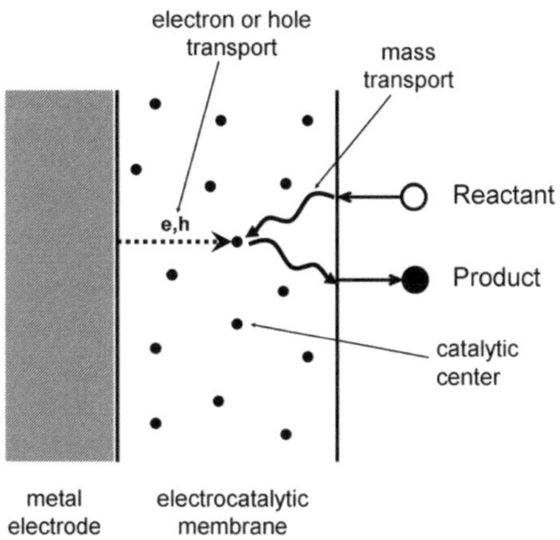

Fig. 10. Model of a three-dimensional (3D) electrocatalytic membrane (Tyczkowski, 2010).

One of the first 3D-electrocatalyst electrodes was tested for the reduction of molecular oxygen. The electrode was made of a conducting matrix, formed by plasma polymerization of acrylonitrile, and cobalt atoms playing the role of active centers, which were introduced to the matrix during the deposition process from bis(acetylacetonate)cobalt(II) precursor (Doblhofer & Dürr, 1980). Just this one example shows how the great possibilities for the construction of electrochemical electrodes are inherent in the plasma deposition technique.

5.1 Catalytic electrodes for fuel cells

Fuel cells are electrochemical membrane reactors that are able to convert chemical energy from a fuel directly to electrical energy through a chemical reaction with oxygen. Hydrogen is the most common fuel, but alcohols like methanol and hydrocarbons such as CH_4 are also used. Although there are many types of fuel cells, all of them consist of two electrodes (negative anode and positive cathode) and an electrolyte (liquid or solid) that allows ionic charges to move between the electrodes. Recently, a lot of effort has gone into improving the quality, reliability, and efficiency of these components by their modification and introduction of new materials (Carrette et al., 2000; Sundmacher, 2010). The cold plasma technology is also widely involved in this process.

For more than a decade, the polymer electrolyte (membrane) fuel cells (PEFC), which can be fed with hydrogen (when proton-exchange membranes are used, such fuel cells are called PEMFC) or, for example, fed directly with methanol (direct methanol fuel cells (DMFC)), are one of the most extensively investigated types of fuel cell worldwide. This results from their high energy-conversion efficiency, relatively low operating temperature, and high power density. The basic catalytic reactions taking place at the electrodes of PEMFC are as follows (Carrette et al., 2000):

$$\text{anode} \qquad H_2 \rightarrow 2H^+ + 2e^- \tag{5}$$

$$\text{cathode} \qquad {}^1/_2\, O_2 + 2H^+ + 2e^- \rightarrow H_2O \tag{6}$$

The anode reaction consists of hydrogen oxidation to protons (hydrogen oxidation reaction - HOR). The protons migrate through the membrane to the cathode. At the cathode, oxygen is reduced (oxygen reduction reaction - ORR) and then recombines with the protons to form water. The electrodes have to be porous to gas diffusion to ensure the supply of the reactant gases to the active zones, where a catalytic material is in contact with the ionic (membrane) and electronic (catalyst substrate) conductors. Similar reactions occur at the electrodes of DMFC (for a proton-exchange membrane) (Carrette et al., 2000):

$$\text{anode} \qquad CH_3OH + H_2O \rightarrow CO_2 + 6H^+ + 6e^- \tag{7}$$

$$\text{cathode} \qquad {}^3/_2\, O_2 + 6H^+ + 6e^- \rightarrow 3H_2O \tag{8}$$

Platinum-based materials (Pt or Pt alloys) are by far the best catalysts for the hydrogen and methanol oxidation (Eqs. (5) and (7)) as well as oxygen reduction reactions (Eqs. (6) and (8)). Unfortunately, Pt is a precious, very expensive metal, which limits the widespread commercialization of Pt-based fuel cells. Besides, the stability of Pt and Pt alloys becomes a serious problem for long-term operation of the cells. Hence, extensive research is underway to overcome these difficulties. The works are going in two directions. Firstly, the new methods to ensure consumption of smaller amounts of platinum and at the same time providing a more stable and effective catalyst are developed. And secondly, the new alternative catalysts that are cheaper than platinum and exhibit at least a comparable catalytic activity are sought (Sundmacher, 2010; Wang, 2005).

As far as the methods of reducing the amount of platinum catalyst are concerned, the most promising appears to be the plasma sputtering technique. By this method, the catalytic electrodes can be prepared with a platinum loading down to 0.005 mg_{Pt}/cm^2, that is drastically lower than that for conventional Pt electrodes (0.5–1.0 mg_{Pt}/cm^2), with no detrimental effect on fuel cell performance. The Pt catalyst is dispersed as nano-clusters of controlled size (sometimes, until less than 2 nm) and controlled crystalline structure that determines the concentration of catalytically active centers. It should be emphasized that such a possibility, in principle, is given only by the application of cold plasma (Caillard et al., 2005; Caillard et al., 2009; Saha et al., 2006; Xinyao et al., 2010). As it was already mentioned, this technology also allows to put the deposited material on virtually any substrate. Thus, the Pt catalyst can be sputtered on both a porous carbon substrate forming the electrode and the surface of a polymer electrolyte (e.g. Nafion), to which the carbon electrode is then pressed.

The main requirement of a good electrode is a three-phase boundary between the fuel supply, the catalyst particle and the ionic (polymer) electrolyte. The catalyst particles also must be in direct contact with the electron conducting electrode (Carrette et al., 2000). To ensure such a contact and at the same time maximize the interphase boundary, co-sputtering or co-deposition of carbon-based material and platinum was used. Materials classified as 3D-electrocatalysts can be obtained in this way. For example, a simultaneous co-sputtering of carbon and platinum on a conventional carbon porous substrate led to high electrodes efficiency for both the hydrogen oxidation and oxygen reduction reactions. The PEMFC tested in this case achieved a specific power of 20 kW per 1 g of platinum, which is one of the best results reported so far (Cavarroc et al., 2009). Carbon and platinum can be also deposited by subsequent sputtering processes. First a porous columnar carbon film (column diameter of 20 nm) is deposited, and then these nanocolumns are decorated by Pt nanoclusters (Rabat & Brault, 2008).

Recently, interesting results have been obtained in this field by combining the plasma polymerization and sputtering methods. For instance, the synthesis of composite thin films made of platinum nanoclusters (3–7 nm) embedded in a porous hydrocarbon matrix was carried out by simultaneous PECVD of pp-ethylene and sputtering of a platinum target. The metal content in the films could be controlled over a wide range of atomic percentages (5–80%) (Dilonardo et al., 2011). Aniline mixed with functionalized platinum nanoparticles as a precursor of PECVD was, in turn, used to prepare a typical 3D-catalyst. The plasma deposition was performed under atmospheric pressure conditions. Plasma polymerized aniline (pp-aniline), which is characterized by both electronic and ionic conductivity, associated with the Pt catalyst in a 3D porous network, without doubt lead to the development of the three-phase boundary (Michel et al., 2010).

Another idea is to deposit on the electrode surface carbon nanofibers (nanotubes) by PECVD and then decorating them by sputtered platinum nanoclusters. Generally, three consecutive plasma deposition steps are carried out to this end: the sputtering of a catalyst used to initiate the growth of the carbon nanofibers (e.g. Fe, Ni, Co), the creation of the nanofibers by PECVD from a mixture of precursors (e.g. CH_4/H_2, CH_4/N_2) and the sputtering of platinum. The plasma produced systems consisting of carbon nanofibers with a diameter of nanometers (13–80 nm) and a length of micrometers (2–20 μm) decorated with 2–5 nm Pt nanoclusters were already the object of research. It was found that such systems prepared on the electrode substrate (carbon cloth or carbon paper) significantly improve the performance of both PEMFC and DMFC compared with the conventional electrodes (Soin et al., 2010). By the way, it should be added that vertically aligned carbon nanotubes as well as graphene layers fabricated by PECVD have recently attracted research interest as supercapacitor electrode materials (Amade et al., 2011; Zhao et al., 2009).

In addition to extensive research into the production of platinum catalysts by the sputtering method, it was also trying to get them through the plasma polymerization method (PECVD). A platinum-containing organic complex, bis(acetylacetonate)platinum(II) ($Pt(acac)_2$), which is characterized by a relatively low sublimation temperature (160–170°C), was used as a precursor of PECVD carried out in an RF discharge. The plasma-polymerized film was then calcined to drive off organic material, leaving behind a catalyst-loaded substrate (Dhar et al., 2005a). The same procedure was used to prepare a composite consisting of ZrO_2 support and Pt catalyst. The support and the catalyst were deposited on a metallic substrate by PECVD as alternate layers from $Zr(acac)_4$ and $Pt(acac)_2$, respectively. It was found that Pt agglomerates were embedded in the zirconia support (Dhar et al., 2005b).

Other potential path of development of fuel cells, in addition to improving the properties of platinum electrodes, is searching for new catalytic materials. This research is mainly focused on the cathode materials, at which the oxygen reduction reactions (ORR) (Eqs. (6) and (8)) constituting the bottleneck in the fuel cell operation proceed (Wang, 2005). The cold plasma technology creates potential and real opportunities in this regard (Brault, 2011). Like the platinum catalyst, also in this case the plasma sputtering technique was used. For example, CoS_2-based thin films were prepared by this method. Electrochemical assessment indicated that the films had significant ORR catalytic activity (Zhu, L. et al., 2008). Similarly, a high ORR catalytic activity showed niobium oxinitride (Nb-O-N) films prepared by plasma (RF) reactive sputtering from a Nb metal plate under various partial pressures of N_2 and O_2 (Ohnishi et al., 2010).

Taking into account the promising electrocatalytic activity for ORR demonstrated by nanoparticles of cobalt oxides (Manzoli & Boccuzzi, 2005), we have undertaken in our laboratory an attempt to produce such a material for the PEMFC electrodes by the plasma polymerization method. The films containing CoO_X were deposited in a parallel plate RF (13.56 MHz) reactor shown in Fig. 2. Cyclopentadienyl(dicarbonyl)cobalt(I) ($CpCo(CO)_2$) was used as a precursor. As a result of the plasma deposition process, very thin films (25–750 nm) composed of a hydrocarbon matrix and amorphous CoO_X were obtained. The amorphousity was determined by the electron diffraction pattern. However, only a moderate thermal treatment was enough to transform the amorphous films into films with nanocrystalline structure of cobalt spinel (Co_3O_4) (Fig. 11). The creation of cobalt spinel nanocrystals was supported by Raman spectroscopy measurements. The electron diffraction and Raman spectroscopy measurements also allowed us to determine the nanocrystals size. It was found that this size can be controlled by parameters of the plasma deposition process. As an example, Fig. 12 shows a dependence of the average size of Co_3O_4 nanocrystals on the flow rate of $CpCo(CO)_2$ vapor through the plasma reactor. This simple example is just enough to show that the cold plasma is a very useful technique, not only for the fabrication of new materials, but also for the precise control of their structure. In the same way as the films with CoO_X, films containing CuO_X were deposited using bis(acetylacetonate)copper(II) ($Cu(acac)_2$). After feeding the reactor by a mixture of $CpCo(CO)_2$ and $Cu(acac)_2$ vapors, thin films of Co–Cu mixed oxides were also fabricated (Tyczkowski, 2011; Tyczkowski et al., 2007).

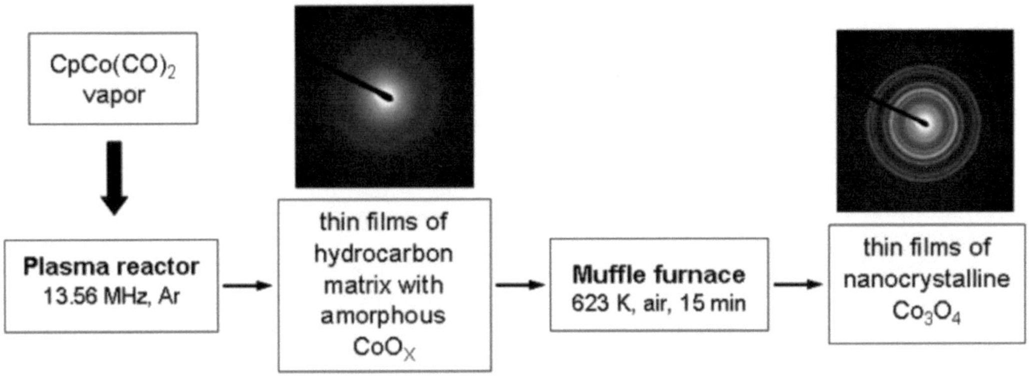

Fig. 11. Outline of the preparation process of nanocrystalline Co_3O_4 films by plasma deposition.

To test the electrocatalytic activity of the above-mentioned films, electrodes for PEMFC were prepared by the deposition of these films on a carbon paper substrate. Then the samples were thermally treated forming anode materials. The opposite electrode (cathode) was prepared from the same carbon paper covered with 10% Platinum on Vulcan XC-72 catalyst (1.56 mg/cm^2) (Kazimierski et al., 2010; Kazimierski et al., 2011). In Fig. 13, preliminary results concerning the current–voltage dependence for the tested fuel cell with various types of anode catalytic materials are shown. Although the characteristics obtained for plasma deposited materials are still far from the model system, in which both electrodes are prepared from Pt (curve f), nevertheless these results are very promising. It is enough to notice that the concentration of catalytic active centers increases with the increase of the deposition time, which is reflected in the improvement of the fuel cell characteristic (curves b, d and e). A simple calculation showed that after 80 min deposition of CoO_X, the whole deposited material loading is only 0.1 mg/cm^2 (moreover, this value is drastically reduced by the annealing) (Kazimierski et al., 2011). However, it seems to be possible to significantly increase the concentration of the centers by optimizing the plasma deposition parameters. Equally important as the centers concentration is the structure of the oxide material. One can see in Fig. 13 that the film composed of Co_3O_4 and CuO_X reveals much higher activity (curve c) than each of these oxides separately (curves a and b). Thus, it is no wonder that further intensive works in this field are planned.

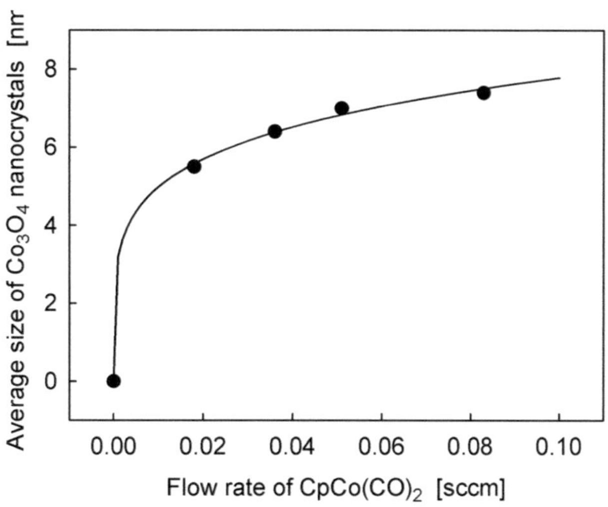

Fig. 12. Dependence of the average size of Co_3O_4 nanocrystals on the flow rate of $CpCo(CO)_2$ vapor through the plasma reactor (Tyczkowski, 2011).

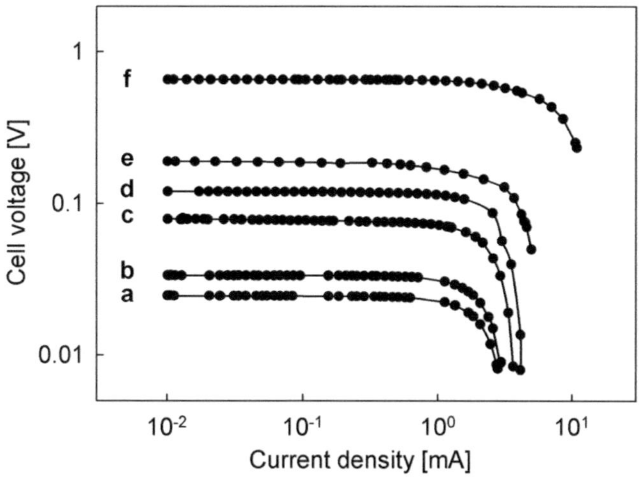

Fig. 13. Fuel cell characteristic for various cathode catalytic materials: (a) - CuO_X (t = 12 min); (b) - Co_3O_4 (t = 12 min); (c) - Co_3O_4+CuO_X (t = 12 min); (d) - Co_3O_4 (t = 60 min); (e) - Co_3O_4 (t = 80 min); (f) - Pt (Tyczkowski, 2011).

5.2 Electrodes for lithium-ion batteries

There has been a significant research work done in the recent past in the development of lithium-ion batteries, which are extensively applied in various electric and portable electronic devices. Although a lot of attractive cathode and anode materials for these batteries are already known, it is still lay much stress on finding new solutions in this area. Increasingly, the cold plasma technology is also used to this end. The first attempts were made in the eighties of the 20th century in Sanyo Electric Co. in Japan, where AF plasma (6.5

kHz) polymerized pyrrole as a conducting cathode layer was use for a Li-ion battery. The pp-pyrrole was deposited on one side of a porous polypropylene separator sheet, a Li layer was vapor deposited on the other site of the sheet, and stainless steel collector layers were formed on both sides by spattering. This battery showed very good properties. Research on the use of polypyrrole (also produced by plasma polymerization) for cathodes of Li-ion batteries is currently being pursued (Cho, S.H. et al., 2007). Another interesting conductive polymer, which was used as the cathode material, is plasma polymerized carbon disulfide (pp-CS_2). These films (approx. 0.5–1 μm) deposited on Pt foil showed satisfactory electrochemical activity in cells vs. Li/Li+. Compared to poly(carbon disulfide) prepared by conventional chemical means, cells having the pp-CS_2 improved cycle life because the plasma polymerized material is more crosslinked and does not depolymerize as readily (Sadhir & Schoch, 1996).

A lot of attention is also paid to entirely new anode materials. Although carbon in nowadays is used as the commercial anode material, it has several shortcomings such as, for instance, low reversible capacity that is usually ranged in 250–300 mAh/g. In order to increase this parameter, silicon with the highest theoretical capacity (e.g., 4000 mAh/g) has been proposed as a new negative electrode material. Unfortunately, this material has also serious drawbacks, which are related to the poor electrical conductivity and drastic volume changes during electrochemical reactions. To solve these problems, silicon and silicide powders with conducting materials such as metals, oxides, and nitrides are used as the composite anodes. The potential of plasma technology for the fabrication of both powders and composite systems are particularly useful in this case. For example, a complex procedure of plasma deposition was used to produce an anode material in the form of copper silicide-coated graphite particles. The graphite particles with mean diameter of 6.0 μm were covered with a very thin film of a-Si:H (30–50 nm) in the PECVD process from SiH_4. Then, copper layer was deposited on the surface of silicon-coated graphite using the next plasma technique, namely RF sputtering. After annealing at a temperature of 300°C, copper silicide was formed. This material used as the anode in Li-ion batteries revealed high capacity properties and good electrical performance (Kim, I.C. et al., 2006).

An example of particles formed by plasma methods for use in the anodes can be the synthesis of monodisperse and non-agglomerated SiO_X nanoparticles by the PECVD method from a mixture of SiH_4 and O_2 in a plasma reactor specially designed for this purpose. It should be emphasized that in this case the nanoparticle size can be controlled by the flow rate of the reactant gases through the reactor. The SiO_X nanoparticles mixed with graphite particles constitutes the anode material (Kim, K. et al., 2010).

More sophisticated materials for Li-ion electrodes were also fabricated by the cold plasma technology. Reactive co-sputtering of Sn and Ru in oxygen plasma has allowed to obtain SnO_2–RuO_2 composite thin films, which reveal unique electrochemical properties (Choi et al., 2004). As the anode, cobalt oxide thin films deposited by reactive sputtering of Co in O_2 plasma were also tested with success. It was found that these films contained Co_3O_4 grains with the size of 4–25 nm (Liao et al., 2006).

6. Other components of electrochemical cells

Plasma deposited thin films are not only very useful for creating electrochemical cell elements where they must be characterized by high electrical conductivity (optoelectronically active materials, electrode materials), but also there, where this conductivity should be as low as possible. Plasma polymers with very low electrical conductivity (see: Fig. 9) are often utilized as a variety of thin-film insulators. In turn, the semi-permeable properties of some plasma polymers allow them to be used as selective membranes. By combining the low electrical conductivity and selective permeability, one can get great barrier materials. Thus, for example, very thin barrier layers for direct methanol fuel cells (DMFC) were produced. For the technical realization of the DMFC, a highly proton conducting polymer electrolyte is necessary. Perfluorosulfonic acid membranes, such as Nafion®, are widely used to this end. Indeed, these membranes have high proton conductivity, but their great disadvantage is too high permeability of methanol molecules that migrate from anode to cathode lowering the cell performance. Deposition of a thin plasma polymer film, produced by PECVD from perfluoroheptane (C_7F_{16}), on the membrane surface decreases the methanol permeability by two-orders of magnitudes (Lue et al., 2007).

The possibility of plasma copolymerization and the production of composite materials allows to design at the molecular level thin films of very low electronic conductivity, but with very high ionic conductivity. Such films can be obtained in the polymer-like form (polymer electrolytes) and as ionic glasses (solid oxide electrolytes). These new solid electrolyte systems enable us to replace the conventional solid electrolytes by the much thinner elements, which in addition have all the other advantages of plasma fabricated materials, for example, selective permeability. Thin films of solid electrolytes produced by cold plasma deposition techniques have been of particular interest recently.

6.1 Solid electrolytes for fuel cells

The most common polymer electrolytes (often called the ion-exchange membranes) for fuel cells are composed of crosslinked macromolecular chains making up a three-dimensional structure on which are distributed some ionizable functional groups giving the membrane its specificity. To maintain the electroneutrality of the material, ionized sites are compensated for by an equivalent number of mobile ions of opposite charge. By jumping between ionized sites, these mobile ions give the membrane its ionic conduction ability. To prepare ion-exchange membranes by plasma polymerization, it is necessary first to choose a precursor with long and flexible chains or, at best, containing spacers in its structure (phenyl groups, for example) and then to initiate the deposition process with a "soft" plasma discharge in order to safely preserve those elements of the precursor likely to constitute the skeleton of the final material. The next criterion, namely, a large quantity of ionizable functional groups favorably distributed in the polymer matrix, requires the selection of a second precursor containing the appropriate ionizable functional group in its structure, which will be embedded in the polymer matrix without defects. A schematic representation of basic processes that occur in the "soft" plasma polymerization of a proton-exchange membrane from styrene and trifluoromethane sulfonic acid (CF_3SO_3H) is shown in Fig. 14.

The first works devoted to the development of plasma-polymerized ion-exchange membranes for fuel cells were carried out in the late 1980's by the Inagaki's group at the Shizuoka University in Japan (Inagaki, 1996). Plasma polymerization of a mixture of fluorinated benzene (C_6F_6, C_6F_5H or $C_6F_4H_2$) and SO_2 gave a Nafion®-like plasma polymer that contains sulfonic acid and sulfonate groups. Such a membrane has the cation-exchange

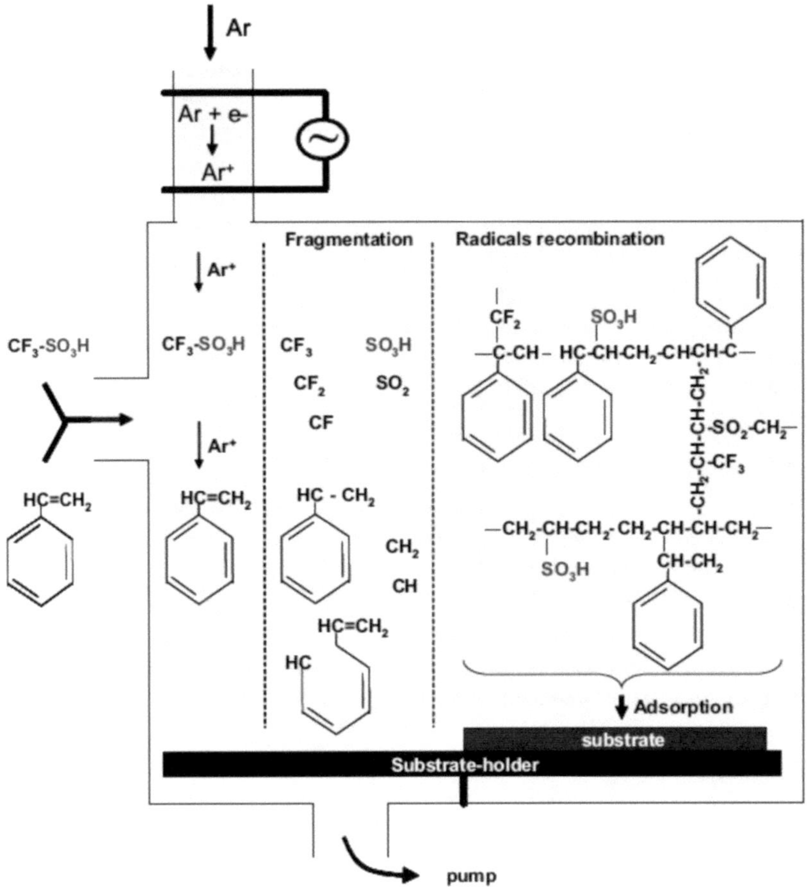

Fig. 14. Schematic representation of the synthesis procedure of proton-exchange plasma polymers (Roualdès et al., 2007).

ability and its main role is to provide the transport of protons from the anode where they are produced by the oxidation of fuel (Eqs. (5) and (7)), to the cathode where they are consumed by the reduction of oxygen into water (Eqs. (6) and (8)). Until recently, the results obtained in this respect were not as good as those for conventional membranes (Roualdès et al., 2007). However, the latest reports provide much more promising results. For example, the mentioned already proton-exchange membranes, which are plasma deposited from styrene and CF_3SO_3H (Fig. 14), can have a higher percentage of proton exchange groups, a higher proton conductivity and a lower fuel permeability, compared with commercially available Nafion® membranes, when appropriate parameters of the plasma process are chosen. Moreover, various plasma procedures are examined in search of these membranes with the best properties. Apart from the typical PECVD method (Ennajdaoui et al., 2010; Roualdès et al., 2007), the remote (after glow) plasma technique (Jiang et al., 2011) and the plasma deposition under high pressure conditions (Merche et al., 2010) have been employed. Other types of proton-exchange membranes prepared by cold plasma deposition, different from those from styrene and CF_3SO_3H, have also been investigated, for example, fluorinated carboxylic membranes from H_2O and C_4F_8 (Thery et al., 2010), phosphorous-doped silicon dioxide membranes from SiH_4, PH_3 and N_2O (Prakash et al., 2008) as well as membranes produced by plasma polymerization from heptylamine ($C_7H_{15}NH_2$) or 1,7-octadiene (C_8H_{14}) and then their treatment by SO_2 plasma (Siow et al., 2009).

The cold plasma technology is also tested for the preparation of anion-exchange membranes, which are presently becoming significant materials for application in alkaline fuel cells, where hydroxyl ions OH^- are the ion charge carriers. Similarly, as in the case of the proton-exchange membranes, the plasma deposition provides formation of very thin crosslinked films with high ion (OH^-) conductivity, high chemical stability and low fuel permeability. The first attempt to obtain such membranes was undertaken only in 2006, when precursors containing tertiary amine groups were plasma polymerized and then the deposited films were quaternized by methyl iodide (Schieda et al., 2006). Very recently, an opposite procedure was applied, namely, the films were plasma polymerized from vinylbenzyl chloride, then the benzyl chloride groups ($-CH_2Cl$) present in the films were quaternized by trimethylamine into $-CH_2N^+(CH_3)_3Cl^-$ groups, and finally these groups were alkalized by KOH into $-CH_2N^+(CH_3)_3OH^-$ groups. This material proves to be an excellent hydroxide ion conductor with great potential for application in alkaline direct alcohol fuel cells (Zhang et al., 2011).

Lastly, we should also mention the solid oxide fuel cells (SOFC), which are currently of great interest. As far as the plasma technology is concerned, the thermal plasma (see: Fig. 1) is particularly relevant in this case (Henne, 2007). However, the cold plasma is trying to use as well. In addition to the reactive sputtering method that is quite justified when we want to obtain thin films of inorganic oxides (e.g. La-Si-O, which is a potential candidate as electrolyte material for intermediate-temperature solid oxide fuel cells (Briois et al., 2007)), attempts to employ the PECVD method have been also made. Popular solid oxide electrolyte material including yttria-stabilized zirconia (YSZ) was prepared by microwave plasma polymerization from (acetylacetonate)zirconium(I) and tris(dipivaloylmethanato)-yttrium(III) as precursors (Itoh & Matsumoto, 1999). Recently performed physicochemical investigations of such deposited solid electrolyte films have shown characteristic nanostructures that are strongly affected by the variation of plasma parameters and the precursor mixture composition. Thus, we can obtain the films with exactly the desired structure and properties, for example, with appropriate ionic conductivity.

6.2 Solid electrolytes for lithium-ion batteries

The polymer electrolytes for Li-ion batteries are fundamentally the same as those used in fuel cells, with the only difference that ions transported in this case are Li^+ ions. These membranes should be very good electronic insulators that separate the anode from the cathode, but very good lithium-ion conductors. They should also have adequate chemical resistance as well as adequate mechanical strength to withstand the pressure changes and stresses of the electrodes during discharge/charge cycling of the battery. All these requirements can be satisfied by plasma polymerized thin films. The first reports on the preparation of such films appeared at the end of the 1980's. The films were deposited from precursors containing alkoxy, siloxane and vinyl groups in one molecule. Then, the films were sprayed with a solution of $LiOCl_4$ to introduce Li^+ into the plasma polymer structure (Ogumi et al., 1989). In subsequent years, more complex systems were prepared. For example, to suppress a reaction between electrodes and the electrolyte, especially to suppress the dendritic growth of lithium during battery charging, a concept of functional gradient solid polymer electrolyte was developed. This electrolyte system was obtained by changing the composition of the mixture of precursors (dimethyl-2-[(2-ethoxyethoxy) ethoxy]vinylsilane and 1,1-didifluoroethylene) during the plasma polymerization process (Ogumi et al., 1997).

The PECVD technique appears to be a unique method that allows for the implementation of Li-ion batteries with particularly sophisticated architecture. A new type of 3D microbatteries with anode or cathode post-arrays has been recently developed in the Tolbert Lab at University of California (Los Angeles). However, such systems require a solid electrolyte in the form of conformal coatings that will evenly cover the high aspect ratio electrodes. Plasma deposited polyethyleneoxide-like electrolyte films, which are electronic insulating and can be intercalated with lithium ions, have been chosen to this end. Currently, these films are intensively investigated (Dudek, 2011).

7. Conclusions and outlook

A huge potential for the production of new materials and control their structure lies in the cold plasma. This technology allows us to produce materials in the form of thin films or nanoparticles, with uniform or gradient construction, and with amorphous or nanocrystalline structure, which can be deposited on substrates of any shape. We can also obtain complex composites in this way. A special place among these composites is occupied by 3D systems. The plasma deposited materials may have very high or very low conductivity (both electronic and ionic), may have very high or very low permeability for a given substance and, in the end, may have surprising catalytic and photocatalytic properties. Without a doubt, the cold plasma technology has strongly consolidated its position in the fabrication of thin-film solar cells. Increasingly, however, it is also employed to produce materials for the components of fuel cells and Li-ion batteries, such as electrodes and solid electrolytes. Many times, these elements showed better properties than those prepared by conventional methods. One can expect that in the near future it will be possible to produce efficient and effective anode–(solid electrolyte)–cathode systems in one continuous process consisting of consecutive acts of plasma deposition. Such a construction will eliminate the problem, inter alia, of ensuring proper contact between the electrodes and solid electrolyte. The plasma technology also opens up prospects for the spectacular solutions, such as microbatteries with high aspect ratio electrodes or asymmetrical supercapacitors. It seems also feasible to produce miniaturized 3D cells in which electrodes are formed from carbon nanotubes decorated with nanoparticles of catalytic material (all fabricated by plasma processes), and covered with a plasma-polymer electrolyte. Indeed, the prospects are very promising.

Another issue, which is only briefly mentioned in this Chapter, is the use of cold plasma for surface modification of conventional materials. We can thus improve the properties of "conventional" elements relevant to the construction of electrochemical cells: electrode substrates, electrodes themselves, separators, etc. Research interest in this field of the cold plasma technology is comparable to that which is focused on entirely new materials produced by plasma deposition techniques. The use of the plasma treatment technique in electrochemical cell engineering is a problem, however, so vast that a separate chapter should be devoted to it.

8. Acknowledgment

I would like to thank all the members of my team: prof. P. Kazimierski, dr. S. Kuberski, dr. J. Sielski, R. Kapica, as well as my doctor students: P. Makowski, W. Redzynia, A. Twardowski, and I. Ludwiczak, for their excellent cooperation. I also thank Ms. K.M. Palinska for her technical assistant in the preparation of this Chapter.

9. References

Amade, R.; Jover, E.; Caglar, B.; Mutlu, T. & Bertran, E. (2011). Optimization of MnO_2/Vertically Aligned Carbon Nanotube Composite for Supercapacitor Application. *J. Power Sources*, Vol. 196, pp. 5779-5783

Artero, V.; Chavarot-Kerlidou, M. & Fontecave, M. (2011). Splitting Water with Cobalt. *Angew. Chem. Int. Ed.*, Vol. 50, pp. 7238-7266

Ayllón, J.A.; Figueras, A.; Garelik, S.; Spirkova, L.; Durand, J. & Cot, L. (1999). Preparation of TiO2 Powder Using Titanium Tetraisopropoxide Decomposition in a Plasma Enhanced Chemical Vapor Deposition (PECVD) Reactor. *J. Mater. Sci. Lett.*, Vol. 18, pp. 1319-1321

Barreca, D.; Devi, A.; Fischer, R.A.; Bekermann, D.; Gasparotto, A.; Gavagnin, M.; Maccato, C.; Tondello, E.; Bontempi, E.; Depero, L.E. & Sada, C. (2011). Strongly Oriented Co_3O_4 Thin Films on MgO(100) and $MgAl_2O_4$(100) Substrates by PE-CVD. *Cryst. Eng. Comm.*, Vol 13, pp. 3670-3673

Battiston, G.A.; Gerbasi, R.; Gregori, A.; Porchia, M.; Cattarin, S. & Rizzi, G.A. (2000). PECVD of Amorphous TiO_2 Thin Films: Effect of Growth Temperature and Plasma Gas Composition. *Thin Solid Films*, Vol. 371, pp. 126-131

Belmonte, T.; Henrion, G. & Gries, T. (2011). Nonequilibrium Atmospheric Plasma Deposition. *J. Therm. Spray Techn.*, Vol. 20, pp. 744-759

Biederman, H. (Ed.). (2004). *Plasma Polymer Films*, Imperial College Press, ISBN 1-86094-467-1, London

Borrás, A.; Sánches-Valencia, J.R.; Garrido-Molinero, J.; Barranco, A. & González-Elipe, A.R. (2009). Porosity and Microstructure of Plasma Deposited TiO_2 Thin Films. *Micropor. Mesopor. Mat.*, Vol. 118, pp. 314-324

Brault, P. (2011). Plasma Deposition of Catalytic Thin Films: Experiments, Applications, Molecular Modeling. *Surf. Coat. Technol.*, Vol. 205, pp. S15-S23

Brenner, J.R.; Harkness, J.B.L.; Knickelbein, M.B.; Krumdick, G.K. & Marshall, C.L. (1997). Microwave Plasma Synthesis of Carbon-Supported Ultrafine Metal Particles. *NanoStructured Mater.*, Vol. 8, pp. 1-17

Briois, P.; Lapostolle, F. & Billard, A. (2007). Investigations of Apatite-Structure Coatings Deposited by Reactive Magnetron Sputtering Dedicated to IT-SOFC. *Plasma Process. Polym.*, Vol. 4, pp. S99-S103

Brudnik, A.; Gorzkowska-Sobaś, A.; Pamuła, E.; Radecka, M. & Zakrzewska, K. (2007). Thin Film TiO_2 Photoanodes for Water Photolysis Prepared by DC Magnetron Sputtering. *Journal of Power Sources*, Vol.173, pp. 774-780

Caillard, A.; Brault, P.; Mathias, J.; Charles, C.; Boswell, R.W. & Sauvage, T. (2005). Deposition and Diffusion of Platinum Nanoparticles in Porous Carbon Assisted by Plasma Sputtering. *Surf. Coat. Technol.*, Vol. 200, pp. 391-394

Caillard, A.; Charles, C.; Ramdutt, D.; Boswell, R. & Brault, P. (2009). Effect of Nafion and Platinum Content in a Catalyst Layer Processed in a Radio Frequency Helicon Plasma System. *J. Phys. D: Appl. Phys.*, Vol. 42, No. 045207

Cao, Y.; Yang, W.; Zhang, W.; Liu, G. & Yue, P. (2004). Improved Photocatalytic Activity of Sn^{4+} Doped TiO_2 Nanoparticulate Films Prepared by Plasma-Enhanced Chemical Vapor Deposition. *New. J. Chem.*, Vol. 28, pp. 218-222

Carlson, D.E. & Wronski, C.R. (1976). Amorphous Silicon Solar Cell. *Appl. Phys. Lett.*, Vol. 28, pp. 671-673

Carrette, L.; Friedrich, K.A. & Stimming, U. (2000). Fuel Cells: Principles, Types, Fuels and Applications. *ChemPhysChem*, Vol. 1, pp. 162-193

Cavarroc, M.; Ennadjaoui, A.; Mougenot, M.; Brault, P.; Escalier, R.; Tessier, Y. & Durand, J. (2009). Performance of Plasma Sputtered Fuel Cell Electrodes with Ultra-Low Pt Loadings. *Electrochem. Commun.*, Vol. 11, pp. 859-861

Cho, J.; Denes, F.S. & Timmons, R.B. (2006). Plasma Processing Approach to Molecular Surface Tailoring of Nanoparticles: Improved Photocatalytic Activity of TiO_2. *Chem. Mater.*, Vol. 18, pp. 2989-2996

Cho, S.H.; Song, K.T. & Lee, J.Y. (2007). Recent Advances in Polypyrrole, In: *Conjugated Polymers*, Skotheim, T.A. & Reynolds, J.R. (Eds.), pp. 243-330, CRC Press, ISBN 978-1-4200-4358-7, Boca Raton

Choi, S.H.; Kim, J.S. & Yoon, Y.S. (2004). Fabrication and Characterization of SnO_2-RuO_2 Composite Anode Thin Film for Lithium Ion Batteries. *Electrochim. Acta*, Vol. 50, pp. 547-552

Conibeer, G.; Green, M.; Corkish, R.; Cho, Y.; Cho, E.C.; Jiang, C.W.; Fangsuwannarak, T.; Pink, E.; Huang, Y.; Puzzer, T.; Trupke, T.; Richards, B.; Shalav, A. & Lin, K. (2006). Silicon Nanostructures for Third Generation Photovoltaic Solar Cells. *Thin Solid Films*, Vol. 511/512, pp. 654-662

Cruz, G.J., Olayo, M.G., López, O.G., Gómez, L.M., Morales, J. & Olayo, R. (2010). Nanospherical Particles of Polypyrrole Synthesized and Doped by Plasma. *Polymer*, Vol. 51, pp. 4314-4318

Ctibor, P. & Hrabovský, M. (2010). Plasma Sprayed TiO_2: The Influence of Power of an Electric Supply on Particle Parameters in the Flight and Character of Sprayed Coating. *J. Eur. Ceram. Soc.*, Vol. 30, pp. 3131-3136

da Cruz, N.C.; Rangel, E.C.; Wang, J.; Trasferetti, B.C.; Davanzo, C.U.; Castro, S.G.C. & deMoraes, M.A.B. (2000). Properties of Titanium Oxide Films Obtained by PECVD. *Surf. Coat. Technol.*, Vol. 126, pp. 123-130

Dang, B.H.Q.; Rahman, M.; MacElroy, D. & Dowling, D.P. (2011). Conversion of Amorphous TiO_2 Coatings into Their Crystalline Forum Using a Novel Microwave Plasma Treatment. *Surf. Coat. Technol.*, Vol. 205, pp. S235-S240

Dhar, R.; Pedrow, P.D.; Liddell, K.C.; Moeller, T.M. & Osman, M.A. (2005a). Plasma-Enhanced Metal-Organic Chemical Vapor Deposition (PEMOCVD) of Catalytic Coatings for Fuel Cell Reformers. *IEEE Trans. Plasma Sci.*, Vol. 33, pp. 138-146

Dhar, R.; Pedrow, P.D.; Liddell, K.C.; Moeller, T.M. & Osman, M.A. (2005b). Synthesis of Pt/ZrO_2 Catalyst on Fecralloy Substrates Using Composite Plasma-Polymerized Films. *IEEE Trans. Plasma Sci.*, Vol. 33, pp. 2035-2045

Dilonardo, E.; Milella, A.; Cosma, P.; d'Agostino, R. & Palumbo, F. (2011). Plasma Deposited Electrocatalytic Films With Controlled Content of Pt Nanoclusters. *Plasma Process. Polym.*, Vol. 8, pp. 452-458

Dittmar, A.; Kosslick, H.; Müller, J.P. & Pohl, M.M. (2004). Characterization of Cobalt Oxide Supported on Titania Prepared by Microwave Plasma Enhanced Chemical Vapor Deposition. *Surf. Coat. Technol.*, Vol. 182 pp. 35-42

Doblhofer, K. & Dürr, W. (1980). Polymer-Metal Composite Thin Films on Electrodes. *J. Electrochem. Soc.*, Vol. 127, pp. 1041-1044

Donders, M.E.; Knoops, H.C.M.; van Kessels, M.C.M. & Notten, P.H.L. (2011). Remote Plasma Atomic Layer Deposition of Co_3O_4 Thin Films. *J. Electrochem. Soc.*, Vol. 158, pp. G92-G96

Dudek, L. (2011). Plasma Polymer Electrolyte Coatings for 3D Microbatteries, In: *IGERT 2011 Poster Competition*, Available from: <www.igert.org/posters2011/posters/25>

Ennajdaoui, A.; Roualdes, S.; Brault, P. & Durand, J. (2010). Membranes Produced by Plasma Enhanced Chemical Vapor Deposition Technique for Low Temperature Fuel Cell Applications. *J. Power Sources*, Vol. 195, pp. 232-238

Fonseca, J.L.C.; Apperley, D.C. & Badyal, J.P.S. (1993). Plasma Polymerization of Tetramethylsilane. *Chem. Mater.*, Vol. 5, pp. 1676-1682

Fujii, E.; Torii, H.; Tomozawa, A.; Takayama, R. & Hirao, T. (1995). Preparation of Cobalt Oxide Films by Plasma-Enhanced Metalorganic Chemical Vapour Deposition. *J. Mater. Sci.*, Vol. 30, pp. 6013-6018

Fujishima, A. & Honda, K. (1972). Electrochemical Photolysis of Water at a Semiconductor Electrode. *Nature*, Vol. 238, pp. 37-38

Garg, D.; Henderson, P.B.; Hollingsworth, R.E. & Jensen, D.G. (2005). An Economic Analysis of the Deposition of Electrochromic WO_3 via Sputtering or Plasma Enhanced Chemical Vapor Deposition. *Mater. Sci. Eng. B*, Vol. 119, pp. 224-231

Goodman, J. (1960). The Formation of Thin Polymer Films in the Gas Discharge. *J. Polym. Sci.*, Vol. 44, pp. 551-552

Gordillo-Vázquez, F.J.; Herrero, V.J. & Tanarro, I. (2007). From Carbon Nanostructures to New Photoluminescences Sources: An Overview of New Perspectives and Emerging Applications of Low-Pressure PECVD. *Chem. Vap. Deposition*, Vol. 13, pp. 267-279

Green, M.A. (2007). Thin-Film Solar Cells: Review of Materials, Technologies and Commercial Status. *J. Mater. Sci.: Mater. Electron.*, Vol. 18, pp. S15-S19

Green, M.A.; Emery, K.; Hishikawa, Y. & Warta, W. (2011). Solar Cell Efficiency Tables (Version 37). *Prog. Photovolt.: Res. Appl.*, Vol. 19, pp. 84-92

Henne, R. (2007). Solid Oxide Fuel Cells: A Challenge for Plasma Deposition Processes. *J. Therm. Spray Techn.*, Vol. 16, pp. 381-403

Huang, C.H.; Tsao, C.C. & Hsu, C.Y. (2011). Study on the Photocatalytic Activities of TiO_2 Films Prepared by Reactive RF Sputtering. *Ceram. Int.*, Vol. 37, pp. 2781-2788

Inagaki, N. (1996). *Plasma Surface Modification and Plasma Polymerization*, Technomic Publ., ISBN 1-56676-337-1, Lancaster

Ingler Jr, W.B.; Attygalle, D. & Deng, X. (2006). Properties of RF Magnetron Sputter Deposited Cobalt Oxide Thin Films as Anode for Hydrogen Generation by Electrochemical Water Splitting. *ECS Trans.*, Vol. 3, pp. 261-266

Itoh, K. & Matsumoto, O. (1999). Deposition Process of Metal Oxide Thin Films by Means of Plasma CVD with β-Diketonates as Precursors. *Thin Solid Films*, Vol. 345, pp. 29-33

James, B.D.; Baum, G.N.; Perez, J. & Baum, K.N. (2009). Report of U.S. DOE Hydrogen Program No GS-10F-009, In: *Technoeconomic Analysis of Photoelectrochemical (PEC) Hydrogen Production*, Available from: < http://205.254.148.40/hydrogenandfuelcells/pdfs//pec_technoeconomic_analysis.pdf>

Jiang, Z.; Jiang, Z. & Meng, Y. (2011). Optimization and Synthesis of Plasma Polymerized Proton Exchange Membranes for Direct Methanol Fuel Cells. *J. Membrane Sci.*, Vol. 372, pp. 303-313

Jiao, F. & Frei, H. (2009). Nanostructured Cobalt Oxide Clusters in Mesoporous Silica as Efficient Oxygen-Evolving Catalysts. *Angew. Chem.*, Vol. 121, pp. 1873-1876

Karthikeyan, J.; Berndt, C.C.; Tikkanen, J.; Reddy, S. & Herman, H. (1997). Plasma Spray Synthesis of Nanomaterial Powders and Deposits. *Mater. Sci. Eng. A*, Vol. 238, pp. 275-286

Kazimierski, P.; Jozwiak, L.; Kapica, R. & Socha, A. (2010). Novel Cu and Co Oxide Based Catalysts for PEMFC Obtained by Plasma-Enhanced Metal-Organic Vapor Deposition PEMOCVD, *Proceedings of 7th ICRP and 63rd GEC*, paper No DTP-093, ISBN 978-4-86348-101-5, Paris, France, October 4-8, 2010

Kazimierski, P.; Jozwiak, L. & Tyczkowski, J. (2011). Cobalt Spinel Catalyst Deposited by Non-Equilibrium Plasma for PEMFC. *Catal. Commun.*, submitted do Editor

Kelly, N.A. & Gibson, T.L. (2006). Design and Characterization of a Robust Photoelectrochemical Device to Generate Hydrogen Using Solar Water Splitting. *Int. J. Hydrogen Energ.*, Vol. 31, pp. 1658-1673

Kim, I.C.; Byun, D.; Lee, S. & Lee, J.K. (2006). Electrochemical Characteristics of Copper Silicide-Coated Graphite as an Anode Material of Lithium Secondary Batteries. *Electrochim. Acta*, Vol. 52, pp. 1532-1537

Kim, K.; Park, J.H.; Doo, S.G. & Kim, T. (2010). Effect of Oxidation on Li-Ion Secondary Battery with Non-Stoichiometric Silicon Oxide (SiO_x) Nanoparticles Generated in Cold Plasma. *Thin Solid Films*, Vol. 518, pp. 6547-6549

Konuma, M. (1992). *Film Deposition by Plasma Techniques*, Springer, ISBN 0-38754-057-1, Berlin

LeComber, P.G. & Spear, W.E. (1979). Doped Amorphous Semiconductors, In: *Amorphous Semiconductors*, Brodsky, M.H. (Ed.), pp. 251-285, Springer, ISBN 3-540-09496-2, Berlin

Li, Z.; Lee, D.K.; Coulter, M.; Rodriguez, L.N.J. & Gordon, R.G. (2008). Synthesis and Characterization of Volatile Liquid Cobalt Amidinates. *Dalton Trans.*, Vol. 2008, pp. 2592-2597

Liao, C.L.; Lee, Y.H.; Chang, S.T. & Fung, K.Z. (2006). Structural Characterization and Electrochemical Properties of RF-Sputtered Nanocrystalline Co_3O_4 Thin-Film Anode. *J. Power Sources*, Vol. 158, pp. 1379-1385

Licht, S.; Ghosh, S.; Tributsch, H. & Fiechter, S. (2002). High Efficiency Solar Energy Water Splitting to Generate Hydrogen Fuel: Probing RuS_2 Enhancement of Multiple Band Electrolysis. *Sol. Energ. Mat. Sol. C.*, Vol. 70, pp. 471-480

Lue, S.J.; Hsiaw, S.Y. & Wei, T.C. (2007). Surface Modification of Perfluorosulfonic Acid Membranes with Perfluoroheptane (C_7F_{16})/Argon Plasma. *J. Membrane Sci.*, Vol. 305, pp. 226-237

Maeda, M. & Watanabe, T. (2005). Evaluation of Photocatalytic Properties of Titanium Oxide Films Prepared by Plasma-Enhanced Chemical Vapor Deposition. *Thin Solid Films*, Vol. 489, pp. 320-324

Manzoli, M. & Boccuzzi, F. (2005). Characterisation of Co-Based Electrocatalytic Materials for O_2 Reduction in Fuel Cells. *J. Power Sources*, Vol. 145, pp. 161-168

Merche, D.; Hubert, J.; Poleunis, C.; Yunus, S.; Bertrand, P.; DeKeyzer, P. & Reniers, F. (2010). One Step Polymerization of Sulfonated Polystyrene Films in a Dielectric Barrier Discharge. *Plasma Process. Polym.*, Vol. 7, pp. 836-845

Michel, M.; Bour, J.; Petersen, J.; Arnoult, C.; Ettingshausen, F.; Roth, C. & Ruch, D. (2010). Atmospheric Plasma Deposition: A New Pathway in the Design of Conducting Polymer-Based Anodes for Hydrogen Fuel Cells. *Fuel Cells*, Vol. 10, pp. 932-937

Mueller, T. (2009). *Heterojunction Solar Cells (a-Si/c-Si)*, Logos, ISBN 978-3-8325-2291-9, Berlin

Nakamura, M.; Aoki, T.; Hatanaka, Y.; Korzec, D. & Engemann, J. (2001). Comparison of Hydrophilic Properties of Amorphous TiO_x Films Obtained by Radio Frequency Sputtering and Plasma-Enhanced Chemical Vapor Deposition. *J. Mater. Res.*, Vol. 16, pp. 621-626

Naseri, N.; Sangpour, P. & Moshfegh, A.Z. (2011). Visible Light Active Au:TiO_2 Nanocomposite Photoanodes for Water Splitting: Sol-Gel vs. Sputtering. *Electrochim. Acta*, Vol. 56, pp. 1150-1158

Nowotny, J.; Bak, T.; Nowotny, M.K. & Sheppard, L.R. (2006). TiO_2 Surface Active Sites for Water Splitting. *J. Phys. Chem. B*, Vol. 110, pp. 18492-18495

Ogumi, Z.; Iwamoto, T. & Teshima, M. (1997). Preparation of Functional Gradient Solid Polymer Electrolyte Thin-Films for Secondary Lithium Batteries, In: *Lithium Polymer Bateries*, Broadhead, J. & Scrosati, B. (Eds.), pp. 4-9, The Electrochemical Society, Inc., ISBN 1-56677-167-6, Pennington

Ogumi, Z.; Uchimoto, Y.; Takehara, Z. & Kanamori, Y. (1989). Preparation of Ultra-Thin Solid-State Lithium Batteries Utilizing a Plasma-Polymerized Solid Polymer Electrolyte. *J. Chem. Soc., Chem. Commun.*, Vol. 21, pp. 1673-1674

Ohnishi, R.; Katayama, M.; Takanabe, K.; Kubota, J. & Domen, K. (2010). Niobium-Based Catalysts Prepared by Reactive Radio-Frequency Magnetron Sputtering and Arc Plasma Methods as Non-Noble Metal Cathode Catalysts for Polymer Electrolyte Fuel Cells. *Electrochim. Acta*, Vol. 55, pp. 5393-5400

Papadopoulos, N.D.; Karayianni, H.S.; Tsakiridis, P.E.; Perraki, M. & Hristoforou, E. (2010). Cyclodextrin Inclusion Complexes as Novel MOCVD Precursors for Potential Cobalt Oxide Deposition. *Appl. Organometal. Chem.*, Vol. 24, pp. 112-121

Pokhodnya, K.; Sandstrom, J.; Olson, C.; Dai, X.; Boudjouk, P.R. & Schulz, D.L. (2009). Comparative Study of Low-temperature PECVD of Amorphous Silicon Using Mono-, Di-, Trisilane and Cyclohexasilane, *Proceedings of Photocoltaic Specialist Conference (PVSC), 2009 34th IEEE*, pp. 001758-001760, doi: 10.1109/PVSC.2009.5411459, Philadelphia, USA, June 7-12, 2009

Prakash, S.; Mustain, W.E.; Park, S. & Kohl, P.A. (2008). Phosphorus-Doped Glass Proton Exchange Membranes for Low Temperature Direct Methanol Fuel Cells. *J. Power Sources*, Vol. 175, pp. 91-97

Raaijmakers, I.J. (1994). Low Temperature Metal-Organic Chemical Vapor Deposition of Advanced Barrier Layers for the Microelectronics Industry. *Thin Solid Films*, Vol. 247, pp. 85-93

Rabat, H. & Brault, P. (2008). Plasma Sputtering Deposition of PEMFC Porous Carbon Platinum Electrodes. *Fuel Cells*, Vol. 08, pp. 81-86

Randeniya, L.K.; Bendavid, A.; Martin, P.J. & Preston, E.W. (2007). Photoelectrochemical and Structural Properties of TiO_2 and N-Doped TiO_2 Thin Films Synthesized Using Pulsed Direct Current Plasma-Activated Chemical Vapor Deposition. *J. Phys. Chem. C*, Vol. 111, pp. 18334-18340

Roualdès, S.; Schieda, M.; Durivault, L.; Guesmi, I.; Gérardin, E. & Durand, J. (2007). Ion-Exchange Plasma Membranes for Fuel Cells on a Micrometer Scale. *Chem. Vap. Deposition*, Vol. 13, pp. 361-369

Sadhir, R.K. & Schoch, K.F. (1996). Plasma-Polymerized Carbon Disulfide Thin-Film Rechargeable Batteries. *Chem. Mater.*, Vol. 8, pp. 1281-1286

Saha, M.S.; Gullá, A.F.; Allen, R.J. & Mukerjee, S. (2006). High Performance Polymer Electrolyte Fuel Cells with Ultra-Low Pt Loading Electrodes Prepared by Dual Ion-Beam Assisted Deposition. *Electrochim. Acta*, Vol. 51, pp. 4680-4692

Schieda, M.; Roualdès, S.; Durand, J.; Martinent, A. & Marsacq, D. (2006). Plasma-Polymerized Thin Films as New Membranes for Miniature Solid Alkaline Fuel Cells. *Desalination*, Vol. 199, pp. 286-288

Schumacher, L.C.; Holzhueter, I.B.; Hill, I.R. & Dignam, M.J. (1990). Semiconducting and Electrocatalytic Properties of Sputtered Cobalt Oxide Films. *Electrochim. Acta*, Vol. 35, pp. 975-984

Searle, T. (Ed.). (1998). *Properties of Amorphous Silicon and its Alloys*, INSPEC, ISBN 0-85296-922-8, London

Siow, K.S.; Britcher, L.; Kumar, S. & Griesser, H.J. (2009). Sulfonated Surfaces by Sulfur Dioxide Plasma Surface Treatment of Plasma Polymer Films. *Plasma Process. Polym.*, Vol. 6, pp. 583-592

Slavcheva, E.; Radev, I.; Bliznakov, S.; Topalov, G.; Andreev, P. & Budevski, E. (2007). Sputtered Iridium Oxide Films as Electrocatalysts for Water Splitting via PEM Electrolysis. *Electrochim. Acta*, Vol. 52, pp. 3889-3894

Soin, N.; Roy, S.S.; Karlsson, L. & McLaughlin, J.A. (2010). Sputter Deposition of Highly Dispersed Platinum Nanoparticles on Carbon Nanotube Arrays for Fuel Cell Electrode Material. *Diam. Relat. Mater.*, Vol. 19, pp. 595-598

Sundmacher, K. (2010). Fuel Cell Engineering: Toward the Design of Efficient Electrochemical Power Plants. *Ind. Eng. Chem. Res.*, Vol. 49, pp. 10159-10182

Täschner, C.; Leonhardt, A.; Schönherr, M.; Wolf, E. & Henke, J. (1991). Structure and Properties of TiC_x Layers Prepared by Plasma-Assisted Chemical Vapour Deposition Methods. *Mater. Sci. Eng. A*, Vol. 139, pp. 67-70

Thery, J.; Martin, S.; Faucheux, V.; Le Van Jodin, L.; Truffier-Boutry, D.; Martinent, A. & Laurent, J.Y. (2010). Fluorinated Carboxylic Membranes Deposited by Plasma Enhanced Chemical Vapour deposition for Fuel Cell Applications. *J. Power Sources*, Vol. 195, pp. 5573-5580

Tyczkowski, J. (1999). Audio-Frequency Glow Discharge for Plasma Chemical Vapor Deposition from Organic Compounds of the Carbon Family. *J. Vac. Sci. Technol. A*, Vol. 17, pp. 470-479

Tyczkowski, J. (2004). Electrical and Optical Properties of Plasma Polymers, In: *Plasma Polymer Films*, Biederman, H (Ed.), pp. 143-216, Imperial College Press, ISBN 1-86094-467-1, London

Tyczkowski, J. (2006). The Role of Ion Bombardment Process in the Formation of Insulating and Semiconducting Plasma Deposited Carbon-Based Films. *Thin Solid Films*, Vol. 515, pp. 922-1927

Tyczkowski, J. (2010). Charge Carrier Transfer: A Neglected Process in Chemical Engineering. *Ind. Eng. Chem. Res.*, Vol. 49, pp. 9565-9579

Tyczkowski, J. (2011). New Materials for Innovative Energy Systems Produced by Cold Plasma Technique. *Funct. Mater. Lett.*, in press

Tyczkowski, J.; Kapica, R. & Łojewska, J. (2007). Thin Cobalt Oxide Films for Catalysis Deposited by Plasma-Enhanced Metal–Organic Chemical Vapor Deposition. *Thin Solid Films*, Vol. 515, pp. 6590-6595

Walsh, A.; Ahn, K.S.; Shet, S.; Huda, M.N.; Deutsch, T.G.; Wang, H.; Turner, J.A.; Wei, S.H.; Yan, Y. & Al-Jassim, M.M. (2009). Ternary Cobalt Spinel Oxides for Solar Driven Hydrogen Production: Theory and Experiment. *Energy Environ. Sci.*, Vol. 2, pp. 774-782

Walter, M.G.; Warren, E.L.; McKone, J.R.; Boettcher, S.W.; Mi, Q.; Santori, E.A. & Lewis, N.S. (2010). Solar Water Splitting Cells. *Chem. Rev.*, Vol. 110, pp. 6446-6473

Wang, B. (2005). Recent Development of Non-Platinum Catalysts for Oxygen Reduction Reaction. *J. Power Sources*, Vol. 152, pp. 1-15

Weber, A.; Poeckelmann, R. & Klages, C.P. (1995). Deposition of High-Quality TiN Using Tetra-Isopropoxide Titanium in an Electron Cyclotron Resonance Plasma Process. *Appl. Phys. Lett.*, Vol. 67, pp. 2934-2935

Wierzchoń, T.; Sobiecki, J.R. & Krupa, D. (1993). The Formation of Ti(OCN) Layers Produced from Metal-Organic Compounds Using Plasma-Assisted Chemical Vapour Deposition. *Surf. Coat. Technol.*, Vol. 59, pp. 217-220

Xinyao, Y.; Zhongqing, J. & Yuedong, M. (2010). Effects of Sputtering Parameters on the Performance of Sputtered Cathodes for Direct Methanol Fuel Cells. *Plasma Sci. Technol.*, Vol. 12, pp. 87-91

Yokoyama, D.; Hashiguchi, H.; Maeda, K.; Minegishi, T.; Takata, T.; Abe, R.; Kubota, J. & Domen, K. (2011). Ta_3N_5 Photoanodes for Water Splitting Prepared by Sputtering. *Thin Solid Films*, Vol. 519, pp. 2087-2092

Zhang, C.; Hu, J.; Nagatsu, M.; Meng, Y.; Shen, W.; Toyoda, H. & Shu, X. (2011). High-Performance Plasma-Polymerized Alkaline Anion-Exchange Membranes for Potential Application in Direct Alcohol Fuel Cells. *Plasma Process. Polym.*, Vol. 8, pp. 1024-1032

Zhao, X.; Tian, H.; Zhu, M;, Tian, K.; Wang, J.J.; Kang, F. & Outlaw, R.A. (2009). Carbon Nanosheets as the Electrode Material in Supercapacitors. *J. Power Sources*, Vol. 194, pp. 1208-1212

Zhu, F.; Hu, J.; Matulionis, I.; Deutsch, T.; Gaillard, N.; Kunrath, A.; Miller, E. & Madan, A. (2009). Amorphous Silicon Carbide Photoelectrode for Hydrogen Production Directly from Water Using Sunlight. *Philos. Mag.*, Vol. 89, pp. 2723-2739

Zhu, L.; Susac, D.; Teo, M.; Wong, K.C.; Wong, P.C.; Parsons, R.R.; Bizzotto, D. & Mitchell, K.A.R. (2008). Investigation of CoS_2-Based Thin Films as Model Catalysts for the Oxygen Reduction Reaction. *J. Catal.*, Vol. 258, pp. 235-242

6

Fuel Cell: A Review and a New Approach About YSZ Solid Oxide Electrolyte Deposition Direct on LSM Porous Substrate by Spray Pyrolysis

Tiago Falcade and Célia de Fraga Malfatti
Federal University of Rio Grande do Sul
Brazil

1. Introduction

Environmental concerns related to energy, already widely held today, will substantially increase in coming years. The energy is one of the main factors to consider in environmentalists discussions, since there is an intimate connection between energy, environment and sustainable development (Stambouli & Traversa, 2002a). In response to a critical need for cleaner energy technology, the solutions have evolved, including energy conservation by improving the efficiency of global energy, a reduction in the use of fossil fuels and an increase in the supply of renewable energies (hydro (Mendez et al., 2006), solar (Akorede et al., 2010), wind (Mendez et al., 2006), biomass (Kirubakaran et al., 2009), geothermal (Fridleifsson, 2001), hydrogen (Louie & Strunz, 2007), etc...). In the context of renewable energy, a good alternative can be found in the development and popularization of fuel cells (Akorede et al., 2010).

A fuel cell is an energy conversion device that generates electricity and heat by combining, electrochemically, fuel gas (hydrogen, for example) and an oxidant gas (oxygen in air). During this process, water is obtained as a reaction product. The fuel cell does not work with charging system such as a battery, it only produces power while the fuel is supplied. The main characteristic of a fuel cell is its ability to convert chemicals directly into electrical energy, with a conversion efficiency much higher than any conventional thermo-mechanical system, thus extracting more electricity from the same amount of fuel to operate without combustion. It is virtually free of pollution and has a quieter operation because there are no moving mechanical parts (Stambouli & Traversa, 2002a).

The initial concept of fuel cells is attributed to the German physical-chemist Friedrich Wilhelm Ostwald in 1894. His idea was to modify the internal combustion engines, eliminating the intermediate stage of combustion and convert chemical energy into electrical energy in a single step. His devices project provided the direct oxidation of natural fuel and oxygen from the air, using the electrochemical mechanism. The device that would perform this direct conversion was called a fuel cell (Wand, 2006).

Ostwald's concepts marked the beginning of a great deal of research in the fuel cells field. Ostwald examined only the theoretical aspect of energy conversion in fuel cells, but completely ignored other practical aspects: the question of whether the electrochemical reactions that involve natural fuels are feasible or not and how they can be efficient. The first experimental studies, conducted after the publication of Ostwald's document, indicated that it was very difficult to build devices for the direct electrochemical oxidation of natural fuels (Wand, 2006).

Ceramic fuel cells came up much later, initially with the discovery of the Nernst solid oxide electrolytes in 1899 [9]. Nearly forty years later, the first ceramic fuel cells began operating at 1000 °C, developed by Baur and Preis in 1937 (Farooque & Maru, 2001). Since 1945, three research groups (USA, Germany and the USSR) made studies on a few main types of generators by improving their technologies for industrial development. This work yielded the current concepts on fuel cells (Wand, 2006).

Nowadays, the development of fuel cells is being mainly driven by environmental reasons mentioned above. Over the last decades, advances in research made possible for a considerable improvement to happen, regarding the characteristics of the cells, in particular their stability and efficiency. Currently, it is possible to classify the fuel cells into two major groups, which differ by basic operational characteristics: the low-temperature and high temperature fuel cells.

1.1 Low-temperature fuel cells

The first large group of fuel cells, the low temperature ones (50 - 250 °C), is characterized by its more focused application on mobile devices and the automotive industry. Within the large group of low-temperature cells, there can also be a division, considering the type of electrolyte used. This classification results in three cell types: proton exchange membrane fuel cell (PEMFC), alkaline fuel cell (AFC); and phosphoric acid fuel cell (PAFC).

1.1.1 Proton exchange membrane fuel cell (PEMFC)

PEM fuel cell uses a solid polymer electrolyte, which is an excellent protonic conductor. In this electrolyte the ion exchange occurs between two porous electrodes. The operating temperature of the fuel cell type PEM is about 100 °C (Ellis et al., 2001).

The advantages of PEM fuel cell are its high charge density and its fast startup time, interesting features for automotive applications. The low temperature makes the technology competitive in the transportation sector and in commercial applications such as laptop computers, bicycles and mobile phones. The main disadvantages of PEM fuel cell are its low operating efficiency (40-45%) and the use of a noble catalyst such as platinum, whose CO intolerance ends up limiting the further popularization of this cell type (Farooque & Maru, 2001).

Two subcategories of PEM fuel cells are currently being widely studied, for allowing the use of other fuels other than hydrogen directly into the cell: direct methanol (DMFC) and direct ethanol (DEFC).

DMFC and DEFC fuel cells use a solid polymer electrolyte for ionic transport. However, they use, respectively, liquid methanol and ethanol as fuel instead of hydrogen. During chemical reactions, the fuel (methanol or ethanol) is directly oxidized in the anode. At the cathode, the reaction occurs with oxygen, producing electricity and water as a byproduct (Ellis et al., 2001; Garcia et al., 2004).

1.1.2 Alkaline fuel cell (AFC)

Initially, it was called Bacon cell, by virtue of its inventor Francis Thomas Bacon. It operates at low temperature around 100 °C and it is able to achieve 60-70% efficiency. It uses an aqueous solution of potassium hydroxide (KOH) as electrolyte solution. This fuel cell has quick startup speed, one of its great advantages. The main disadvantage is that it is very sensitive to CO_2 (Farooque & Maru, 2001). It needs an external system to remove CO_2 from the air. Furthermore, the use of a liquid electrolyte is also a disadvantage because it reduces the cell lifetime and makes the assembly handling and transport more difficult.

1.1.3 Phosphoric acid fuel cell (PAFC)

The phosphoric acid fuel cell (PAFC) operates at around 175-200 °C. This range of operating temperature is almost twice as high as the PEM's. It uses phosphoric acid as electrolyte. Unlike PEMFC and AFC, the PAFC is very tolerant to impurities in reforming hydrocarbons. The chemical reaction involved in this type of fuel cell is the same as PEMFC, where hydrogen is used as fuel input, however PAFC is more tolerant to CO_2 (Farooque & Maru, 2001). Cogeneration is also possible due to its relatively high operating temperature. The disadvantage of the PAFC is the same as the PEM's, its cost also increases due to the use of platinum as a catalyst (O'Sullivan, 1999).

1.2 High temperature fuel cells

The second major group of fuel cells, the high temperature ones (650 - 1000 °C) has as its main feature the high efficiency, since the high operating temperature facilitates the reactions. One application of this type of fuel cell is stationary generation, such as primary or secondary source of energy. Two cell types in this group are molten carbonate fuel cell (MCFC) and solid oxide fuel cell (SOFC).

1.2.1 Molten carbonate fuel cell (MCFC)

The molten carbonate fuel cell (MCFC) operates at high temperature, which is about 600-700 °C. It consists of two porous conductive electrodes in contact with an electrolyte of molten carbonate. This type of cell allows the internal reform. The main advantage of the MCFC is its high efficiency (50-60%) without external reformer and metal catalyst, due to the high operating temperature (Farooque & Maru, 2001). This cell is intolerant to sulfur and its launching is slow, these are its main disadvantages.

1.2.2 Solid oxide fuel cell (SOFC)

The solid oxide fuel cell (SOFC) is the cell that operates at the highest temperatures (800-1000 °C). It uses a solid electrolyte, which consists of a dense ceramic material with high ionic conductivity. In this cell, oxygen ion is transported through the electrolyte and, in the interface with the anode, it combines with hydrogen to create water and energy. The main advantages of the SOFC are that it produces electricity with high efficiency of 50-60% and does not require an external reformer to extract hydrogen from fuel due to its ability to internal reform. The waste heat can be recycled to produce additional electricity in the operation of cogeneration. The high temperature, which provides satisfactory characteristics in the cell as already mentioned, is also responsible for its major disadvantages, such as the problems of material selection, high heat detritions and the impossibility of using metallic materials, which cost much less than the ceramic materials currently used (Aruna & Rajam, 2008; Farooque & Maru, 2001; Srivastava et al., 1997).

1.3 Evolution of SOFC

The studies on solid oxide fuel cells have promoted considerable improvements on their characteristics and properties. Several layouts were developed to improve the characteristics of SOFC, considering the current needs and costs of production and efficiency.

1.3.1 Tubular layout

The solid oxide fuel cell is highly influenced by the temperature variation during operation, constantly suffering from thermal stress, which requires excellent compatibility between the thermal expansion coefficients of the cell components. The cylindrical shape of tubular SOFC contributes significantly to minimize the difference in the coefficients of thermal expansion, thus avoiding the formation of cracks and delamination. This model also makes unnecessary the use of sealant gases. On the other hand, the efficiency is impaired, since the path made by the electric current is increased, causing ohmic losses (Minh, 1993).

1.3.2 Planar layout

This design consists of flat electrodes and electrolyte, separated by thin interconnects. The components can be manufactured separately, giving the production simplicity. It has higher energy density than the tubular. A disadvantage is the long time needed for heating and cooling of the cell, used to prevent the formation of cracks (Minh, 1993).

Inside the planar design of SOFC cells, the development of new materials and techniques of production allowed an evolution of the configurations of planar cells.

The first generation of SOFC fuel cells (1G-SOFC) operated at temperatures of around 1000 °C. The system consisted of electrolyte support and the mechanical stability of the cell was given by the thickness of the electrolyte. In this design the anode and cathode were quite thin (around 50 μm), while the electrolyte had thickness of 100 μm to 200 μm. However, the operating temperature of the cell was a limiting factor in popularizing this type of energy source, which led to the development of new designs for the cell (Wang, 2004).

At high temperature, the thickness of the electrolyte around 200 μm did not provide a problem in the ionic conductivity. However, the reduction of cell operating temperature to 700 °C - 800 °C would result in a drastic reduction of the electrolyte conductivity, so, the second generation of planar SOFC (2G-SOFC) was developed aiming to reduce the electrolyte thickness. The mechanical stability of the cell can no longer be granted by the electrolyte, which now is less than 20 μm. For that, the second-generation cells are anode support type, where the anode is responsible for mechanical stability, with thickness between 300 μm and 1500 μm. In this generation, the cathode thickness was of 50 μm (Wang, 2004).

However, the costs of the cell are directly related to the costs of obtaining and producing ceramic materials present in their components, and an anode or cathode with very high thickness results in a substantial increase in production costs. This motivation led to the development of third generation SOFC (3G-SOFC) interconnect support type, allowing a reduction in the ceramic components thickness, accompanied by a reduction in operating temperature of the cell (800 °C) and the use of metallic interconnects. In this design, the electrolyte thickness has less than 20 μm and both, anode and cathode thickness, has about 50 μm (Wang, 2004). Figure 1 shows a schematic configuration of planar SOFCs.

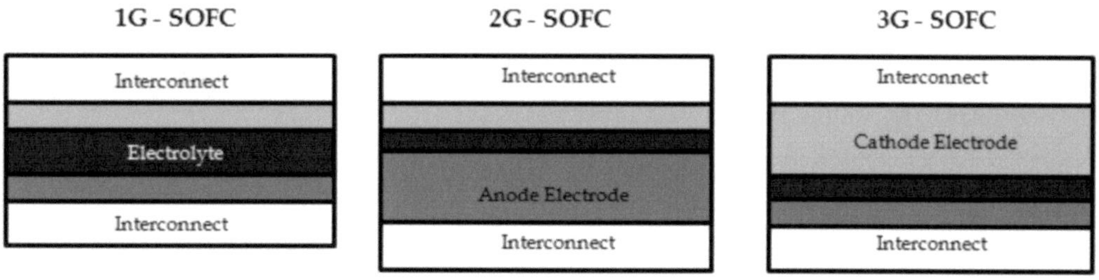

Fig. 1. Representation of the configuration for the three-generations of planar SOFCs.

1.4 Components of SOFC

1.4.1 Anode

The anode of a fuel cell is the interface between the fuel and electrolyte. The main functions of the anode are:

- Provide sites for electrochemical reactions of combustible gas catalytic oxidation with ions from the electrolyte.
- Allow the diffusion of fuel gas for the reactive sites of the electrode/electrolyte interface and the removal of byproducts.
- Transport of generated electrons to the interconnect (external circuit).

The anode material must possess, under the operating conditions of the fuel cell: good physical and chemical stability, chemical and structural compatibility with the electrolyte and interconnect, high ionic and electronic conductivity and catalytic activity for fuel oxidation (Ralph et al., 2001). The thermal stability is an important aspect to maintain the structural integrity throughout the temperature variations at which this component is subjected.

In general, the performance of the anode is defined by its electrical and electrochemical properties and therefore has a strong dependence on its microstructure. Thus, the control parameters, such as composition, size and distribution of particles and pores, are very important for optimizing the performance of the anode material of a solid oxide fuel cell.

Ceramic-metal composites, typically Ni-based, have been commonly used. Among them, the NiO-YSZ composite is the material of conventional fuel cells. Ni is also used because it has good electrical, mechanical and catalytic properties (Ralph et al., 2001). More recently, mixed conductors based on ceria (NiO-GDC) and transition metal perovskites (such as Fe, Mn, Cr and Ti) are being studied as potential candidates for anode materials for solid oxide fuel cells. However, to date, these materials do not present, in reducing atmospheres, values of electronic conductivity high enough for high performance fuel cell (Gong et al., 2010). The oxides of transition metals can have different oxidation states that can induce electron transport, and usually increase the catalytic activity. Some compositions can be highlighted, such as: $Zr_{1-x-y}Ti_xY_yO_2$ (Tao & Irvine, 2002), $La_{1-x}Sr_xA_{1-y}M_yO_3$ (A: Cr ou Fe; M: Ru, Cr ou Mn) (Sauvet & Fouletier, 2001), $Sr_{1-x}Y_xTiO_3$ (Hui & Petric, 2001), e $La_{1-x}Sr_xTiO_3$ (Canales-Vázquez et al., 2003).

1.4.2 Cathode

The cathode of a fuel cell is the interface between air (or oxygen) and the electrolyte. Its main functions are to catalyze the reaction of oxygen reduction and to transport the generated electrons to the interconnect (external circuit).

In the same way of all the materials used in solid oxide fuel cells, the cathode must present certain general characteristics:

- Low cost and ease of fabrication.
- Minor differences between the thermal expansion coefficients of various components in the cell.
- Phase and microstructure stability during the operation.
- Chemical stability.
- High electrical conductivity, both electronic and ionic
- Porous microstructure during the entire operation of the cell
- High catalytic activity for oxygen reduction and stability in highly oxidizing atmospheres.

The materials, perovskite-type ABO_3, commonly used as cathodes in fuel cells are solid oxide ceramics based on lanthanum manganite ($LaMnO_3$) substituting A for Sr ions. This material fills most of the requirements for its use as cathode for ceramic fuel cells operating at temperatures around 1000 °C. The ionic conductivity of materials based on $LaMnO_3$ is significantly smaller than the ionic conductivity of YSZ electrolytes, but the ionic conductivity increases significantly substituting Mn for Co. The diffusion coefficients of oxygen ions in lanthanum cobaltite can reach 4-6 orders of magnitude higher when compared to those of lanthanum manganites with similar doping (Carter, 1992). Other materials have been used as cathode in SOFC, such as: Lanthanum strontium ferrite (LSF), $(LaSr)(Fe)O_3$, Lanthanum strontium cobaltite (LSC), $(LaSr)CoO_3$, Lanthanum strontium cobaltite ferrite (LSCF), $(LaSr)(CoFe)O_3$, Lanthanum strontium manganite ferrite (LSMF), $(LaSr)(MnFe)O_3$, Samarium strontium cobaltite (SSC), $(SmSr)CoO_3$, Lanthanum calcium cobaltite ferrite (LCCF), $(LaCa)(CoFe)O_3$, Praseodymium strontium manganite (PSM), $(PrSr)MnO_3$, Praseodymium strontium manganite ferrite (PSMF), $(PrSr)(MnFe)O_3$ (Stambouli & Traversa, 2002b).

1.4.3 Electrolyte

The design of fuel cells with solid oxide electrolyte must be based on the concept of oxygen ion conduction through the electrolyte, with ions O^{2-} migrating from the cathode to the anode, where they react with the fuel (H_2, CO, etc..) generating an electrical current.

The materials that have been studied the most are: yttria stabilized zirconia, doped ceria with gadolinium and lanthanum gallate doped with strontium and magnesium. The solid oxide fuel cells can, in principle, operate in a wide temperature range between 500 °C and

1000 °C. Thus, they can be divided into two types: operating at high temperatures (> 750 °C) and intermediate temperatures (500 °C to 750 °C). One of the determinant factors for the operating temperature is the characteristic of the solid electrolyte. The ohmic losses associated with the electrolyte are important for the cell performance. In order to reduce the operating temperature of SOFC, aiming the use of more conventional steel alloys as interconnects at temperatures around 700 °C (Horita et al., 2008; Perednis & Gauckler, 2004), it is necessary to employ electrolytes with high oxygen ionic conductivity at low temperatures.

YSZ is so far the most widely used solid electrolyte for application in high temperature SOFC. For many years, the zirconium oxide is already known as a conductor of oxygen ions.

The yttria addition to the zirconia-yttria solid solution has two functions: to stabilize the cubic structure type fluorite and to form oxygen vacancies in concentrations proportional to the yttria content. These vacancies are responsible for high ionic conductivity. Yttria stabilized zirconia is a suitable ionic conductor at temperatures above 800 °C, since thin dense membranes (less than 20 μm) can be manufactured. These membranes should be free of impurities. The stabilized zirconia is chemically inert to most reactive gases and electrode materials.

In view of the limitations encountered in using other types of ceramic conductors than yttria stabilized zirconia, its efficiency at low temperatures had to be improved. To reduce the operating temperature of the cell without affecting the efficiency of oxygen ion conduction, the electrolyte has to be as thin as possible in order to compensate the increase in ohmic losses (Huijsmans, 2001). Other advantages of fuel cells with thin electrolytes are reduction in material costs and improvement in the characteristics of the cells (Perednis & Gauckler, 2004). Therefore, the yttria stabilized zirconia is still a material with great prospects in the application as electrolyte in solid oxide fuel cells. Research about this type of material is aimed to improve its characteristics in order to adapt the needs of current applications.

1.5 Techniques to obtain SOFC materials

The methods employed in the deposition of thin films of oxides can be divided into two major groups based on the nature of the deposition processes. Physical methods of deposition: physical vapor deposition (PVD) (Kueir-Weei et al., 1997), ion beam (Xiaodong et al., 2008) and sputtering (Haiqian et al., 2010). The chemical methods of deposition, which can be subdivided as to the nature of the precursor: gas phase and solution. The gas phase methods: chemical vapor deposition (CVD) (Bryant, 1977) and atomic layer epitaxy (ALE) (Suntola, 1992). The solution methods: spray pyrolysis (Chamberlin & Skarman, 1966), sol-gel (Brinker et al., 1990) and electrodeposition.

The table 1 shows de main advantages and disadvantages of some technique that can used to obtain SOFC materials.

The technique of spray pyrolysis can be used to obtain both, dense or porous oxide films, and to produce ceramic coatings and powders. Compared to other deposition techniques, spray pyrolysis is a simple method for operational control. It is also cost-effective, especially regarding the cost of system implementation. Furthermore, deposition in multi-layers can be easily obtained by this versatile technique.

Technique	Advantage	Disadvantage
CVD	Applicable to ceramic coatings	Thin, non-uniform coatings, high cost
Screen printing	Simple	Non-uniform, porous coating
Spray pyrolysis	Simple, applicable to ceramic coatings, low cost	Many experimental variables
Sol-gel	Simple, applicable to ceramic coatings	Thin, non-uniform coatings
Electrodeposition	Simple, effective for complex shapes	Complex for ceramic materials

Table 1. Advantages and disadvantages of some techniques used to obtain SOFC materials.

1.5.1 Spray pyrolysis

A typical setup for spray pyrolysis consists of an atomizer, the precursor solution, the heated substrate and a temperature controller. Three types of atomizers are commonly used in spray pyrolysis: compressed air (the solution is exposed to a beam of air) (Balkenende et al., 1996); ultrasound (short wavelengths are produced by ultrasonic frequencies, generating a very fine spray) (Arya & Hintermann, 1990); electrostatic (the solution is exposed to a high electric field) (Chen et al., 1996).

Films prepared by spray pyrolysis have been used in several devices, such as solar cells, sensors, anti-reflective coatings, thermal barriers, solid oxide fuel cells, among others.

The deposition of thin films via spray pyrolysis involves spraying a metallic salt solution on a heated substrate. The solution droplets reach the substrate surface, where solvent evaporation and the decomposition of the metal salt occurs, forming a film. The film morphology and thickness depend on the volume of solution sprayed and the substrate temperature. The film formed is usually a metal salt, which is converted to oxide in the heated substrate.

Many processes occur sequentially or simultaneously during the formation of a thin film by spray pyrolysis: atomization of the solution, transport and evaporation of solution drops, solution spread on the substrate, evaporation of the solvent and, finally, drying and decomposition of the precursor salt. Understanding these processes helps to improve the quality of the films obtained, facilitating the production scale and reproducibility of the process.

The initial parameter in the preparation of a film with certain desired characteristics, using the technique of spray pyrolysis, is the definition of the precursor solution to be used. This solution should provide the ions, needed to form the desired film, and should contain a solvent, with adequate evaporation rate and stability in the process conditions.

In the aerosol, the drops of solution are transported and eventually evaporate. The gravitational, electrical or thermophoretic forces, have influence on the trajectory of droplets and their evaporation, so the modeling of film growth should be taken into account.

When the formation of dense films is wanted, it is important that the maximum amount of drops reaches the substrate without the formation of particles along the path. Sears et al.

investigated the mechanism of SnO_2 film growth (Sears & Gee, 1988). In case of excessively rapid evaporation of the solvent along the way, the droplet size decreases and precipitates precursor salt on the edges of the drop, causing the deposition of precipitates on the substrate surface. This phenomenon is extremely deleterious to obtain dense and homogeneous films, since the particles formed in the atomizer-substrate path add to the substrate surface, forming a porous crust (Perednis, 2003).

On the other hand, if the drops are sprayed against the substrate with sufficiently high strength, spread lightly, maintain an evaporation rate equivalent to the solute precipitation rate, the solute nucleates and precipitates homogeneously, creating a dense and continuous film (Yu & Liao, 1998).

A model of the possible transport situations of aerosol from the atomizer, toward the heated substrate can be seen in Figure 2. In zone I, the droplet is too large, has a very slow solvent evaporation rate and results in the formation of a brittle precipitate. In the second case, the drops have size and strength suitable for spraying, forming homogeneously aggregates of precipitates (zone II). And finally, in zone III, the drops are too small and not strong enough to reach the substrate, causing particles to appear before reaching the substrate.

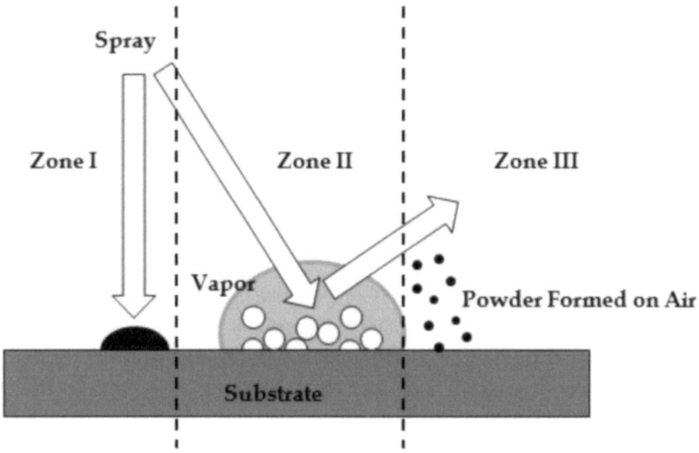

Fig. 2. Aerosol transport model.

The substrate temperature can be considered the most important factor to determine the solvent evaporation rate. It is directly linked to the time the solvent takes to spread over the surface of the substrate and the speed at which it evaporates after being spread. In addition, the substrate temperature should be high enough so that the salt decomposition reactions occur, in order to form the desired final material.

It is known that low substrate temperatures provide an excellent scattering of the solution, however, the film layer that is formed is very rich in solvent, taking too much time to evaporate. After stopping the solution spraying, the film is still wet and the local rise in temperature causes the solvent to evaporate, diminishing the volume and contributing to the formation of tension in the film. The strong adhesion between film and substrate prevents free contraction of the film, promoting the formation of cracks (Neagu et al., 1981).

However, if the substrate temperature is high, it will reduce the time of solvent evaporation to a point when the solvent partially evaporates before touching the substrate, inhibiting the spread. In cases where the temperature is too high, the droplets of solution not even touch the substrate, all the solvent is evaporated on the way between the atomizer and the surface and only particles are deposited on the substrate (Perednis & Gauckler, 2004).

The choice of an intermediate temperature is necessary to obtain a film with the desired characteristics. This should consider the solvent used, so it evaporates after a light scattering of the solution on the substrate.

1.6 Research directions

In recent years, research has been aimed at further reducing the operating temperature of the cell, leading to the development of a new class of planar SOFC, the intermediate temperature (IT-SOFC). This configuration provides an even greater reduction in operation temperature, allowing a large-scale use of metal interconnects. However, the ionic conductivity of the electrolyte is greatly affected by reducing the temperature. In this context, there are two ways to develop electrolytes for IT-SOFC: by changing the electrolyte to a material that has a high conductivity; or by reducing of the electrolyte thickness. In IT-SOFC cells, the electrolyte thickness should be as thin as possible (less than 10 μm has been suggested). In this case, the cell can operate at temperatures between 600 °C and 800 °C (Cooper et al., 2008).

In recent years, research has been aimed at further reducing the operating temperature of the cell, leading to the development of a new class of planar SOFC, the intermediate temperature (IT-SOFC). This configuration provides an even greater reduction in operation temperature, allowing a large-scale use of metal interconnects. However, the ionic conductivity of the electrolyte is greatly affected by reducing the temperature. In this context, there are two ways to develop electrolytes for IT-SOFC: by changing the electrolyte to a material that has a high conductivity; or by reducing the electrolyte thickness. In IT-SOFC cells, the electrolyte thickness should be as thin as possible (less than 10 μm has been suggested). In this case, the cell can operate at temperatures between 600 °C and 800 °C.

For the case of electrolyte change, materials with better ionic conductivity than YSZ have been studied to replace it. LSGM has been successful when applied with a metallic Ni anode, presenting ionic conductivity in temperatures around 400 °C (Sasaki et al., 2008; Tuker, 2010). The problem with such material is its low chemical stability. Another exhaustly studied material for SOFC electrolyte is gadolinium-doped ceria (CGO). This material has better ionic conductivity than YSZ and can be used in lower temperatures. However, at 600 °C the Ce reduction occurs. This reduction confers electric conduction to the material and the cell can suffer a short circuit (Tuker, 2010; Yuan et al., 2010).

The greatest challenge nowadays is to reach fuel cells operating around 700 °C without efficiency losses. The operation temperature can be lowered when the electrolyte thickness is reduced and the densification optimized. Several researches aim the obtaining of thin and dense electrolytes (Gaudon et al., 2006).

In the present study, some results about the elaboration of the YSZ thin film by spray pyrolysis process will be presented.

2. Materials and methods

The spray pyrolysis setup consisted mainly of the following parts: a spraying unit, a liquid feeding unit, and a temperature control unit (Figure 3). The spray unit consisted of an airbrush (Campbell Hausfeld) using an air blast atomizer. The liquid feeding unit is the precursor solution, constituted by yttrium chloride ($YCl_3.6H_2O$) (Aldrich Chemicals) and zirconium acetylacetonate ($Zr(C_6H_7O_2)_4$) (Aldrich Chemicals) dissolved in three different solvents: (1) mixture of ethanol (C_2H_5OH) (FMaia) and propylene glycol ($C_3H_8O_2$) (Proton) (1:1 vol.%); (2) mixture of ethanol and 2-methoxy, 1-propanol ($C_4H_{10}O_2$) (Aldrich Chemicals) (1:1 vol.%); (3) mixture of ethanol and diethylene glycol monobutyl ether ($C_8H_{18}O_3$) (Aldrich Chemicals) (1:1 vol.%). The Table 2 shows the boiling temperatures of any individual solvents used in this work.

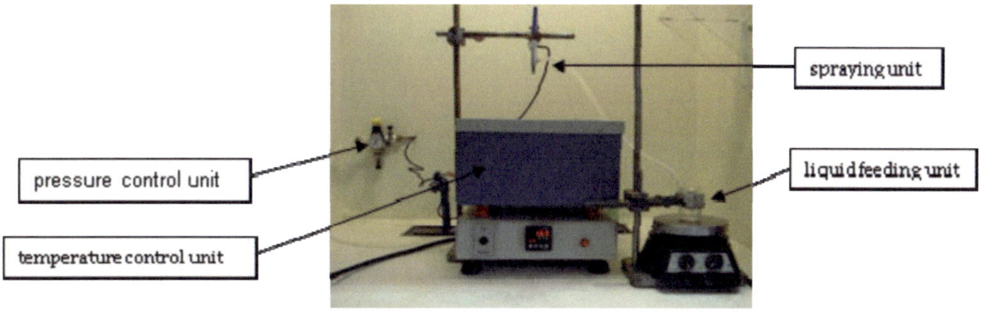

Fig. 3. Spray pyrolysis experimental apparatus.

Solvent	Boiling Point [°C]
ethanol	78.4
propylene glycol	188.2
2-methoxy, 1-propanol	120.0
diethylene glycol monobutyl ether	230.4

Table 2. Boiling temperatures of the solvents used.

The solutions were prepared according to the stoichiometry required to the films $(ZrO_2)_{0.92}(Y_2O_3)_{0.08}$ (Perednis & Gauckler, 2004) and adopting a final concentration of salts in solution of 0.1 $mol.L^{-1}$. The precursor solution was maintained under stirring and heating at 50 °C in a hotplate stirrer (Fisaton), in order to obtain the complete dissolution of the salts and decrease the heat loss of the substrate. Finally, the temperature control unit consisted in a hotplate, used for heating the substrate. A thermostat controlled the hotplate temperature and the substrate temperature was monitored by an infrared pyrometer. The precursor solution was sprayed on the heated LSM porous substrate, in order to obtain the YSZ films.

The substrate temperature is determining in the morphology of films obtained, since it is directly related to the solvent evaporation rate, and a quick evaporation of the solvent promotes the formation of particles instead of forming a continuous film. On the other hand, the slow evaporation of the solvent, promotes crack formation in the film. Previous studies were made to determine the optimum temperature to obtain continuous films. For this reason, in this chapter, a single temperature was studied, 350 °C.

Three other important parameters influence in the solvent evaporation rate and in the attainment of dense and continuous films: solution flow rate, nozzle distance and air pressure,. In this study, these parameters were kept constant, and their values were, respectively, 35 $mL.h^{-1}$, 250 mm and 3 $kgf.cm^{-2}$.

The YSZ films are amorphous, after deposition, and a heat treatment was required in order to stabilize the zirconia cubic phase. This heat treatment was performed at 700 °C for two hours in a furnace with a constant heating rate of 2 $°C.min^{-1}$ and slow cooling.

Two protocols were used for the film deposition: one-step deposition and multi-layer deposition. The first protocol consists in a deposition of 50 mL of precursor solution followed by heat treatment and the other protocol consists in a deposition of 150 mL of precursor solution in three sequential steps with intermediate heat treatment after each deposition.

The microstructure and the morphology of the films were evaluated by Fourier Transform Infrared Spectroscopy (FT-IR), X-ray diffraction (XRD) and Scanning Electron Microscopy (SEM).

3. Results and discussion

3.1 Microstructural characterization

3.1.1 X-ray diffraction (XRD)

The spectrum of X-ray diffraction (Figure 4) shows that after the deposition, the film is amorphous.

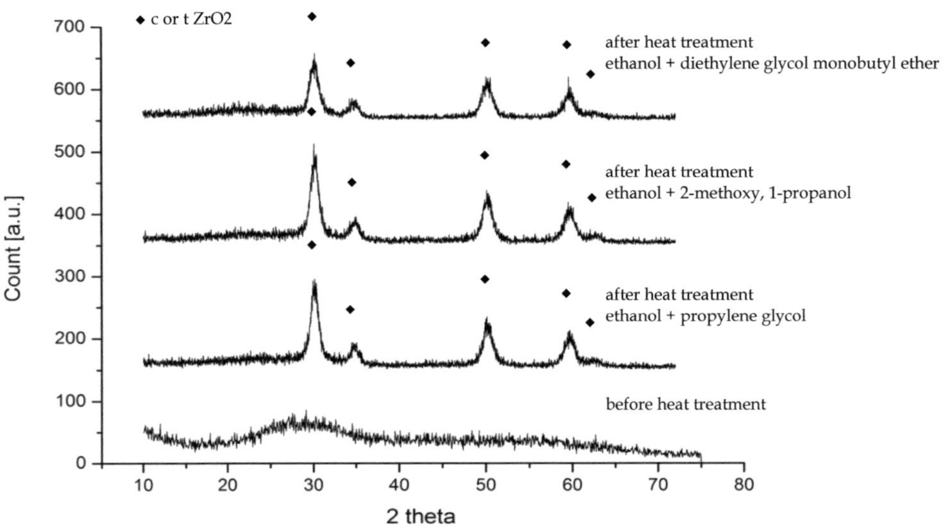

Fig. 4. XRD spectrum of YSZ obtained from different precursor solutions before and after heat treatment at 700 °C for 2 hours.

However, after the heat treatment at 700 °C for 2 hours, the crystallization of zirconia was observed for all the solutions tested. There was no influence of the solvent used in the stabilization of the zirconia phase. The overlapping of the tetragonal and cubic zirconia peaks impedes the determination of the predominant stabilized phase (Wattnasiriwech et al., 2006). Given the importance of determining the stable phase, other techniques can be used to complement the analysis of YSZ X-ray diffraction. In this work Fourier Transform Infrared Spectroscopy (FT-IR) was used to determinate the zirconia phase stabilized.

3.1.2 Fourier Transform Infrared Spectroscopy (FT-IR)

The FT-IR spectrum (Figure 5) shows a very pronounced peak around 471 cm^{-1} (zone IV). Comparing it to FT-IR spectra for cubic zirconia, presented in the literature (Khollam et al., 2001), it is possible to see the presence of vibrational frequencies resulting from metal-oxygen bonds characteristic of this phase around 471 cm^{-1}, revealing that the heat treatment at 700 °C allowed the stabilization of the zirconia cubic phase for the films prepared. The band located at 3455 cm^{-1} (zone I) can be attributed to bonds O - H, possibly due to the presence of solvent excess adsorbed in the film. The bands presented around 2330 cm^{-1} (zone II) correspond to adsorbed atmospheric CO_2, according to (Andrade et al., 2006). In the region from 1630 to 1619 cm^{-1} (zone III), the bands correspond to asymmetric and symmetric stretch of COO-group (Farhikhte, 2010).

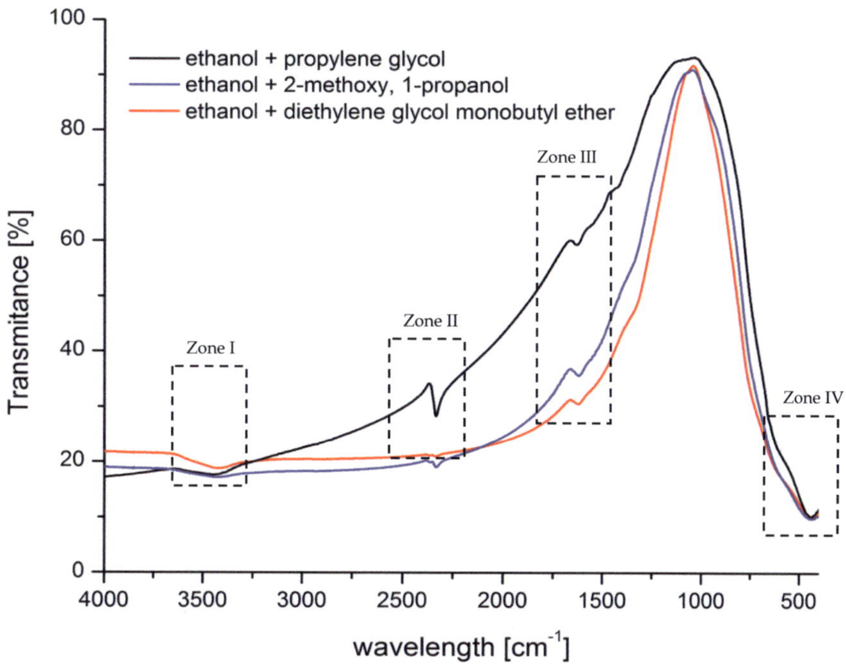

Fig. 5. FT-IR spectrum Of YSZ obtained from different precursor solutions after heat treatment at 700 °C for 2 hours.

3.2 Morphology characterization

3.2.1 One-step deposition (influence of solvent)

Variations of the solvent used in the precursor solution cause changes in the solution characteristics. A solvent mixture that has, for example, a lower boiling point, leads to a faster evaporation of the solvent for the same temperature. This stage of the study aimed to observe the morphological changes associated to the use of different solvents for the same temperature.

Figure 6 shows the film obtained at 350 °C, using ethanol and propylene glycol as solvent in the precursor solution. The morphology obtained showed a large amount of cracked plates. Locally, the plates are fairly homogeneous, with little porosity, as evidenced in detail showed in Figure 6b. This morphology may be associated to the high viscosity of the solvent, which hinders the spread of the solution on the substrate surface, resulting in a thicker solution film rich in solvent. When the solvent evaporates, the cohesion and adhesion to the substrate throughout the film generate a lot of residual stresses, causing the oxide film to crack.

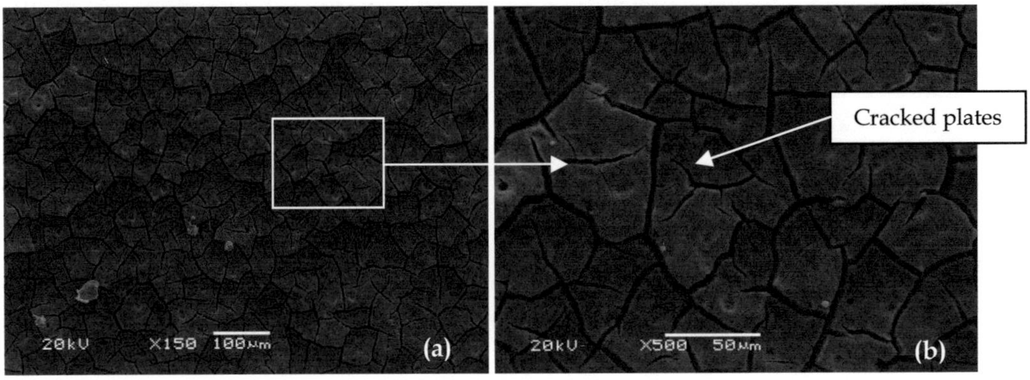

Fig. 6. (a) SEM image of the film obtained from ethanol and propylene glycol on the substrate at 350 °C, without heat treatment; (b) magnification of (a).

The films deposited from ethanol and propylene glycol solution presented clefts (Figure 6) and after heat treatment, the cracks increased, not only in number but also in intensity (Figure 7). The plates, which were well bonded to the surface, suffered a major influence of the contraction during the zirconia crystallization, contributing to the increase in cracking and detachment of the film as shown in Figure 7. This mechanism has been proposed by (Φstergård, 1995).

Depositions from the solution of ethanol and 2-methoxy, 1-propanol showed, at first, the formation of an apparently continuous film instead of plaque formation. This can be seen in Figure 8a. However, as this solution has lower boiling point and greater fluidity, there is a very high scattering on the surface. This phenomenon and the rapid evaporation of the solvent after the scattering, lead to a cracked film.

It is also possible to observe the deposition of some precipitates distributed over the layer of the film (Figure 8). These are related to the evaporation during the transport of the droplets toward the substrate, forming YSZ powder that clings to the surface of the film.

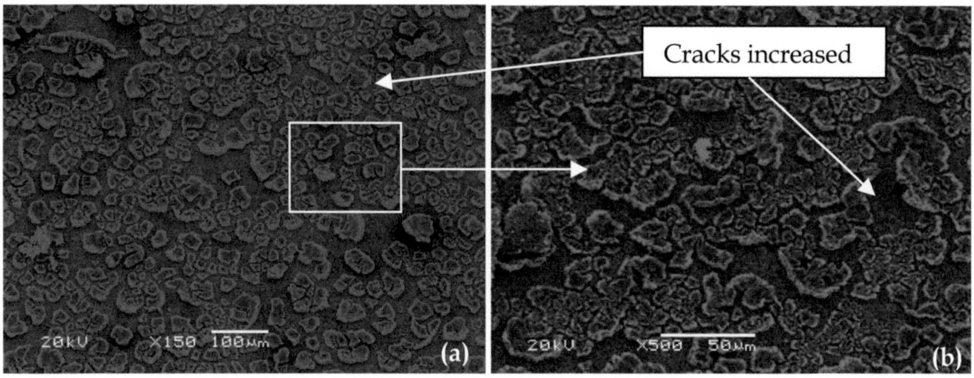

Fig. 7. (a) SEM image of the film obtained from ethanol and propylene glycol on the substrate at 350 °C, after heat treatment at 700 °C for 2 hours; (b) magnification of (a).

On the other hand, the film formed (Figure 8b), is not continuous and suggests the overlapping of non-homogeneous rough boards, which are formed during the rapid evaporation of the solvent. This type of morphology is undesirable to use in SOFC electrolyte, because it forms deep cracks and discontinuities.

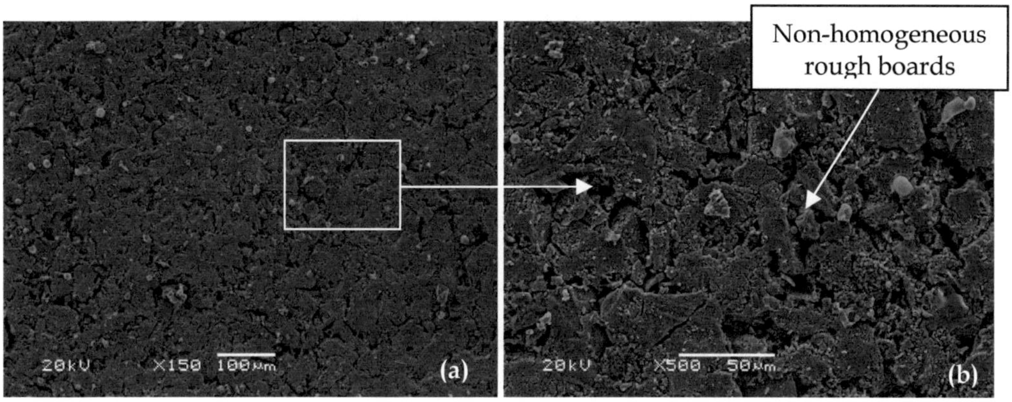

Fig. 8. (a) SEM image of the film obtained from ethanol and 2-methoxy, 1-propanol on the substrate at 350 °C, without heat treatment; (b) magnification of (a).

After heat treatment (Figure 9), there seems to be a softening of the cracks seen after the deposition, possibly related to the contraction during the zirconia crystallization. Which, in this case, tends to reduce overlapping plates, softening the final morphological structure. However, this effect is not enough to homogenize the surface, and the cracks are not completely eliminated (Figure 9b).

The films obtained from ethanol and diethylene glycol monobutyl ether showed a considerable reduction in the amount of cracks and discontinuities (Figure 10a). After a soft scattering of the solution onto the substrate surface, the solvent evaporates properly, thus reducing internal stresses in the film and reducing the number of cracks, making it more homogeneous (Neagu et al., 2006).

However, the magnification of the image (Figure 10b) revealed the presence of small cracks distributed throughout the film. These cracked regions jeopardize the homogeneity of the

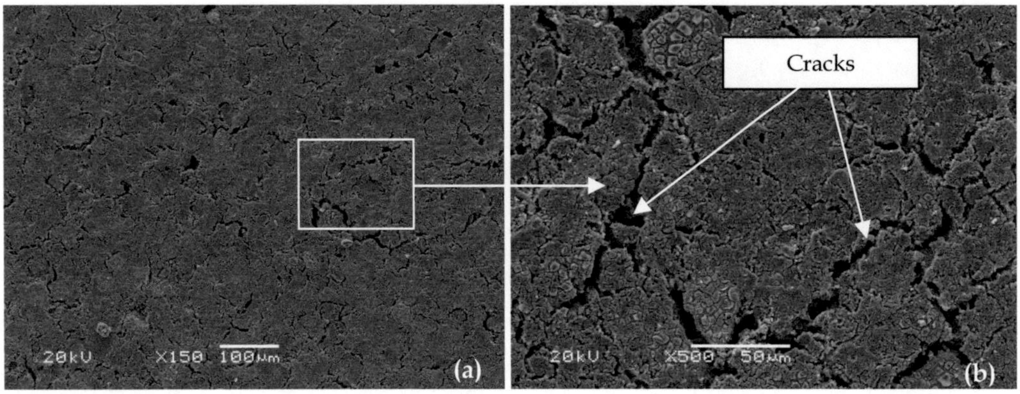

Fig. 9. (a) SEM image of the film obtained from ethanol and 2-methoxy, 1-propanol on the substrate at 350 °C after heat treatment at 700 °C for 2 hours; (b) magnification of (a).

film and show that the use of this solvent alone is not enough for complete densification of the deposited layer.

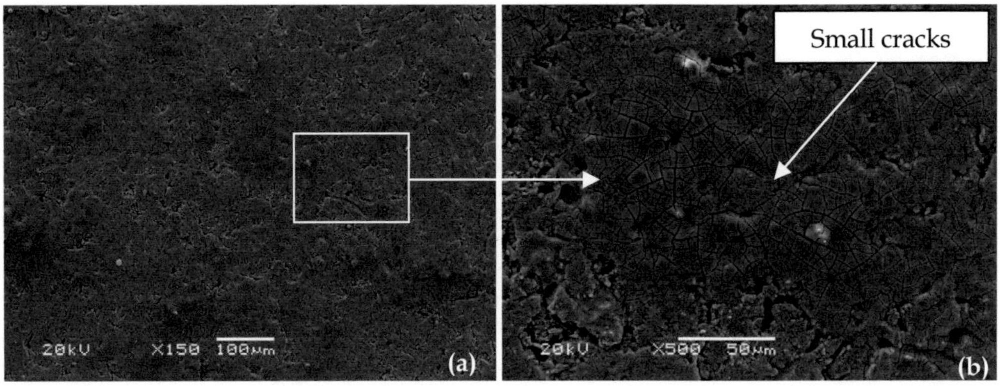

Fig. 10. (a) SEM image of the film obtained from ethanol and diethylene glycol monobutyl ether on the substrate at 350 °C, without heat treatment; (b) magnification of (a).

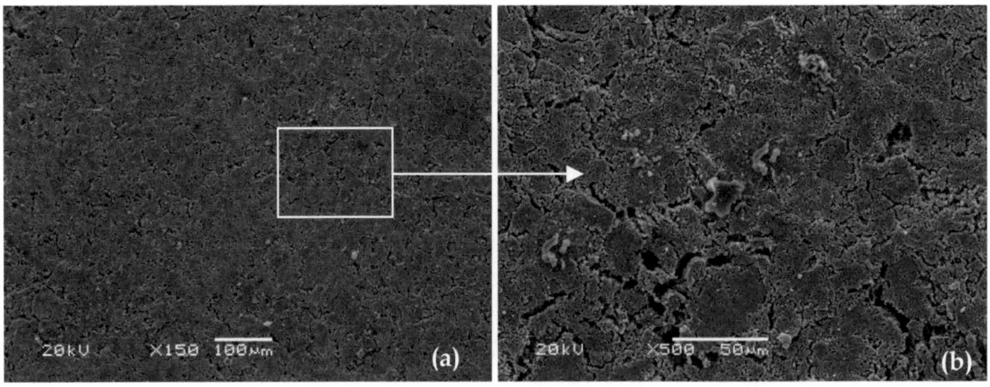

Fig. 11. (a) SEM image of the film obtained from ethanol and diethylene glycol monobutyl ether on the substrate at 350 °C, after heat treatment at 700 °C for 2 hours; (b) magnification of (a).

The films have undergone little morphological changes after heat treatment (Figure 11a). There was a slight increase of the cracks, which can be best shown in Figure 11b, that may be associated to the changes in the volume of the film during the crystallization of the zirconia cubic phase.

3.2.2 Multi-layer deposition

From the results obtained in depositions with different solvents, the solution (ethanol and diethylene glycol monobutyl ether) was chosen to use in the protocol of multi-layer depositions with intermediate heat treatment.

Figure 12 shows the film after the first layer deposition, where the surface is completely covered. After heat treatment (Figure 12b) there is a homogenization of the surface morphology, however, this homogenization is accompanied by the appearance of some discontinuities distributed throughout the film.

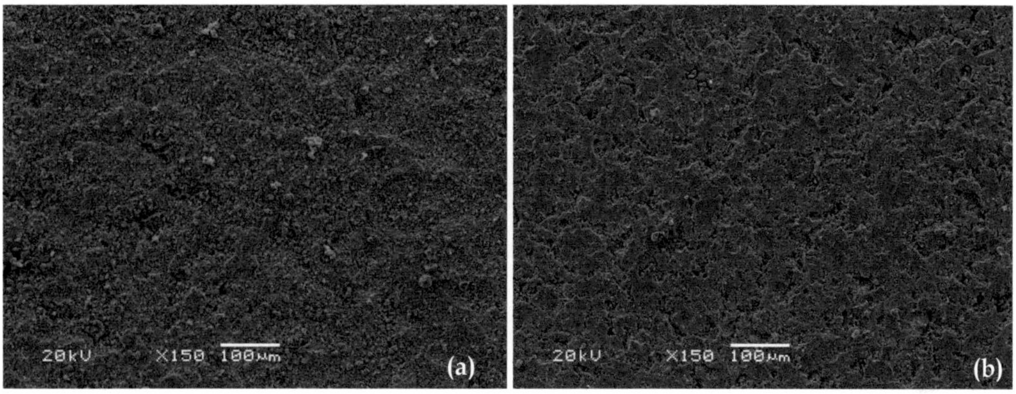

Fig. 12. SEM images of the film obtained from intermittent deposition at substrate temperature of 350 °C: (a) after the first deposition; (b) after the first heat treatment at 700 °C for 2 hours.

Fig. 13. SEM images of the film obtained from intermittent deposition at substrate temperature of 350 °C: (a) after the second deposition; (b) after the second heat treatment at 700 °C for 2 hours.

After the second layer deposition, the film covered most of these irregularities, achieving the aimed goal of the intermediate deposition protocol. However, the second and first layer overlapping, led to the formation of an even more irregular layer, both before (Figure 13a) and after heat treatment (Figure 13b).

The same behavior was observed after the last deposition. This layer apparently had some leveling effect on the film, slightly reducing the surface irregularities (Figure 14a). There were no significant morphological changes after the final heat treatment (Figure 14b). For the protocol of multi-layer depositions, with intermediate heat treatment, a crack-free film was obtained, but the surface presented a roughness increase.

Fig. 14. SEM images of the film obtained from intermittent deposition at substrate temperature of 350 °C: (a) after the final deposition; (b) after the final heat treatment at 700 °C for 2 hours.

4. Conclusions

Results showed that YSZ films can be obtained directly on porous LSM substrate by spray pyrolysis technique. A dense and homogeneous film can be obtained by multi-layers deposition with intermediate heat treatment.

The heat treatment at 700 °C allowed the stabilization of the zirconia cubic phase, which is the phase of interest for application as electrolyte in solid oxide fuel cells, as indicated by x-ray diffraction analysis and confirmed by the Fourier transform infrared spectroscopy analysis.

The type of solvent used influences in the morphology of the films obtained. Solvents that have very low boiling point lead to the uncontrolled evaporation of the solvent, forming a quite brittle film. The increase in the solvent boiling temperature helps to a proper evaporation, resulting in films with more satisfactory morphological characteristics. In all cases of one-layer deposition, heat treatment caused an increase of surface discontinuities, demonstrating its influence on the films morphology.

5. Acknowledgments

This work was supported by CAPES, the Brazilian Government agency for the development of human resources. The authors also thank the financial support of CNPq.

6. References

Akorede, M.F., Hizam, H. & Pouresmaeil, E. (2010). Distributed energy resources and benefits to the environment. *Renewable and Sustainable Energy Reviews*. Vol 14, pp. 724.

Andrade, I. M. de. (2006). Zircônia Estabilizada com a adição de Cério e Neodímio. *17º CBECIMat - Congresso Brasileiro de Engenharia e Ciência dos Materiais*, Foz do Iguaçu, Brazil.

Aruna, S.T. & Rajam, K.S. (2008). A study on the electrophoretic deposition of 8YSZ coating using mixture of acetone and ethanol solvents. *Materials Chemistry and Physics*. Vol. 111, pp. 131.

Arya, S.P.S. & Hintermann, H.E. (1990). Growth of Y-Ba-Cu-O Superconducting Thin Films by Ultrasonic Spray Pyrolysis. *Thin Solid Films*. Vol. 193, pp. 841.

Balkenende, A.R., Bogaerts, A., Scholtz, J.J., Tijburg, R.R.M. & Willems, H.X. (1996). Thin MgO Layers for Effective Hopping Transport of Electrons. *Philips Journal of Research*. Vol. 50, pp. 365.

Brinker, C.J., Hurd, A.J., Frye, G.C., Ward, K.J. & Ashley, C.S. (1990). Sol-Gel Thin Film Formation. *Journal of Non-Crystalline Solids*. Vol. 121, pp. 294.

Bryant, W.A. (1977). The fundamentals of chemical vapour deposition. *Journal of Materials Science*. Vol. 12, pp. 1285.

Canales-Vázquez, J., Tao, S. W. & Irvine, J. T. S. (2003). Electrical properties in La2Sr4Ti6O19−δ: a potential anode for high temperature fuel cells. *Solid State Ionics*. Vol. 159, pp. 159.

Carter, S., Selcuk, A., Chater, R. J., Kajda, J., Kilner, J. A. & Steele, B. C. H. (1992). Oxygen transport in selected nonstoichiometric perovskite-structure oxides. *Solid State Ionics*. Vol. 53, pp. 597.

Chamberlin, R.R. & Skarman, J.S. (1966). Chemical Spray Deposition Process for Inorganic Films. *Journal of the Electrochemical Society*. Vol. 113, pp. 86.

Chen, C.H., Kelder, E.M., Van der Put, P.J.J.M. & Schoonman, J. (1996). Morphology control of thin LiCoO2 films fabricated using the electrostatic spray deposition (ESD) technique. *Journal of Materials Chemistry*. Vol. 6, pp. 765.

Cooper, L., Benhaddad, S., Wood, A. & Ivey, D.G. (2008). The effect of surface treatment on the oxidation ferritic stainless steels used for solid oxide fuel cell interconnects. *Journal of Power Sources*. Vol. 184, pp. 220.

Ellis, M. W., Von Spakovsky, M.R. & Nelson D.J. (2001). Fuel cell systems: efficient, flexible energy conversion for the 21st century. *IEEE Proceedings*. Vol. 89, pp. 1808.

Farhikhte, S., Maghsoudipour, A. & Raissi, B. (2010). Synthesis of nanocrystalline YSZ (ZrO2-8Y2O3) powder by polymerized complex method. *Journal of Alloys and Compounds*. Vol. 491, pp. 402.

Farooque, M. & Maru, H. C. (2001). Fuel cells-the clean and efficient power generators. *IEEE Proceedings*. Vol. 89, pp. 1819.

Fridleifsson, I.B. (2001) Geothermal energy for the benefit of the people. *Renewable and Sustainable Energy Reviews*. Vol 5, pp. 299.

Garcia, B. L., Sethuraman, V.A., Weidner, J. W., White, R.E. & Dougal, R. (2004). Mathematical Model of a Direct Methanol Fuel Cell. *Journal of Fuel Cell Science and Technology*. Vol. 1, pp. 43.

Gaudon, M., Laberty-Robert, Ch., Ansart, F. & Stevens, P. (2006). Thick YSZ films prepared via modifited sol-gel route: Thickness control (8-80 μm). *Journal of European Ceramic Society*. Vol. 26, pp. 3153.

Gong, M., Bierschenk, D., Haag, J., Poeppelmeier, K.R., Barnett, S. A., Xu, C., Zondlo, J.W. & Liu, X. (2010). Degradation of LaSr2Fe2CrO9−δ solid oxide fuel cell anodes in phosphine-containing fuels. *Journal of Power Sources*. Vol. 195, pp. 4013.

Haiqian, W., Weijie, J., Lei, Z., Yunhui, G., Bin, X., Yousong, J. & Yizhou, S. (2010). Preparation of YSZ films by magnetron sputtering for anode-supported SOFC. *Solid State Ionics*. Article in press.

He, Z., Yuan, H., Glasscock, J.A., Chatzichristodoulou, C., Phair, J.W., Kaiser, A. & Ramousse, S. (2010). Densification and grain growth during early-stage sintering of $Ce_{0.9}Gd_{0.1}O_{1.95-\delta}$ in reducing atmosphere. *Acta Mater.* Vol 58, pp. 3860.

Horita, T., Kishimoto, H., Yamaji, K., Xiong, Y., Sakai, N., Brito, M.E. & Yokokawa, H. (2008)Evaluation of Laves-phase forming Fe–Cr alloy for SOFC interconnects in reducing atmosphere. *Journal of Power Sources*. Vol. 176, pp. 54.

Hui, S. & Petric, Q. A. (2001). Electrical Properties of Yttrium-Doped Strontium Titanate under Reducing Conditions. *Journal of Electrochemical Society*. Vol. 149, pp. 1.

Huijsmans, J.P.P. (2001). Ceramics in solid oxide fuel cells. *Cur. Opinion in Solid State and Materials Science*. Vol. 5, pp. 317.

Khollam, Y.B., Deshpande, A. S., Patil, A. J., Potdar, H. S., Deshpande, S. B. & Date, S. K. (2001). Synthesis of yttria stabilized cubic zirconia (YSZ) powders by microwave-hydrothermal route. *Materials Chemistry and Physics*. Vol. 71, pp. 235.

Kirubakaran, V., Sivaramakrishnan, V., Nalini, R., Sekar, T., Premalatha, M. & Subramanian, P. (2009). A review on gasification of biomass. *Renewable and Sustainable Energy Reviews*. Vol. 13, pp. 179.

Kueir-Weei, C., Jong, C. & Ren, X. (1997). Metal-organic vapor deposition of YSZ electrolyte layers for solid oxide fuel cell applications. *Thin Solid Films*. Vol. 304, pp. 106.

Louie, H. & Strunz, K. (2007) Superconducting Magnetic Energy Storage (SMES) for Energy Cache Control in Modular Distributed Hydrogen-Electric Energy Systems. *IEEE Transactions on Applied Superconductivity*. Vol. 17, pp. 2361.

Mendez, V.H., Rivier, J., de la Fuente, J.I., Gómez, T., Arceluz, J., Marín, J. & Madurga, A. (2006). Impact of distributed generation on distribution investment deferral. *International Journal of Electrical Power and Energy Systems*. Vol 28, pp. 244.

Minh, N. Q. (1993). Ceramic Fuel Cells. *Journal of American Ceramic Society*. Vol. 76 pp. 563.

Neagu, R., Perednis, D., Princivalle, A. & Djurado, E. (2006). Influence of the process parameters on the ESD synthesis of thin film YSZ electrolytes. *Solid State Ionics*. Vol. 177, pp. 1981.

O'Sullivan, J. B. (1999). Fuel cells in distributed generation. *Power Engineering Society Summer Meeting*, Vol. 1, pp. 568.

Perednis, D. & Gauckler, L. J. (2004). Solid oxide fuel cells with electrolytes prepared via spray pyrolysis. *Solid State Ionics*. Vol. 166, pp. 229.

Perednis, D. (2003). Thin Film Deposition by Spray Pyrolysis and the Application in Solide Oxide Fuel Fuel Cells. *Docotorate Thesis - Suiss Federal Institute of Technology*, Zurich, Switzerland.

Ralph, J. M., Schoeler, A.C. & Krumpelt, M. (2001). Materials for lower temperature solid oxide fuel cells. *Journal of Material Science*. Vol. 36, pp. 1161.

Sasaki, K., Muranaka, M., Suzuki, A. & Terai, T. (2008). Synthesis and characterization of LSGM thin films electrolyte by RF magnetron spputering for LT-SOFCS. *Solid State Ionics*. Vol. 179, pp. 1268.

Sauvet, A.-L. & Fouletier, J. (2001). Catalytic properties of new anode materials for solid oxide fuel cells operated under methane at intermediary temperature. *Journal of Power Sources*. Vol. 101, pp. 259.

Sears, W.M. & Gee, M.A. (1988). Mechanics of Film Formation During the Spray Pyrolysis of Tin Oxide. *Thin Solid Films*. Vol. 165, pp. 265.

Srivastava, P.K., Quach, T., Duan, Y.Y., Donelson, R., Jiang, S.P., Ciacchi, F.T. & Badwal, S.P.S. (1997). Electrode supported solid oxide fuel cells: Electrolyte films prepared by DC magnetron sputtering. *Solid State Ionics*. Vol. 99, pp. 311.

Stambouli, A.B. & Traversa E. (2002a). Fuel cells, an alternative to standard sources of energy. *Renewable and Sustainable Energy Reviews*. Vol 6, pp. 297.

Stambouli, A.B. & Traversa, E. (2002b). Solid oxide fuel cells (SOFCs): a review of an environmentally clean and efficient source of energy. *Renewable and Sustainable Energy Reviews*. Vol 6, pp. 433.

Suntola, T. (1992). Atomic layer epitaxy. *Thin Solid Films*. Vol. 216, pp. 84.

Tao, S. & Irvine, J.T.S. (2002). Optimization of Mixed Conducting Properties of Y2O3–ZrO2–TiO2 and Sc2O3–Y2O3–ZrO2–TiO2 Solid Solutions as Potential SOFC Anode Materials. *Journal of Solid State Chemistry*. Vol. 165, pp. 12.

Tucker, M.C. (2010). Progress in metal supported solid oxide fuel cells: a review. *Journal of Power Sources*. Vol. 195, pp. 4570.

Wand, G. (2006). *Fuel Cell History*, Part 1.

Wang, C.Y. (2004). Fundamental Models for Fuel Cell Engineering. *Chemical Reviews*. Vol. 104, pp. 4727.

Wattnasiriwech, D. Wattnasiriwech, S. & Stevens R. (2006). A sol-powder coating technique for fabrication of yttria stabilized zirconia. *Materials Research Bulletin*. Vol. 41, pp. 1437.

Xiaodong, H., Bin, M., Yue, S., Bochao, L. & Mingwei, L. (2008). Electron beam physical vapor deposition of YSZ electrolyte coatings for SOFCs. *Applied Surface Science*. Vol. 254, pp. 7159.

Yu, H. F. & Liao, W.H. (1998). Evaporation of solution droplets in spray pyrolysis. *International Journal of Heat and Mass Transfer*. Vol. 41, pp. 993.

Φstergård, M.J.L. (1995). Manganite-zirconia composite cathodes for SOFC: Influence of structure and composition. *Eletrochimica Acta*. Vol. 40, pp. 1971.

7

Investigations of Intermediate-Temperature Alkaline Methanol Fuel Cell Electrocatalysis Using a Pressurized Electrochemical Cell

Junhua Jiang[1] and Ted Aulich[2]
[1]*Illinois Sustainable Technology Center, University of Illinois at Urbana-Champaign,*
[2]*Energy and Environmental Research Center, University of North Dakota,*
USA

1. Introduction

Direct methanol fuel cells (DMFCs) possess obvious advantages over traditional hydrogen fuel cells in terms of hydrogen storage, transportation, and the utilization of existing infrastructure. However, the commercialization of this fuel cell technology based on the use of proton-conductive polymer membranes has been largely hindered by its low power density owing to the sluggish kinetics of both anode and cathode reactions in acidic media and high cost owing to the use of noble metal catalysts. These could be potentially addressed by the development of alkaline methanol fuel cells (AMFCs). In alkaline media, the polarization characteristics of the methanol electrooxidation and oxygen electroreduction are far superior to those in acidic media (Yu et al., 2003; Prabhuram & Manoharan, 1998). Another obvious advantage of using alkaline media is less-limitations of electrode materials. The replacement of Pt catalysts with non-Pt catalysts will significantly decrease the cost of catalysts. Recently, the AMFCs have received increased attention (Dillon et al., 2004). However, these fuel cells are normally operated at temperature lower than 80 °C. In this low temperature range, both methanol electrooxidation and oxygen electroreduction reactions are not sufficiently facile for the development of high performance AMFCs. Considerable undergoing efforts are now focused on the development of highly active catalysts for accelerated electrode reactions.

Alternatively, increasing temperature has been proven as an effective way to accelerate electrode reactions. The changes of the reaction rates with increasing temperature are strongly determined by the values of activation energy, as described by the Arrhenius equation. More obvious changes are expected for the methanol electrooxidation in alkaline media than in acidic media since reported values of the activation energy are higher in alkaline media (Cohen et al, 2007). Additionally, increasing temperature may decrease concentration polarization, Ohmic polarization, and CO poisoning of the catalysts. All these advantages can contribute to the performance improvement of the AMFCs.

Further finding of increasing temperature for the methanol oxidation is that methanol can be efficiently converted with water in the aqueous phase over appropriate heterogeneous catalysts at temperatures near 200 °C to produce primarily H_2 and CO_2 (Huber et al., 2003;

Cortright et al., 2002). The aqueous-phase reforming (APR) process eliminates the need to vaporize both water and the oxygenated hydrocarbon, which reduces the energy requirements for producing hydrogen. Moreover, the formation of CO could be minimized since the APR occurs at temperatures and pressures where the water-gas shift reaction is favorable. The APR of methanol on supported Pt over 200~265 °C results in the production of H_2 at a usually high selectivity of around 99% as follows (Davda et al., 2005):

$$CH_3OH + H_2O \rightarrow CO_2 + 3H_2 \tag{1}$$

During the methanol APR, trace amount of methane is the side product and the use of more basic/neutral catalyst favors H_2 production. The methanol APR indicates that sluggish methanol oxidation reaction which has plagued low temperature AMFCs, could become highly facile in alkaline/neutral media at temperatures close to 200 °C where the APR of methanol is triggered. Substantially accelerated electrooxidation of methanol would make it possible to achieve low anode overpotentials. Therefore, the investigations of the electrooxidation of methanol in an intermediate-temperature range over the methanol-boiling temperature (around 80 °C) and the triggering temperature of the methanol APR (about 200 °C) would be of academic and practical importance for the development of high performance AMFC technology. This intermediate temperature range has been rarely used because of the limitation of boiling points of both methanol and water.

Our research efforts have been focused on the investigations of fuel cell electrocatalysis in alkaline media in the intermediate-temperature range of 80 to 200 °C for accelerated anode and cathode reaction kinetics, and the development of high performance AMFCs. In this work, we have successfully developed a pressurized electrochemical cell by modifying a Parr autoclave which can be operated at pressure up to 2000 psi and at temperature up to 200 °C. An Ag/AgCl electrode has been identified as a suitable internal reference electrode with good stability in this intermediate temperature range. It has been found that the methanol oxidation and oxygen reduction reactions can be significantly accelerated in aqueous alkaline media with increasing temperature. The former is characterized by an onset overpotential of less than 0.1 V at 150 °C for substantial methanol electrooxidation at Pt. Furthermore, highly facile methanol oxidation and oxygen reduction reactions have been also achieved at non-Pt electrodes. This accelerated kinetics of both the methanol oxidation and oxygen reduction reactions provides fundamental support for the development of novel methanol fuel cells. Accordingly, high performance intermediate-temperature alkaline methanol fuel cells using Pt and non-Pt electrocatalysts have been successfully demonstrated.

2. Experimental

2.1 Pressurized electrochemical cell

The pressurized electrochemical cell was constructed by modifying a 300 ml Parr autoclave equipped with a 200 ml glass liner, as shown in Fig. 1 (A) and (B). The working electrode was prepared by mechanically depositing high-surface-area catalysts onto a gold disk electrode of 0.5 mm in diameter, following a powder-rubbing procedure (Kucernak & Jiang, 2003), Fig. 1 (C). The mass of the catalyst layer was measured by dissolving the rubbed layer in boiling aqua regia, followed by removing the excess acid and measuring inductively coupled plasma–atomic emission spectroscopy (ICP-AES) spectra. A 15 cm length Pt wire of 0.5 mm in diameter was used as the counter electrode. Three kinds of Ag-based reference electrodes were evaluated as the internal reference electrode under our conditions by monitoring the changes of both hydrogen- and oxygen-electrochemistry on measured voltammograms for a Pt electrode in 0.5 mol dm^{-3} potassium hydroxide (KOH) at varying temperature. It was found that the Ag/AgCl reference provides the most reproducible results while both Ag wire quasi-reference electrode and Ag/Ag_2O electrode were inapplicable. The Ag/AgCl reference electrode was introduced into a glass tube containing 0.1 mol dm^{-3} HCl which was separated from KOH solution in the working electrode chamber by a microporous ceramic pellet fixed at the top end of the glass tube. The potentials of the Ag/AgCl were measured versus a reversible hydrogen electrode (RHE) at varying temperature in a hydrogen atmosphere. In the following sections, all potentials reported were referred to the RHE unless otherwise stated.

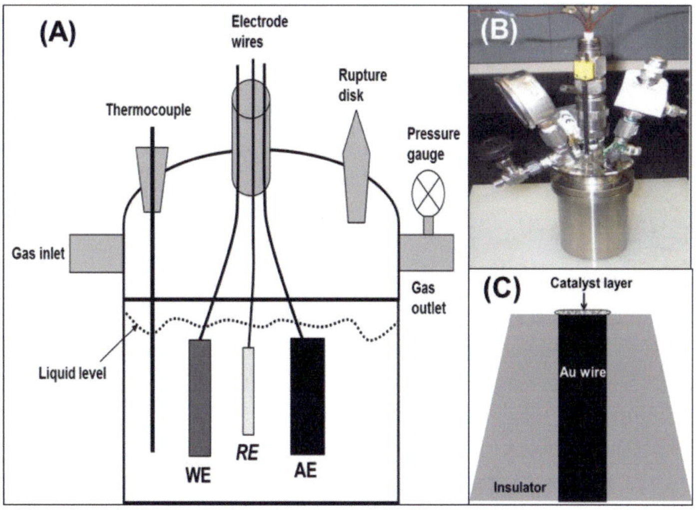

Fig. 1. A pressurized electrochemical cell based on a modified Parr autoclave. Schematic diagram of the cell (A); Image of the cell (B); and Schematic structure of a working electrode (C).

2.2 Instruments and materials

Both voltammetric and chronoamperometric measurements were performed using Autolab general purpose electrochemical system (Ecochemie, Netherland). A lab-constructed single-cell system with temperature control, gas flow rate and pressure control, and liquid flow rate and pressure control constructed was used for the measurements of fuel cell performance.

All chemicals and materials were used as received without further purification. All solutions were prepared from methanol (>99.9%, Aldrich) or KOH (>85%, Aldrich) with deionized water (18.2 MΩ cm). Unsupported electrocatalysts were used in voltammetric and chronoamperometric measurements, including Pt black (Alfa, S.A. typically 27 m^2 g^{-1}), Pd black (Alfa, S.A. typically 20 m^2 g^{-1}) and Ag nanopowder (Alfa, 20-40 nm). Carbon supported Pt (Fuel Cell Store, 60 wt% Pt/C) and carbon supported Pd (Sigma-Aldrich, 30 wt% Pd/C) were used for the preparation of gas diffusion electrodes with carbon-cloth as the diffusion layer. All gases used were of research grade.

2.3 Electrochemical measurements

During the measurements of base voltammograms and methanol electrooxidation, high-purity nitrogen was introduced into the electrochemical cell to inhibit vaporization of the liquid phase at elevated temperature, and gas-phase pressure was set at 300 psi unless otherwise stated. Background voltammograms were measured in 0.5 mol dm^{-3} KOH at 50 mV s^{-1}. All methanol oxidation experiments were carried out in 0.5 mol dm^{-3} KOH + 0.5 mol dm^{-3} methanol. Steady-state voltammograms for the methanol oxidation were recorded at a scan rate of 10 mV s^{-1}. The measurements of the chronoamperograms were performed by stepping potential from -1.1 V where no methanol oxidation occurs to a given value where methanol is oxidized. For the purpose of comparison, the oxidation of hydrogen was investigated under similar conditions by introducing high-purity hydrogen into the electrochemical cell to equilibrate with the solution of 0.5 mol dm^{-3} KOH solution at 200 psi. All oxygen reduction experiments were carried out in 0.5 mol dm^{-3} KOH solution equilibrated with 300 psi high purity O_2. To investigate the electrochemical oxidation of CO, high purity CO was introduced into the electrochemical cell with the pressure set at 300 psi.

3. Results and Discussion

3.1 Ag/AgCl reference electrode

Base cyclic voltammograms for a high-surface-area Pt-coated Au disk in 0.5 mol dm^{-3} KOH solution as a function of temperature are shown in Fig 2. Analogous to literature results, three characteristic zones corresponding to hydrogen electrochemistry, double layer and oxygen electrochemistry are observed. The peaks associated with hydrogen adsorption/desorption in the hydrogen electrochemistry zone over −1.15 to around −0.60 V are negatively shifted as the temperature is increased, accompanied by obvious decrease in the peak currents at higher temperature. The decrease in these peak currents and the double layer currents observed in a narrow potential range over -0.60 V to -0.50 V suggest that the electrochemical surface area of the electrodes is decreased at higher temperature. This could be caused by the accelerated adsorption of electrolyte components and/or impurities onto the electrode surface, resulting in the blocking of some electrode surfaces. In the oxygen electrochemistry zone over −0.5 to approximately 0.2 V, the potential of the cathode peak corresponding to the reduction of surface oxides is slightly shifted as the temperature is increased from 20° to 150°C. All these facts are indicative of the possibility of using the Ag/AgCl electrode as the internal reference electrode in aqueous KOH solution at elevated temperature.

The standard potential of the Ag/AgCl electrode in aqueous solution containing chloride has been measured in a wide temperature range over 0 to approximately 300 °C, and its temperature dependence has been analyzed by several groups (Öijerholm et al., 2009; Greeley et al., 1960). Based on the standard potential of the Ag/AgCl electrode of around 0.23 V vs. standard hydrogen electrode (SHE) and the reversible hydrogen potential in 0.5 mol dm^{-3} KOH solution (−0.81 V vs. *SHE*) at 20 °C, the value of the onset potential for hydrogen evolution would be around −1.04 V vs. Ag/AgCl if the hydrogen evolution overpotential on Pt is neglected. This value is very close to our measured value (−1.1 V). This agreement between the calculated value and measured value suggests that the Ag/AgCl electrode satisfactorily functions as the internal reference electrode in hydroxide solution. This is also supported by the characters of the hydrogen and oxygen electrochemistry for Pt electrode at elevated temperature.

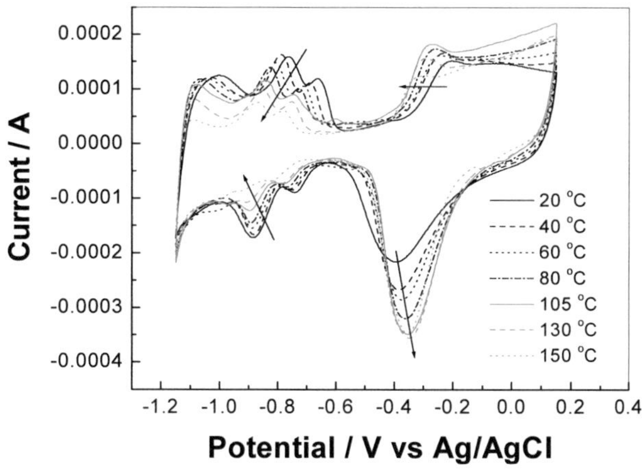

Fig. 2. Cyclic voltammograms for a high-surface Pt-coated Au disk electrode of 0.5 mm diameter in 0.5 mol dm^{-3} KOH at a scan rate of 50 mV s^{-1} as a function of reaction temperature.

To further evaluate the suitability of using the Ag/AgCl electrode as the internal reference in the wide temperature range of our interest, the cyclic voltammograms for a high-surface-area Pt-coated Au disk in 0.5 mol dm^{-3} KOH solution equilibrated with 200 psi H_2 are measured and shown in Fig. 3. The positive-going scans are characteristic of fast-rising currents at lower overpotentials, followed by slow-rising currents and limiting currents at higher overpotentials. These characteristics are similar to literature results for H_2 oxidation at Pt electrodes in aqueous base solution (Bao & Macdonald, 2007; Schmidt et al., 2002). Increasing reaction temperature clearly increases the limiting currents. These changes are mainly caused by the concentration changes of dissolved H_2 as a function of the temperature. The potential value at zero current ($E_{H2,I=0}$) measured in Fig. 3 is negatively shifted by around 40 mV as the temperature is increased from 20 to 150 °C.

The reversible potential of hydrogen reaction depends upon the reaction temperature and the pH value of the reaction medium as follows:

$$E^{o}_{H_2} = \frac{-2.303RT}{F} pH \tag{2}$$

where T is in K. Substituting pH=13.69 (corresponding to 0.5 mol dm^{-3} OH^-) and the values of R and F into Equation 2 results in its simplification as follows:

$$E^{o}_{H_2} = -2.7 \times 10^{-3} T \tag{3}$$

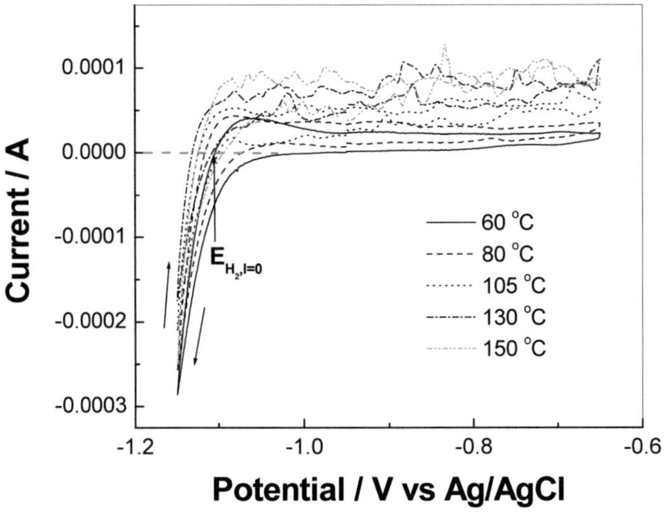

Fig. 3. Temperature dependence of cyclic voltammograms for a high-surface Pt-coated Au disk electrode of 0.5 mm diameter in 0.5 mol dm^{-3} KOH equilibrated with 200 psi H_2 at a scan rate of 10 mV s^{-1}.

According to Equation 3, the reversible hydrogen potential (*RHE*) should be negatively shifted with increasing temperature. At 20°C, the value is −0.81 V, and it is expected to be −1.14 V at 150°C. This means that a negative reversible potential shift should be around 0.33 V as the reaction temperature is increased from 20° to 150°C if the potential of the Ag/AgCl reference electrode is a constant independent upon the temperature. Because the value of $E_{H2,I=0}$ is a signature of the reversible hydrogen potential, its shift with increasing temperature should be theoretically similar to that of the *RHE* if the Ag/AgCl reference electrode potential remains constant. However, Fig. 3 demonstrates that increasing temperature from 20° to 150°C produces a negative shift of only around 40 mV, much lower than 0.33 V. Therefore, the small $E_{H2,I=0}$ shift strongly indicates that the potential of the Ag/AgCl reference electrode would have a temperature dependence similar to that of the *RHE*. This consistence provides a big convenience to investigate the kinetics of fuel cell reactions in a wide temperature range using an internal Ag/AgCl reference electrode since the Ag/AgCl reference electrode could be used as an equivalent of the reversible hydrogen electrode.

The potential difference between the Ag/AgCl electrode and the *RHE* is experimentally measured in a H_2 atmosphere. Fig. 4 shows the temperature dependence of the potential difference. It is clearly seen that the difference is less dependent upon the reaction temperature in comparison to the individual *RHE*. A potential difference of around 80 mV as the temperature is increased from 20 to 150 °C. The temperature dependence of their potential difference could be linearly approximated as follows:

$$\Delta E = E_{Ag/AgCl} - E_{RHE} = -4.4 \times 10^{-4} T - 0.980 \qquad (4)$$

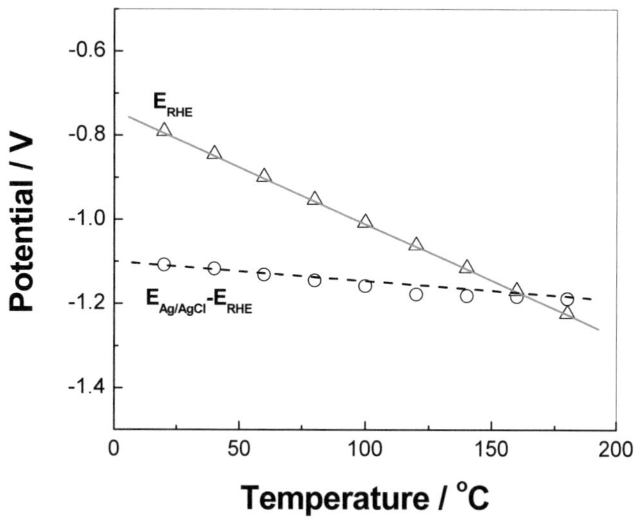

Fig. 4. Temperature dependence of calculated *RHE* potential and measured potential difference between Ag/AgCl electrode and *RHE*.

In Equation 4, the first term is much smaller than the second term being a constant. This supports that the temperature dependence of the Ag/AgCl electrode is similar to that of the RHE. Therefore, the Ag/AgCl electrode can be approximated as an equivalent of the RHE in preliminary investigations. The variations of measured electrode potentials with temperature are signature of the changes in the reaction kinetics. For the kinetic analysis, all potentials can be corrected to the RHE scale according to Equation 4.

3.2 Methanol electrooxidation

Cyclic voltammograms for a high-surface-area Pt-coated Au disk electrode in 0.5 mol dm^{-3} KOH solution containing 0.5 mol dm^{-3} methanol as a function of temperature (Fig. 5) clearly show that the onset potentials of substantial methanol electrooxidation are negatively shifted by increasing temperature. Because this prominent negative potential shift with increasing temperature is not caused by the potential shift of the reference electrode, it is reasonably believed that the electrooxidation of methanol is substantially accelerated. To assess the kinetics of methanol electrooxidation, we have compared the onset potentials for methanol oxidation and H_2 oxidation (Fig. 3) under similar conditions. The value of their onset potential difference is significantly decreased with increasing temperature. At 150 °C, a typical value is approximately 60 mV. It is well accepted that the hydrogen oxidation is highly facile on Pt, this value indicates that highly facile electrooxidation of methanol can be achieved on single-element Pt electrocatalyst in aqueous alkaline solution in the intermediate-temperature range.

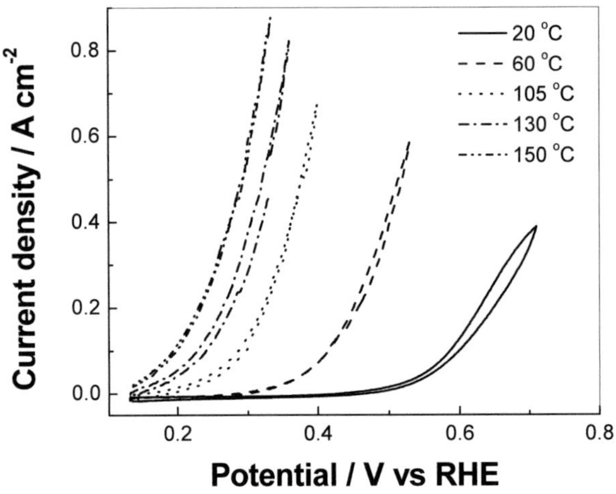

Fig. 5. Temperature dependence of cyclic voltammograms for a high-surface Pt-coated Au disk electrode of 0.5 mm diameter in 0.5 mol dm^{-3} KOH + 0.5 mol dm^{-3} CH_3OH at a scan rate of 10 mV s^{-1}.

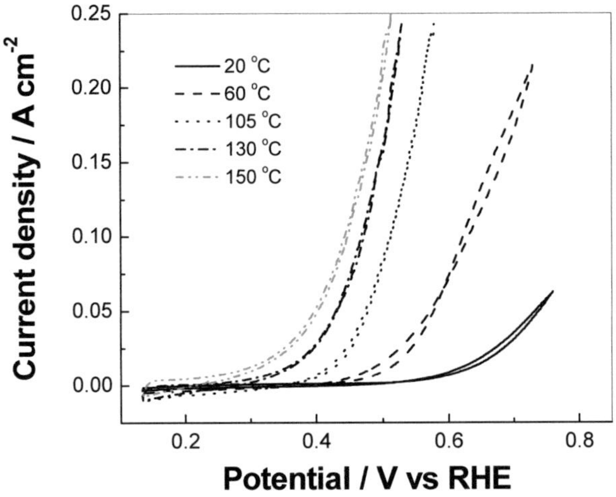

Fig. 6. Temperature dependence of cyclic voltammograms for a high-surface Pd-coated Au disk electrode of 0.5 mm diameter in 0.5 mol dm^{-3} KOH + 0.5 mol dm^{-3} CH_3OH at a scan rate of 10 mV s^{-1}.

Cyclic voltammograms for a high-surface-area Pd-coated Au disk electrode in 0.5 mol dm^{-3} KOH solution containing 0.5 mol dm^{-3} methanol as a function of temperature are shown in Fig. 6. Similar temperature dependence is observed. Increasing temperature significantly shifts the onset overpotential for methanol electrooxidation to more negative potentials. The value of the onset overpotential is decreased from 0.52 V to approximately

0.20 V when the temperature is increased from 20 °C to 150 °C. Although these values are higher compared to those measured at Pt electrode under the similar conditions, the value of 0.20 V would suggest that Pd is highly active toward the methanol oxidation. It is therefore prospective to replace Pt with Pd in alkaline methanol fuel cells for decreased cost since Pd is normally three times cheaper than Pt. Moreover, the activity of Pd-based catalysts could be further improved by introducing metal oxides to Pd or alloying Pd with other metal elements.

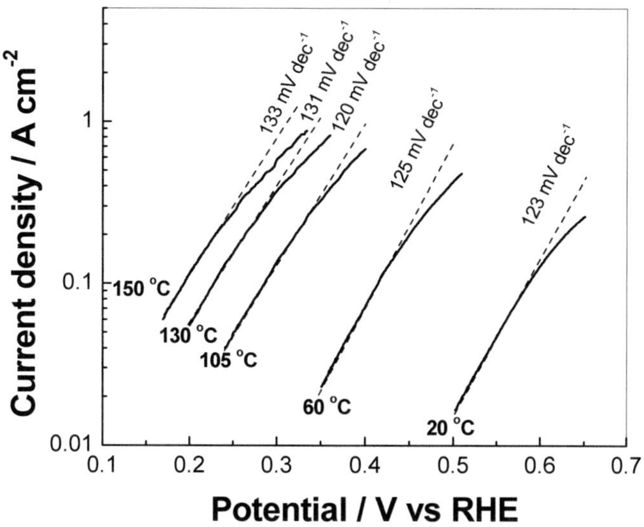

Fig. 7. Tafel plots for methanol electrooxidation at a high-surface Pt-coated Au disk electrode of 0.5 mm diameter in 0.5 mol dm^{-3} KOH + 0.5 mol dm^{-3} CH_3OH as a function of reaction temperature with data extracted from the positive-scans in Fig. 5.

Tafel plots for the methanol electroxidation at the high-surface-area Pt electrode in a solution of 0.5 mol dm^{-3} KOH and 0.5 mol dm^{-3} CH_3OH as a function of temperature are shown in Fig. 7. The values of measured *Tafel* slopes range from 120 mV dec^{-1} to 133 mV dec^{-1} in the intermediate-temperature range over 20 to approximately 150 °C. These values are in agreement with literature results obtained under similar conditions at a platinized Pt electrode and single crystal Pt(110) and Pt(111) electrodes (Tripković et al., 1998 & 2002). It has been proposed that the chemical reaction between the surface intermediate HCO_{ad} and OH_{ad} is the rate-determining step and the overall rate equation for the methanol electrooxidation could be written as follows (13):

$$j = Ac_{CH_3OH}^{0.5}c_{OH}^{0.5}\exp(\frac{\alpha F}{RT}\eta) \tag{5}$$

where A is a constant and other terms have their normal meanings. From this expression, a *Tafel* slope ranging from 120 to 164 mV dec^{-1} as the temperature is increased from 20 to 150 °C could be obtained, which fits with the experimental data shown in Fig. 7.

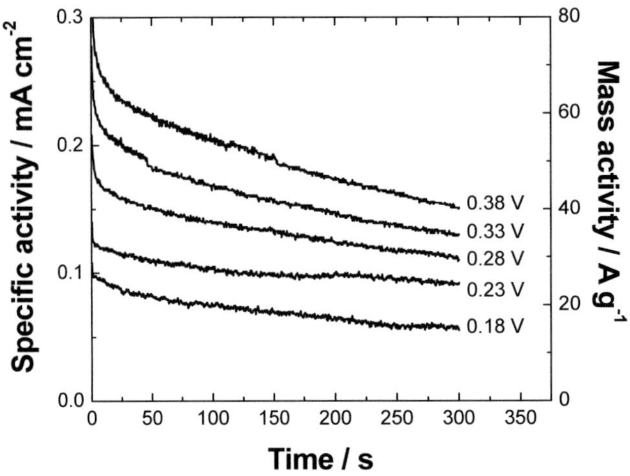

Fig. 8. Dependence of chronoamperometric curves at a controlled potential at 150 °C for a high-surface Pt-coated Au disk electrode of 0.5 mm diameter in 0.5 mol dm^{-3} KOH + 0.5 mol dm^{-3} CH_3OH upon polarization time.

The activity of high-surface-area Pt toward the methanol electrooxidation in aqueous KOH solution at 150 °C is evaluated using chronoamperometry. The current-time transients at controlled potentials are shown in Fig. 8. The mass activity was estimated from measured pseudo steady-state chronoamperometric current upon a polarization of 300 s. This figure shows that pseudo current densities can be attained even at 0.18 V vs RHE. This indicates that methanol can be dominantly oxidized to non-poisoning products in alkaline media even at low overpotentials. The activity of Pt is rather high, characterized by a mass activity of 14.9 A g^{-1} and a specific area activity of 0.05 mA cm^{-2} at an overpotential of 0.18 V. This fact is important for the improvement of the anodic reaction kinetics. In this figure, the activity decays are slow at low overpotentials and they become fast at high overpotentials. The slow decays may be caused by the surface blocking of Pt electrode owing to the adsorption of surface poisons (Matsuoka et al., 2005). Analogous to methanol oxidation in acidic media, CO has been also proposed as the predominant surface poison formed in the methanol oxidation in alkaline media. Additionally, progressive carbonation of the solution caused by CO_2 produced by oxidation reaction may decrease the pH value of the solution close to the electrode surface, leading to a decrease in reactivity. It is reasonable to attribute the progressive carbonation for the fast activity decays.

3.3 CO electrooxidation

Adsorption and oxidation of CO on Pt surfaces in aqueous electrolytes has been extensively studied, primarily because of the importance of CO as a catalyst poison and also as a reaction intermediate (Garcia & Koper, 2011; Spendelow et al., 2006). The onset of CO oxidation on Pt occurs at lower overpotentials alkaline media than in acidic media. However, all these studies have been performed at low temperature. Here we have investigated the oxidation of dissolved CO in alkaline media in the intermediate temperature range. Fig. 9 shows the temperature dependence of cyclic voltammograms for a high-surface Pt-coated Au disk electrode of 0.5 mm diameter in 0.5 mol dm^{-3} KOH equilibrated with 300 psi CO. The voltammetric characters for the oxidation of dissolved CO in the alkaline media are significantly different from those reported for adsorbed CO. The CO oxidation commences at around 0.23 V vs RHE at 60 °C. The oxidation current initially increases with increasing potential until a maximum current is reach at approximately 0.65 V. Further increase in the potential results in sudden oxidation current decrease. These facts interestingly indicate that the oxidation of CO proceeds on oxide-free Pt surface and the formation of the surface oxide inhibits the CO oxidation since the formation of the surface oxide on Pt commences at around 0.65 V (Fig. 2). These are quite different from literature results for the oxidation of adsorbed CO on Pt in acidic and alkaline media since the latter is normally triggered by the formation of surface oxide.

Increasing temperature substantially increases the CO oxidation at the oxide-free Pt surface and diminishes the onset overpotential. At temperature higher than 130 °C, the onset overpotential is approximately 0.15 V. This value is lower than those potentials used for the chronoamperometric measurements for the methanol oxidation on Pt (Fig. 8). Therefore, the current decays are less likely to be caused by the formation of surface CO as a catalyst poison.

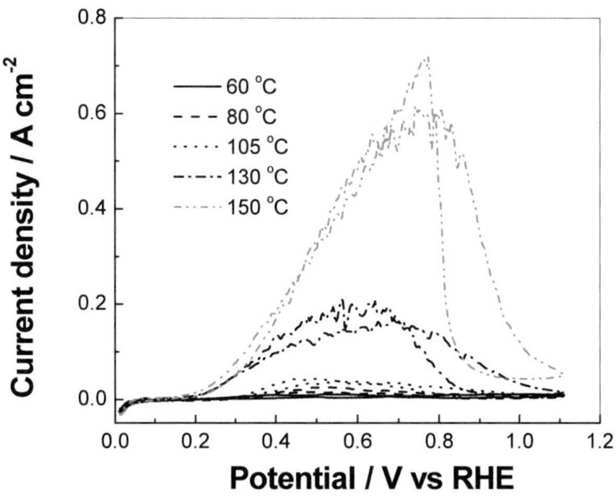

Fig. 9. Variation of cyclic voltammograms for a high-surface Pt-coated Au disk electrode of 0.5 mm diameter in 0.5 mol dm^{-3} KOH equilibrated with 300 psi CO as a function of reaction temperature.

We have also investigated the oxidation of dissolved CO at the Pd electrode. Fig. 10 shows the temperature dependence of cyclic voltammograms for a high-surface Pd-coated Au disk electrode of 0.5 mm diameter in CO-saturated 0.5 mol dm^{-3} KOH. The temperature dependence of the CO oxidation is more complicated at the Pd electrode than at the Pt electrode. At temperature lower than 105 °C, the oxidation of dissolved CO is very slow. A broad oxidation wave can be seen in the double layer region. However, the CO oxidation is inhibited at more positive potentials. This behaviour becomes more pronounced at 130 °C. The onset of the oxidation occurs at 0.20 V. Further increasing temperature causes substantial changes in the voltammograms. At 150 °C, the onset potential for the CO oxidation is shifted negative to approximately 0 V vs RHE. This value is much lower than the onset potential (0.2 V) for the methanol oxidation under similar conditions (Fig. 6). Moreover, the oxidation current is increased with increasing potentials even in the oxide formation region. These strongly suggest that the CO oxidation is more facile than the methanol oxidation at the Pd electrode at 150 °C. Therefore, it is highly likely that methanol can be oxidized to non-poisoning products rather than surface poison CO under our conditions.

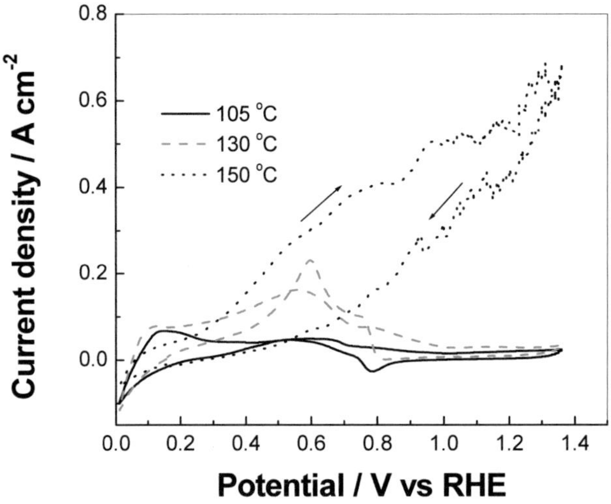

Fig. 10. Variation of cyclic voltammograms for a high-surface Pd-coated Au disk electrode of 0.5 mm diameter in 0.5 mol dm^{-3} KOH equilibrated with 300 psi CO as a function of reaction temperature.

It is well accepted that the oxidation of the adsorbed CO at Pt and Pd follows a Langmuir-Hinshelwood mechanism with the reaction between adsorbed CO and surface OH as the rate-determining step as follows (Spendelow et al., 2004):

$$OH^- - e^- \rightarrow OH_{ad} \tag{6}$$

$$CO_{ad} + OH_{ad} \rightarrow COOH_{ad} \tag{7}$$

This mechanism suggests that the substantial oxidation of adsorbed CO occurs in the oxide-formation potential region. However, cyclic voltammograms in Figs. 9 and 10 show that the CO oxidation occurs at the oxide-free electrodes and is inhibited by the oxide formation. The CO oxidation is also likely to proceed via the following mechanism:

$$CO_{ad} + OH^- - e^- \rightarrow COOH_{ad} \tag{8}$$

$$COOH_{ad} + OH^{-} - e^{-} \rightarrow CO_2 + H_2O \tag{9}$$

A theoretical *Tafel* slope of around 167 mV dec^{-1} will be expected at 150 °C if the electrochemical formation of $COOH_{ad}$ is the rate-determining step. This value is close to our experimental value of 138 mV dec^{-1} measured from the polarization region over 0.2 to 0.35 V shown in Fig. 9 at the same temperature.

3.4 Oxygen electroreduction

Pt is the most used and active catalyst for the oxygen reduction reaction (*orr*) and all of the Pt-group metals reduce oxygen in alkaline media according to the 4-electrode process (Lima & Ticianelli, 2004). Silver has been studied as a potential replacement of Pt due to its high activity for the *orr* and its low cost. The *orr* occurs with the participation of 2 and 4-electron processes, depending on its oxidation state and electrode potential (Kotz & Yeager, 1980) Moreover, the size of the Ag particles affects the different catalytic activity for these two processes (Demarconnay et al., 2004).

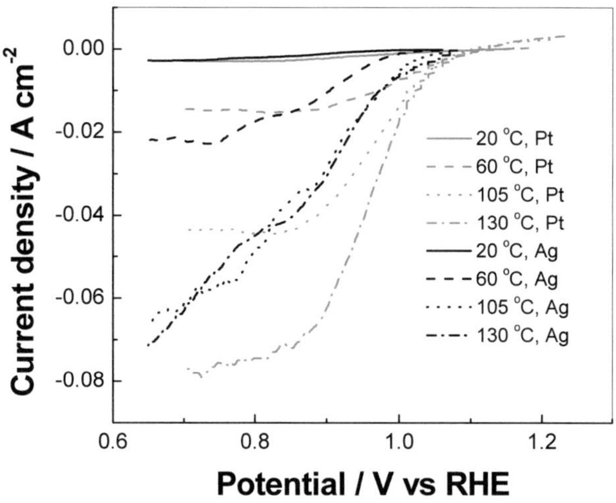

Fig. 11. Temperature dependence of polarization curves at a scan rate of 10 mV s^{-1} for a high-surface Pt (in gray) and Ag (in black) coated Au disk electrode of 0.5 mm diameter in 0.5 mol dm^{-3} KOH equilibrated with 300 psi O_2.

We have compared the activities of Pt and Ag for the *orr* in alkaline media in the intermediate temperature range. Fig. 11 shows the temperature dependence of voltammograms for high surface area Ag and Pt coated gold disk electrode of 0.5 mm in diameter in 0.5 M KOH solution equilibrated by 300 psi O_2. At both Pt and Ag electrodes, the values of the current density at lower overpotentials are substantially increased with increasing temperature with obvious positive shift of the onset potential. At higher overpotentials, a limiting current plateau is observed and it is increased with increasing temperature. The increase of the limiting current is probably caused by higher O_2 diffusion efficient and concentration in the aqueous solution at higher temperature.

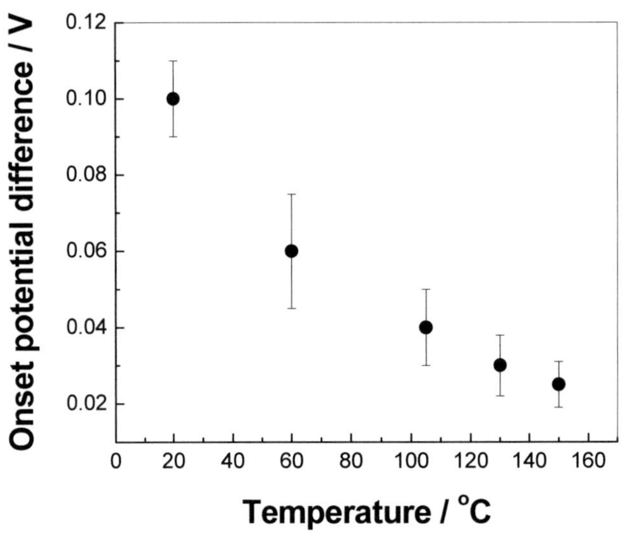

Fig. 12. Temperature dependence of onset polarization potential difference for the *orr* on Pt and Ag with data taken from Fig. 11.

The onset potential difference for the *orr* at Ag and Pt as a function of reaction temperature is shown in Fig. 12. Their potential difference is obviously decreased with increasing temperature. At room temperature, the difference is approximately 0.10 V. It falls in a range of 0.02 to 0.03 V when the temperature is higher than 130 °C. This indicates that Ag is a very promising electrocatalyst for the *orr* in the intermediate temperature range. Although the stability of Ag in alkaline media is questioned, a few strategies have been proven efficient to this issue.

3.5 Single cell performance

Based on the above fundamental studies, an intermediate-temperature alkaline methanol fuel has been developed and its performance is measured using a single-cell system with temperature control, gas flow rate and pressure control, and liquid flow rate and pressure control.

The performance of the novel fuel cell utilizing commercial Pt/C as both anode and cathode catalysts under optimized operating conditions is demonstrated in Fig. 13. The peak power density seen at around 280 mA cm^{-2} reaches 90 mW cm^{-2}. This value is much higher than a typical value of around 50 mW cm^{-2} of the state-of-the-art DMFCs using Nafion®-based proton-conducting membranes and PtRu anode catalysts (Dillon et al, 2004).

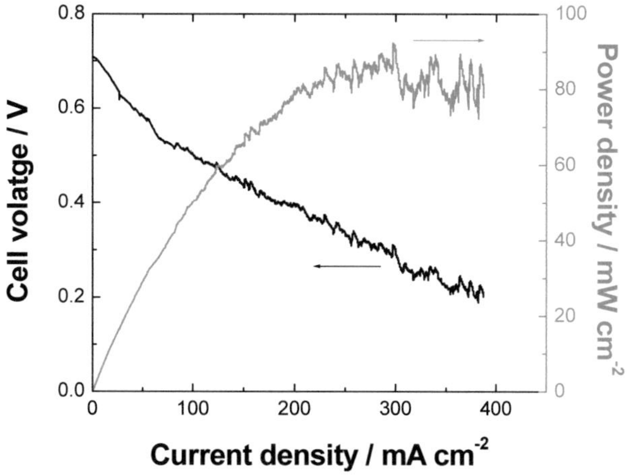

Fig. 13. Dependence of cell voltage and power density on current density for an intermediate-temperature alkaline methanol fuel using Pt/C for both anode and cathode and operated at 120 °C. Anode feed: 2 mol dm^{-3} CH_3OH + 2 mol dm^{-3} KOH; cathode feed: 120 SCCM O_2.

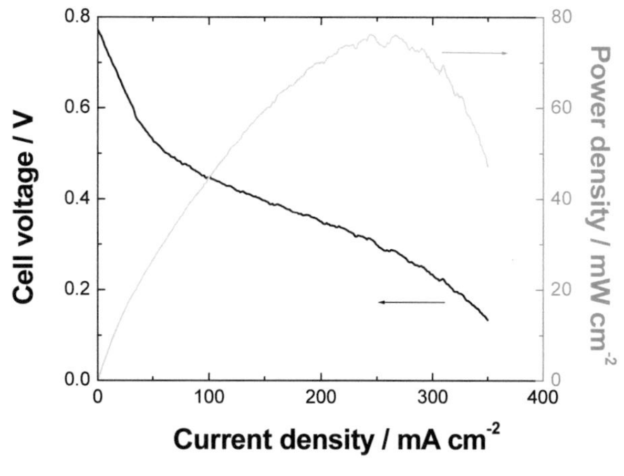

Fig. 14. Dependence of cell voltage and power density on current density for an intermediate-temperature alkaline methanol fuel using Pd/C for anode and Ag/C for cathode operated at 140 °C. Anode feed: 2 mol dm^{-3} CH_3OH + 2 mol dm^{-3} KOH; Cathode feed: 120 SCCM O_2.

To evaluate the potential of using non-Pt catalysts in the novel fuel cell, the performance of the fuel cell utilizing Pd/C anode and Ag/C cathode has been measured and is shown in Fig. 14. It is characterized by a peak power density of around 75 mW cm^{-2} seen at around 260 mA cm^{-2}. This performance is comparable to that of the fuel cell utilizing Pt catalysts. Therefore, significant cost reduction would be expected for the novel fuel cell without the cost of the performance because Pd and Ag are much cheaper than Pt. Further cost reduction will be dependent up the advance of the non-noble catalysts for the methanol oxidation and the *orr*.

4. Conclusion

Increasing temperature has been proven as an effective way to accelerate fuel cell reactions. To investigate the fuel cell reactions in the intermediate temperature of 80 to 200 °C, we have developed a pressurized electrochemical cell based on a modified commercial Parr autoclave which can be pressurized up to 2000 psi. In this electrochemical cell, aqueous electrolyte solutions and liquid fuels can exist in their liquid forms in the intermediate-temperature range at varying balance pressure. This makes the investigations of intermediate-temperature fuel cell electrocatalysis possible. To further address experimental challenges, we have evaluated three kinds of Ag-based electrodes as an internal reference electrode in basic media and have found that the Ag/AgCl electrode could act as the internal reference electrode with satisfactory stability. To facilitate the investigations of the fuel cell reactions on high-surface-area electrocatalysts, a powder-rubbing procedure has been used to mechanically deposit the electrocatalysts onto a gold substrate of 0.5 mm in diameter. Based upon these efforts, well-developed cyclic voltammograms and chronoamperograms for the electrochemical methanol oxidation and oxygen reduction in the intermediate temperature range have been obtained.

It is encouragingly found that the methanol electrooxidation in alkaline media on high-surface-area Pt and Pd electrodes can be substantially accelerated by increasing temperature, characterized by obvious decrease in the onset overpotential with increasing temperature. Moreover, CO could be oxidized at lower onset potentials than methanol in the intermediate temperature range under similar conditions. This strongly indicates that CO is no longer a poison limiting the methanol oxidation. Replacement of Pt with Pd produces no substantial decrease in the activity towards the methanol oxidation. For the electrochemical oxygen reduction, silver demonstrates a high activity in the intermediate temperature range which is comparable to that of high surface area Pt. This indicates the possibility of using non-Pt catalysts in intermediate-temperature alkaline methanol fuel cells. Our preliminary fuel cell studies demonstrate the high performance of intermediate-temperature alkaline methanol fuel cell with Pt and non-Pt catalysts.

5. References

Yu, H., Scott, K. & Reeve, W. (2003). A Study of the Anodic Oxidation of Methanol on Pt in Alkaline Solutions. *Journal of Electroanalytical Chemistry*, Vol.547, No.1, (April 2003), pp. 17-24, ISSN: 00220728

Prabhuram, J. & Manoharan, R. (1998). Investigation of Methanol Oxidation on Unsupported Platinum Electrodes in Strong Alkali and Strong Acid. *Journal of Power Sources*, Vol.74, No.1, (July 1998), pp. 54-61, ISSN: 03787753

Dillon, R., Srinivasan, S., Aricò, S. & Antonucci, V. (2004). International Activities in DMFC R&D: Status of Technologies and Potential Applications. *Journal of Power Sources*, Vol.127, No. 1-2, (March 2004), pp. 112-126, ISSN: 03787753

Cohen, J., Volpe, D. & Abruña, H. (2007). Electrochemical Determination of Activation Energies for Methanol Oxidation on Polycrystalline Platinum in Acidic and Alkaline Electrolytes. *Physical Chemistry Chemical Physics*, Vol.9, No.1, (January 2007), pp. 49-77, ISSN: 1463-9076.

Huber, W., Shabaker, W. & Dumesic, A. (2003). Raney Ni-Sn Catalyst for H_2 Production from Biomass-derived Hydrocarbons. *Science*, Vol.300, No.5628, (June 2003), pp. 2075-2077, ISSN: 00368075

Cortright, D., Davda, R. & Dumesic, A. (2002). Hydrogen from Catalytic Reforming of Biomass-derived Hydrocarbons in Liquid Water. *Nature*, Vol.418, No.6901, (August 2002), pp. 964-967, ISSN: 00280836

Davda, R., Shabaker, W., Huber, W., Cortright, D. & Dumesic, A. (2005). A Review of Catalytic Issues and Process Conditions for Renewable Hydrogen and Alkanes by Aqueous-phase Reforming of Oxygenated Hydrocarbons over Supported Metal Catalysts. *Applied Catalysis B: Environmental*, Vol. 56, No. 1-2, (March 2005), pp. 171-186, ISSN: 09263373

Kucernak, A. & Jiang, J. (2003). Mesoporous Platinum as a Catalyst for Oxygen Electroreduction and Methanol Electrooxidation. *Chemical Engineering Journal*, Vol.93, No.1, (May 2003), pp. 81-90, ISSN: 13858947

Öijerholm, J., Forsberg, S., Hermansson, P. & Ullberg, M. (2009). Relation between the SHE and the Internal AgAgCl Reference Electrode at High Temperatures. *Journal of the Electrochemical Society*, Vol.156, No.3, pp. P56-P61, ISSN: 00134651

Greeley, S., Smith Jr., T., Stoughton, W. & Lietzke, H. (1960). Electromotive Force Studies in Aqueous Solutions at Elevated Temperatures. I. The Standard Potential of the Silver-silver Chloride Electrode. *Journal of Physical Chemistry*, Vol.64, No.5, pp. 652-657, ISSN: 00223654

Bao, J. & Macdonald, D. (2007). Oxidation of Hydrogen on Oxidized Platinum. Part I: The Tunneling Current. *Journal of Electroanalytical Chemistry*, Vol.600, No.1, (February 2007), pp. 205-216, ISSN: 00220728

Schmidt, J., Ross Jr., N. & Markovic, M. (2002). Temperature Dependent Surface Electrochemistry on Pt Single Crystals in Alkaline Electrolytes: Part 2. The Hydrogen Evolution/Oxidation Reaction. *Journal of Electroanalytical Chemistry*, Vol.524-525, No.3, (May 2002), pp. 252-260, ISSN: 00220728

Tripković, V., Popović, D., Momčilović & D., Dražić, M. (1998). Kinetic and Mechanistic Study of Methanol Oxidation on a Pt(110) Surface in Alkaline Media. *Electrochimica Acta*, Vol.44, No.6-7, (November 1998), pp. 1135-1145, ISSN: 00134686

Tripković, V., Popović, D., Momčilović, D. & Drazić, M. (1998). Kinetic and Mechanistic Study of Methanol Oxidation on a Pt(100) Surface in Alkaline Media. *Journal of Electroanalytical Chemistry*, Vol.448, No.2, (May 1998), pp. 173-181, ISSN: 00220728

Matsuoka, K., Iriyama, Y., Abe, T., Matsuoka, M. & Ogumi, Z. (2005). Electro-oxidation of Methanol and Ethylene Glycol on Platinum in Alkaline Solution: Poisoning Effects and Product Analysis. *Electrochimica Acta*, Vol.51, No.6, (November 2005), pp. 1085-1090, ISSN: 00134686

García, G. & Koper, M. (2011). Carbon Monoxide Oxidation on Pt Single Crystal Electrodes: Understanding the Catalysis for Low Temperature Fuel Cells. *ChemPhysChem*, Vol.12, No.11, (August 2011), pp. 2064-2072, ISSN: 14394235

Spendelow, S., Goodpaster, D., Kenis, A. & Wieckowski, A. (2006). Mechanism of CO Oxidation on Pt(111) in Alkaline Media. *Journal of Physical Chemistry B*, Vol.110, No.19, (May 2006), pp. 9545-9555, ISSN: 15206106

Spendelow, S., Lu, Q., Kenis, A. & Wieckowski, A. (2004). Electrooxidation of Adsorbed CO on Pt(1 1 1) and Pt(1 1 1)/Ru in Alkaline Media and Comparison with Results from Acidic Media. *Journal of Electroanalytical Chemistry*, Vol.568, No.1-2, (July 2004), pp. 215-224, ISSN: 00220728

Lima, B. & Ticianelli, A. (2004). Oxygen Electrocatalysis on Ultra-thin Porous Coating Rotating Ring/Disk Platinum and Platinum-Cobalt Electrodes in Alkaline Media. *Electrochimica Acta*, Vol.49, No.24, (September 2004), pp. 4091-4099, ISSN: 00134686

Kötz, R & Yeager, E. (1980). Raman Studies of the Silver/Silver Oxide Electrode. *Journal of Electroanalytical Chemistry*, Vol.111, No.1, (July 1980), pp. 105-110, ISSN: 00220728

Demarconnay, L., Coutanceau, C. & Léger, M. (2004). Electroreduction of Dioxygen (ORR) in Alkaline Medium on Ag/C and Pt/C Nanostructured Catalysts - Effect of the Presence of Methanol. *Electrochimica Acta,* Vol.49, No.25, (October 2004), pp. 4513-4521, ISSN: 00134686

8

Electrochemical Cells with Multilayer Functional Electrodes for NO Decomposition

Sergey Bredikhin[1] and Masanobu Awano[2]

[1]Institute of Solid State Physic Russian Academy of Sciences, Chernogolovka,
[2]Advanced Manufacturing Research Institute, National Institute of Advanced Industrial Science and Technology (AIST), Shimo-shidami, Moriyama-ku, Nagoya
[1]Russia
[2]Japan

1. Introduction

Striking progress has recently been made in understanding the central role of nitrogen oxide radicals, NO_x, in atmospheric processes (Lerdau et al., 2000). NO_x is implicated in the formation of acid rain and a tropospheric ozone (the principal toxic component of smog and a greenhouse gas) (Finlayson-Pitts, B.J. & Pitts, J.N., 1997; Lerdau et al., 2000). The major known source of NO_x is fuel combustion and biomass burning. Air pollution by nitrogen oxides (NO_x) in combustion waste causes serious environmental problems in urban areas. The reduction of nitrogen oxide emissions has become one of the greatest challenges in environment protection (Libby, 1971; Nishihata et al., 2002). This is why the different methods of NOx decomposition are intensely studied by numerous groups from academic as well as industrial research laboratories (Garin, 2001; Parvulescu et al., 1998).

The main activities of scientific groups working in the field of NO decomposition are concentrated on the reduction of the NOx in the presence of NH_3, CO, H_2 or hydrocarbons. These scientific groups have been tested a large number of categories of catalysts with a different ways of NO decomposition reactions. The main directions of the research can be described as follows.

First, the selective catalytic reduction of NO with ammonia, typical for chemical industrial plants and stationary power stations (Bosch & Janssen, 1988; Janssen & Meijer, 1993). The main step is the reduction on NO or NO_2 to N_2 and H_2O. Generally, liquid ammonia is injected in the residual gas before the catalytic reaction takes place.

$$4NO+4NH_3+O_2 \longrightarrow 4N_2+6H_2O$$

$$6NO+4NH_3 \longrightarrow 5N_2+6H_2O$$

Second, the catalytic reduction of NO in the presence of CO and/or hydrogen. These reactions are typical for the automotive pollution control. The use of CO or H_2 for catalytic reduction was one of the first possibilities investigated in view of eliminating NO from automotive exhaust gas (Baker & Doerr, 1964, 1965; Klimisch & Barnes, 1972; Nishihata et al., 2002; Roth & Doerr, 1961).

$$NO+CO \longrightarrow CO_2+1/2\,N_2$$

$$NO+H_2 \longrightarrow 1/2\,N_2+H_2O$$

Third, the selective catalytic reduction of NO in the presence of hydrocarbons and more particularly methane, a method which has not yet reached industrial use but can be applied both for automotive pollution control and in various industrial plants (Armor, 1995; Hamada et al., 1991; Iwamoto, 1990; Libby, 1971; Miura et al., 2001; Sato et al., 1992).

Fourth, the direct decomposition of NO. The decomposition of NO would represent the most attractive solution in emission control, because the reaction does not require that any reactant be added to NO exhaust gas and could potentially lead to the formation of only N_2 and O_2 (Garin, 2001; Lindsay et al., 1998; Miura et al., 2001; Rickardsson et al., 1998).

The goal of this paper is to represent a **fifth** direction of an intense research effort focused on electrochemical cells for the reduction of NO_x gases due to the need to design an effective method for the purification of the exhaust gases from lean burn and diesel engines.

2. Traditional type of electrochemical cells for NO decomposition

Electrochemical cells have become an important technology, which contributes to many aspects of human life, industry and environment. Now it is understandable that the reduction of NOx emission can be achieved not only by catalytic NOx decomposition but also by electrochemical decomposition, where the removal of oxygen by a gaseous reducing reagent is replaced by the more effective electrochemical removal. Additional reducing reagents such as hydrocarbons, CO, H_2 or ammonia can lead to the production of secondary pollutants like oxygenated hydrocarbons, CO, CO_2, N_2O or ammonia or, even, as was often reported in the past, cyanate and isocyanate compounds.

Without coexisting oxygen the successful decomposition of NO gas into oxygen and nitrogen in a primitive electrochemical cell (Fig.1) was first demonstrated over 25 years ago (Gur & Huggins, 1979; Pancharatnam et al., 1975). In 1975 *Pancharatnam et al.* (Pancharatnam et al., 1975) proposed to use for NO gas decomposition an electrochemical cell represented by the following cell arrangement

$$\text{Pt(Cathode)}\,|\,\text{YSZ}\,|\,\text{Pt(Anode)} \tag{1}$$

On applying a voltage to such cells NO gas is directly reduced at the triple-phase boundary *(tpb)* (cathode - yttrium-stabilized zirconia (YSZ) - gas) forming gaseous N_2 and solid –phase oxygen ions:

$$2NO + 4e^- + 2\,V_O(ZrO_2) \rightarrow N_2 + 2O^{2-}(YSZ) \tag{2}$$

Under the external voltage the oxygen ions are transported through the solid electrolyte from cathode to anode and gaseous O_2 is evolved at the anode.

Unfortunately, excess O_2 in the combustion exhaust gas is adsorbed and decomposed at the *tpb* in preference to the NO gas (Fig.1):

$$O_2 + 4e^- + 2\,V_O(ZrO_2) \rightarrow 2O^{2-}(YSZ) \tag{3}$$

As a result, the additional ionic current though the cell associated with the oxygen ions produced due to this unwanted reaction (Eq.(3)) far exceeds the current associated with the desired reaction (Eq.(2)). In 1997 *Hibino et al.* (Hibino et al., 1997) has shown that at first stage the electrochemical oxygen pumping is carried out without NO decomposition, and that NO decomposition began at corresponding currents after the electrochemical oxygen pump is complete. As illustration Fig.2 shows the dependence of NO conversion on the value of the current passing through the two chambers cell at 1000ppm of NO without oxygen (Curve 1) and at 2% of Oxygen (Curve 2) in He (the balance) at gas flow rate 50ml/min. It is seen that in the presence of oxygen the decomposition of NO take place only when all oxygen should be pumped away from the near electrode area.

Recently, many attempts to improve the properties of electrochemical cells operating in the presence of excess oxygen have been carried out by using different catalysts as the cathode material (Hibino, 2000a, 2000b; Marwood & Vayenas, 1997; Nakatani et al., 1996; Walsh & Fedkiw, 1997). *Walsh* (Walsh & Fedkiw, 1997) proposed substitute dense Pt electrodes to the porous platinum and to use a mixture of ionic (CeO) and electronic (Pt) conductors as a porous cathode. It is well known that substitution of dense electrode to the porous should increase gas penetration to the *tpb* on the surface of the YSZ-disc solid electrolyte and using of the mixture of ionic (CeO) and electronic (Pt) conductors should lead to the increase of the *tpb* surface area inside the cathode. As the result both oxygen and nitrogen oxide decomposition take place in such cells and for effective NO adsorption and decomposition the *tpb* should be free from the adsorbed oxygen. This conclusion agrees well with a fact that the NO decomposes after the oxygen pumping is completed.

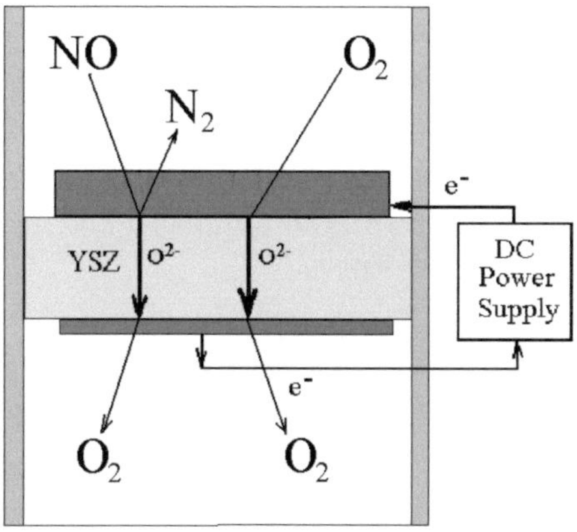

Fig. 1. Conceptual representation of the electrochemical cell for NO decomposition.

To improve the selectivity for NO gas adsorption and decomposition in the presence of the oxygen excess *K. Iwayama* (Iwayama & Wang, 1998; Washman et al., 2000) proposed to coat Pt cathode by different metals or metal oxide. Decomposition activity was measured on metal oxide/Pd(cathode)/YSZ/Pd(anode) at 773–973 K and 3.0V of applied voltage in a flow of 50 ml/min containing 1000 ppm of NO and 6% of O_2 in helium. Coating of various metal oxides onto the cathode electrode greatly changed the decomposition activity; the order was RuO_2>>Pt>Rh_2O_3>Ni>none>Ag>WO_3. The activity of the system modified by RuO_2 has been investigated as a function of the kind of electrode, the applied voltage, and the reaction temperature. The cell of RuO_2/Ag(cathode)/YSZ/Pd(anode) was found to show the most excellent activity among the cells examined.

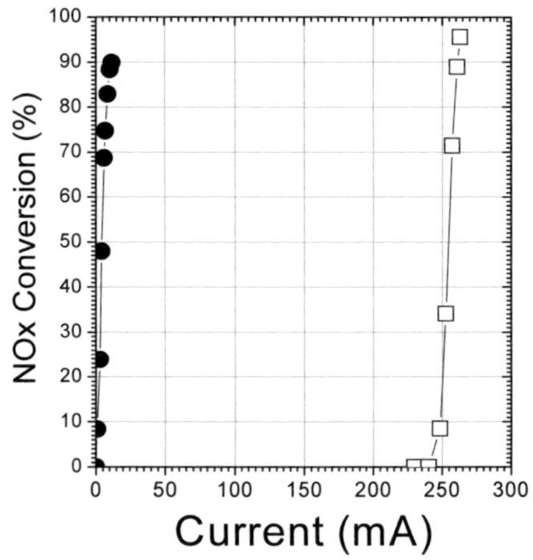

Fig. 2. Dependence of NO conversion on the value of the current passing through the two chambers cell at 1000 ppm of NO without oxygen (-•- Curve 1) and at 2% of Oxygen (-□- Curve 2) in He (the balance) at gas flow rate 50 ml/min.

Later (Hwang et al., 2001; Matsuda et al., 2001; Washman et al., 2000) a series $La_{1-x}A_xBO_3$ perovskite were prepared and systematically evaluated for substitution of the Pt or Pd electrodes. A major target of all these researches was the promotion of NO reduction by F – center type defects in the YSZ surface or inside perovskite type cathodes (Hwang et al., 2001; Matsuda et al., 2001; Washman et al., 2000).

An important characteristic of the efficiency of electrochemical cell is the value of the current efficiency coefficient (η). Current efficiency (η) can be defined from the value of the oxygen ionic current (I_{NO}) due to the oxygen from decomposed NO gas (see eq.2) and a total ionic current flux through the cell ($I = I_{NO} + I_{O2}$) as:

$$\eta = I_{NO} / (I_{NO} + I_{O2}) \qquad (4)$$

As illustration current efficiency for NO decomposition against current is plotted in Fig.3 for LSC|YSZ|Pt and LSCP|YSZ|Pt cells (Hwang et al., 2001). It is seen that relatively high value of current efficiency, *ca.* 1.5% can be obtained between 300 and 350 mA. This result shows that the unwanted reaction (Eq. (3)) of oxygen gas adsorption and decomposition fare exceed the desirable reaction of NO gas adsorption and decomposition.

In 2007 *Simonsen et al.* (Simonsen et al., 2007) studied spinels with composition $CoFe_2O_4$, $NiFe_2O_4$, $CuFe_2O_4$, and Co_3O_4 as electro-catalyst for the electrochemical reduction of nitric oxide in the presence of oxygen. It was shown that spinels are active for the reduction of both nitric oxide and oxygen. The composition $CuFe_2O_4$ shows the highest activity for the reduction of nitric oxide relative to the reduction of oxygen. However now information was given on the characteristics of the electrochemical cells based on these cathode materials.

*K. Kammer (*Kammer, 2005) reviewed the investigations in the field of electrochemical reduction of nitric oxide. He has shown that the electrochemical reduction of nitric oxide in

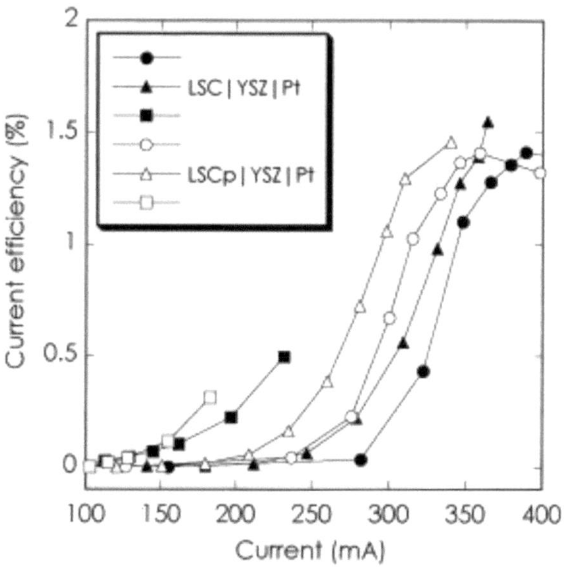

Fig. 3. Current efficiency vs. current curves of LSC u YSZ u Pt and LSPC u YSZ u Pt electrochemical cells between 600 and 800°C. (-•-,-○-, 800°C), (-▲-,-Δ- 700°C), (-■-,-□- 600°C).

several types of all solid-state electrochemical cells is possible, and proposed, that in order to reduce nitric oxide in an atmosphere containing excess oxygen further development of cathode materials are needed.

In accordance with above we can conclude that all known cathode materials used up now for electrochemical reduction of nitric oxide show a low selectivity for NO reduction in the presence of the excess oxygen and can't be used for practical application.

3. Electrochemical reactors with multi–layer functional electrode

To solve the problem of effective electrochemical reduction of nitric oxide in the presence of the excess oxygen *S.Bredikhin et al.* (Awano et al., 2004; Bredikhin et al., 2001a, 2001b) proposed the concept of artificially designed multilayer structure which should operate as an electrode with high selectivity. At present time a new type of electrochemical reactor with a functional multi-layer electrode has been successfully designed in *National Institute of Advanced Industrial Science and Technology (AIST), Nagoya, Japan* (Awano et al., 2004; Bredikhin et al., 2004). The typical values of current efficiency in such electrochemical reactors are of the order of 10% - 20% at gas composition: 1000 ppm NO and 2% O_2 balanced in He and at gas flow rate 50 ml/min. The value of current efficiency depends on the functional multi-layer electrode composition, structure and operating temperature. Such electrochemical reactors show the value of NO selectivity (ν_{sel}) with respect to oxygen gas molecules $\nu_{sel} > 5$. This means that the probability for NO gas molecules to be adsorbed and decomposed is at least 5 times higher than for oxygen gas molecules.

The arrangement of the electrochemical reactor with a functional multi-layer electrode is illustrated schematically in Fig.4. An YSZ disc with a thickness of 500μm and a diameter of

20mm was used as the solid electrolyte. The composite Pt(55vol%)-YSZ(45vol%) paste was screen–printed with an area of 1.77 cm^2 on one surface of the YSZ disk as the cathode, and then calcined at 1673 K for 1 hour, to produce a dense Pt(55vol%)-YSZ(45vol%) composite with a thickness of about 3 μm and diameter of 15 mm (Bredikhin et al., 2004). A dense Pt collector was connected with a cathode. The NiO-YSZ paste was screen-printed with an area of 2 cm^2 over the cathode and sintered at 1773 K for 4 hours to produce a nano-porous NiO-YSZ electro-catalytic electrode with a diameter of 16mm and a thickness of about 5-6 μm (Aronin et al., 2005; Awano et al., 2004a; Bredikhin et al., 2006; Hiramatsu, 2004). The nano-porous YSZ layer with a thickness of about 2 μm was deposited over the electro-catalytic electrode as a covering layer (Awano et al., 2004b). The commercial TR-7070 (Pt-YSZ) paste was screen-printed with an area of 1.77 cm^2 on to the other surface of the YSZ disk as the anode, and then calcined at 1473 K for 1 hour. Platinum mesh and wire were attached to the cathode and the anode, for connection with the power supply unit.

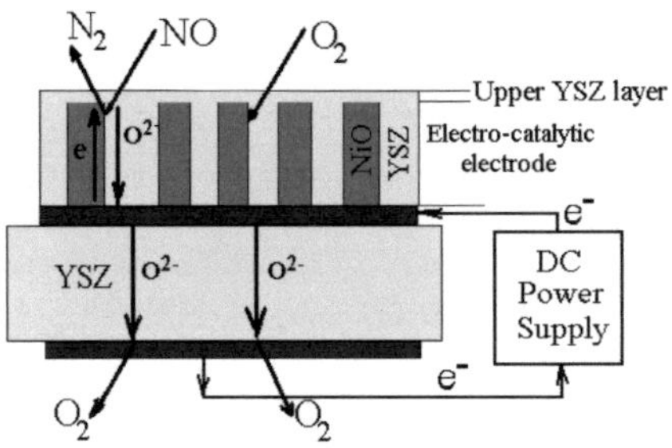

Fig. 4. Conceptual representation of the electrochemical cell with multilayer electro-catalytic electrode.

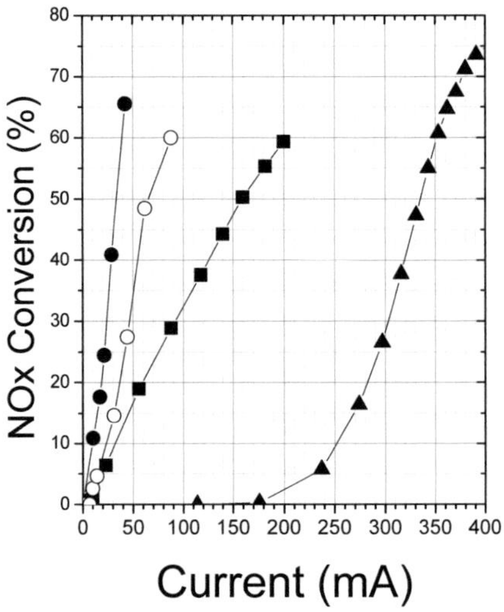

Fig. 5. The dependence of NO conversion on the value of the current for electrochemical reactors with functional multi-layer electrodes (-●- 2% and -○- 10% of oxygen) and for a reactor with a monolayer electro-catalytic electrode (-■- 2% of oxygen) and a traditional type of electrochemical cell with Pt-YSZ types of cathode (-▲- 2% of oxygen).

The electrochemical reactor was set in a quartz house and connected to a potensio – galvanostat (SI1267 and 1255B, SOLARTRON). The applied voltage and current dependence of NO decomposition behavior was investigated. The range of the applied voltage to the electrochemical cell was from 0 to 3V. The electrochemical decomposition of NO was carried out at 573 - 873 K by passing a mixed gas of 500-1000 ppm of NO and 2-10% of O_2 in He (balance) at a flow rate v = 50 ml/min. The concentrations of NO and of N_2 in the outlet gas ($[NO]_{out}$) were monitored using an on-line NOx (NO, NO_2 and N_2O) gas analyser (Best Instruments BCL-100uH, BCU-100uH) and a gas chromatograph (CHROMPACK Micro-GC CP 2002), respectively.

S. Bredikhin et al. (Awano et al., 2004b; Bredikhin et al., 2004; Hiramatsu et al., 2004) and *K. Hamamoto et al.* (Hamamoto et al., 2006, 2007) have shown that electrochemical reactors with multi-layer electro-catalytic electrode effectively operate even at low concentration of NOx (300-500ppm) and at the high concentration of oxygen (10%). From Fig.5 it is seen that the efficiency of NO decomposition by electrochemical reactors with the functional multi-layer electrode far exceeds the efficiency of the traditional type of electrochemical cells.

4. Microstructure and properties of functional layers of multi-layer electrode

The electrochemical reactors for selective NOx decomposition can be represented by the following reactor arrangements:

(Covering layer | Electro-catalytic electrode | Cathode) | YSZ | (Anode) (5a)

(Covering layer | Cathode | Electro-catalytic electrode) | YSZ | (Anode) (5b)

Let us consider in detail the arrangement of the electrochemical reactor with a functional multi-layer electrode and the properties of each functional layer. The cross-section view of the functional multi-layer electrode is shown in Fig.6.

Fig. 6. The cross-section view of the multi-layer electro-catalytic electrode.

The cathode is a dense Pt(55vol%)-YSZ(45vol%) composite with a thickness of about 2-3 μm. The nano- porous NiO-YSZ electro-catalytic electrode with a thickness of about 6-8 μm was deposited over the cathode. The porous YSZ layer with a thickness of about 2-3 μm was deposited over the cathode. It is seen that the multi-layer electrode consists from three main functional layers: 1. Cathode; 2. Electro-catalytic electrode; 3. Covering layer.

4.1 Microstructure and the properties of composite cathode

The external voltage in the electrochemical reactor with a functional multi-layer electrode is applied between the cathode and anode. This voltage leads to the polarization of the YSZ-disk and to the generation of a high concentration of oxygen vacancies in the near cathode region (Fig. 4). Due to the gradient in the concentration of the oxygen ions between the near-cathode region of the YSZ disk and the electro-catalytic electrode, diffusion of oxygen ions from the electro-catalytic electrode to the YSZ disk takes place. Since the oxygen ions are a charged species, their diffusion from the electro-catalytic electrode to the YSZ disk leads to an equal flux of electrons from the cathode to the electro-catalytic electrode (Aronin et al., 2005; Awano et al., 2004a; Bredikhin et al., 2006; Hiramatsu, 2004). As a result of electroneutrality, a decrease in the flux of the electrons should lead to the same decrease in the flux of the oxygen ions and the diffusion of oxygen ions should be stopped when the transport of electrons is blocked. In accordance with this consideration we can conclude that for effective reactor operation the cathode should be an electronic and oxygen ionic current conductor with high electronic conductivity along the cathode plane and high oxygen ionic conductivity from the electro-catalytic electrode through the cathode to the YSZ solid electrolyte (Awano et al., 2004b).

In 2004 *S.Bredikhin et al.* (Bredikhin et al., 2004) studied the correlation between the efficiency of NO decomposition by electrochemical cells with electro-catalytic electrode and the YSZ-Pt cathode compositions. In this study, *S.Bredikhin et al.* (Bredikhin et al., 2004) examined $YSZ_{(X)}$-$Pt_{(1-X)}$ composite as a cathode for electrochemical cells with a functional multilayer electrode. Electrochemical cells with electro-catalytic electrode and $YSZ_{(X)}$-$Pt_{(1-X)}$ composite cathode with 0, 15.0, 24.8, 35.2, 45.4, 49.9, 55.0, and 64 vol. % of YSZ were obtained. Investigation of the current-voltage (*I-V*) characteristics of the electrochemical cells with multilayer electrode has shown a strong dependence on the composition of the YSZ-Pt cathode (Bredikhin et al., 2004). The best performance was observed for electrochemical cells with YSZ volume contents slightly lower than 50 vol %. To investigate this behavior in detail the value of the current has been plotted as a function of the volume content of YSZ in the YSZ-Pt composite cathode for different values of the electrochemical cell operating voltage. These experimental data are shown in Fig. 7. From this figure, it is seen that an increase of the YSZ content from 0 to 49.9% leads to a four to five times increase in the value of the current through the cell at the same value of the cell operating voltage. At the same time a small change in the composition of the YSZ-Pt cathode by increasing the YSZ content to more than 50 vol % leads to an abrupt decrease in the value of the current passed through the cells (Fig. 7) and at 55 vol % of YSZ current higher than 1 mA cannot be passed even at applied voltages to the cells of more than 3 V. To describe these phenomena let us consider the processes of electronic and ionic transport through the composite YSZ-Pt cathode. The geometry of the employed electrochemical cell means that electronic transport take place along the cathode through the network of Pt particles and from the cathode through the three-dimensional network of pathways from NiO or Ni particles to three-phase boundary

(TPB) on the surface of the pores inside the electro-catalytic electrode. Oxygen ionic transport takes place perpendicular to the YSZ-Pt cathode plane from the electro-catalytic electrode through the network of YSZ particles to the YSZ disk.

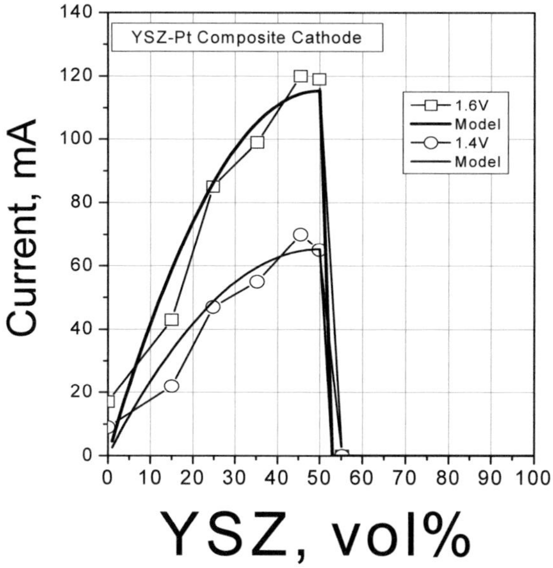

Fig. 7. The experimental dependence of the current on the volume fraction of YSZ in the YSZ-Pt cathode for applied voltages of 1.4 Volts and 1.6 Volts, and comparison with calculated dependencies.

Figure 8 shows the cross-sectional view of the electrochemical cells for different YSZ-Pt cathode compositions. It is seen that the addition of YSZ particles to the cathode leads to the formation of oxygen-conducting YSZ bridges through the electronically conducting Pt cathode and that the number of such bridges increases with an increasing amount of YSZ in the cathode. From Fig. 8 it is seen that the average size of the electronically insulating (YSZ) particles and electronically conducting (Pt) particles are of the order of 2-3 mm and are the same as the thickness of the cathode layers. These observations give us the possibility to conclude that the YSZ-Pt cathode is a quasi two-dimensional system and that the two-dimensional percolation model can be used to describe the electronic conductivity along the plane of the YSZ-Pt cathode. In accordance with this model a sharp transition in electronic conductivity along the cathode should be observed at 50 vol % of the electronically conducting Pt phase (Bredikhin et al., 2004). This means that the electronically conducting Pt phase is continuous when the volume fraction of the electronically insulating YSZ phase is less than 50 vol % and the Pt phase becomes disconnected when the volume fraction of insulating YSZ phase is greater that 50 vol %. This two-dimensional percolation model prediction is in good agreement with the experimentally observed sharp threshold of the value of the current through the electrochemical cell as a function of the volume content of YSZ in the YSZ-Pt composite cathode (Fig. 7) (Bredikhin et al., 2004).

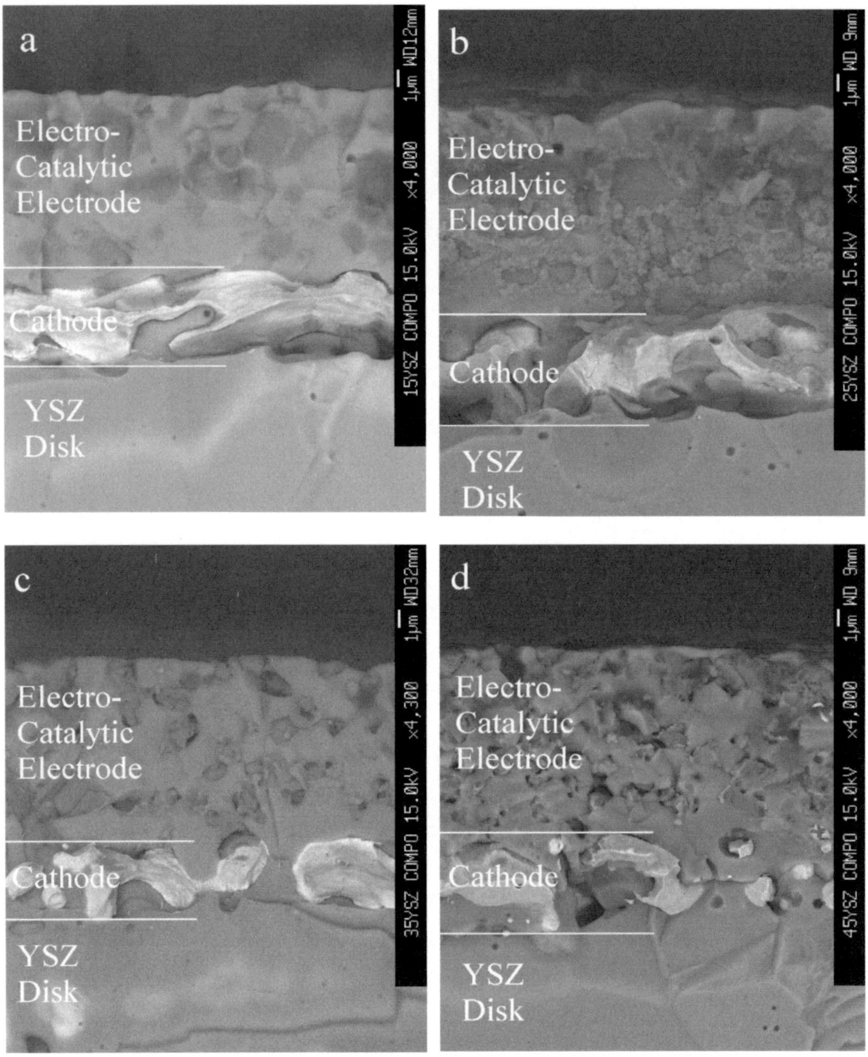

Fig. 8. SEM images of the cross-section of the multi-layer cells with different YSZ-Pt cathode compositions (a - 15vol% of YSZ, b - 25vol% of YSZ, c - 35vol% of YSZ, d - 45vol% of YSZ).

As follows from Figs. 7 and 8 an increase of the YSZ content leads to an increase in the oxygen ionic current through the cell at a given value of the applied external voltage. At the same time when the YSZ content exceeds the percolation threshold (50 vol %) the Pt phase becomes disconnected, and the flux of electrons from the cathode to the electro-catalytic electrode is blocked. As a result the oxygen ionic diffusion from the electro-catalytic electrode to the YSZ-disk is stopped. In accordance with the above results it is seen that the most critical place for oxygen ion diffusion from the electro-catalytic electrode to the YSZ-disk is the oxygen transport through the composite YSZ-Pt cathode, and that compositions with a volume fraction of YSZ slightly lower than the two-dimensional percolation threshold (1/2) should have the highest efficiency for charge transport through the electrochemical cell (Bredikhin et al., 2004).

4.2 Electro-catalytic electrode

a. Peculiarity of ambipolar conductivity of the electro-catalytic electrode

The external voltage (V_0) in the electrochemical reactor is applied between the cathode and the anode (Fig. 4), and the voltage drop is equal to the sum of the polarizing voltage (V_{pol}) and to the Ohms voltage ($V_{Ohm} = R_{YSZ}\times I_{ox}$) drop through the YSZ disk

$$V_0 = V_{pol} + V_{Ohm} = V_{pol} + R_{YSZ}\times I_{ox}, \quad (6)$$

where R_{YSZ} is the resistance of the YSZ disk and I_{ox} is the value of oxygen ionic current through the cell. The formation of the gradient in the concentration of the oxygen ions in the YSZ disc under the DC polarization voltage V_{pol} can be described in accordance with the well-known equation (Hamamoto et al., 2007; Kobayashi et al., 2000; Schoonman, as cited in Chowdari & Radakrishna, 1998; Wagner, as cited in Delahay & Tobias, 1966)

$$\mu_{cathode} = \mu_{anode} + 2e\times V_{pol} = \mu_o + \tfrac{1}{2}\, kT \ln(P_{O2}/P_o) + 2e\times V_{pol}, \quad (7)$$

where μ_o and P_o are standard values and where the chemical potential of the oxygen at the surface of the electro-catalytic electrode (μ_{CE}) and in the anode region of YSZ disc (μ_{anode}) is fixed by the oxygen gas pressure (P_{O2}) ($\mu_{CE} = \mu_{anode} = \mu_o + \tfrac{1}{2}\, kT \ln(P_{O2}/P_o)$). Then the difference in the chemical potential between the near cathode region and the surface of the electro-catalytic electrode should result in the Nernst potential formation ($\Delta\phi$) through the electro-catalytic electrode

$$\Delta\phi = V_{pol} = (V_0 - R_{YSZ}\times I_{ox}) \quad (8)$$

The value of the oxygen ionic current through the electro-catalytic electrode depends on the value of the Nernst potential and on the value of the electro-catalytic electrode ambipolar conductivity (σ_{amb}) as

$$I_{ox} = \Delta\phi\times\sigma_{amb} \quad (9)$$

As follows from equations (8) and (9) the value of ionic current through the electrochemical cell depends on the value of ambipolar conductivity of electro-catalytic electrode (σ_{amb}) and on the value of oxygen ionic conductivity (R_{YSZ}) of YSZ disc as

$$I_{ox} = (V_0\times\sigma_{amb})/(1+R_{YSZ}\times\sigma_{amb}). \quad (10)$$

It is seen that when $R_{YSZ}\times\sigma_{amb} < 1$, there is a linear dependence between ionic current through the cell and the value of ambipolar conductivity

$$I_{ox} \approx (V_0\times\sigma_{amb}). \quad (11)$$

In the year 2001 *S.Bredikhin et al.* (Bredikhin et al., 2001a, 2001b, 2001c) have shown that for electrochemical cells with nano-porous electro-catalytic electrode there is a linear dependence between the value of NO conversion (ΔNO) and value of oxygen ionic current passed through the cell

$$\Delta NO = (1/(F\times\nu\times n))\times \eta\times I_{ox}, \quad (12)$$

where n=2 is the charge of the oxygen ions, F- is a Faraday constant, ν is a total gas flow rate and η is current efficiency. From equations (11) and (12) it is follows that the rate of NO decomposition depends on the external voltage applied to the cell and on the value of ambipolar conductivity as

$$\Delta NO = (1/(F \times \nu \times n)) \times \eta \times V_0 \times \sigma_{amb} \tag{13}$$

It is obvious that for optimization of the electrochemical cell for NO decomposition it is necessary to design the cell with the highest value of the ambipolar conductivity. To analyze the specific features of the ambipolar conductivity of the NiO-YSZ composite electrode the calculated value of σ_{amb} (see Eq.11) has been plotted as a function of the volume content of NiO in the NiO-YSZ composite electrode. These data are shown in Fig.9.

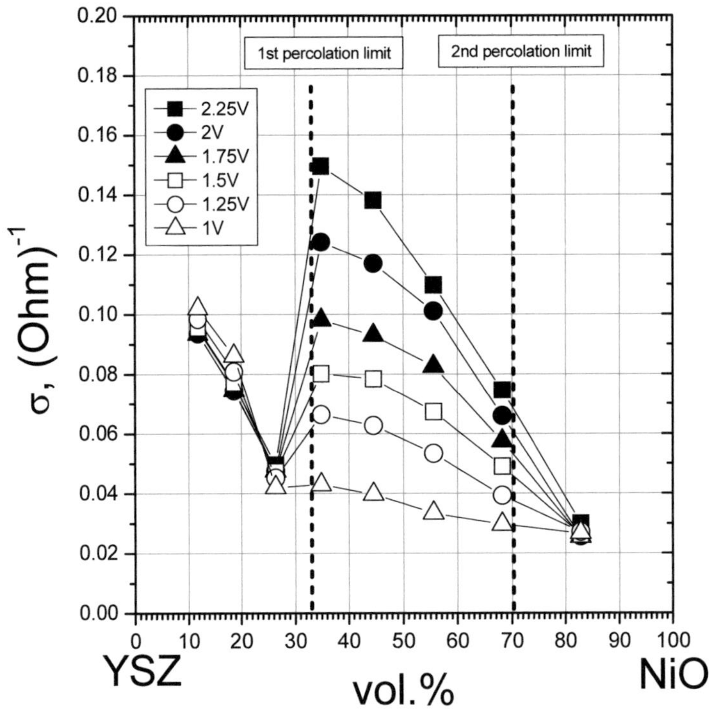

Fig. 9. The dependence of the value of the ambipolar conductivity on the volume fraction of NiO in Ni-YSZ electro-catalytic electrode

It is seen that when the external voltage is higher than 1.5 volt, the two percolation transitions of electronic and ionic conductivities lead to a composition range from 1/3 to 2/3, in which the ambipolar conductivity is much higher than those in the other regions. This suggests that in order to achieve high value of ambipolar conductivity, the volume fractions of each phase have to be within 1/3 ~ 2/3 and external voltage should be higher then 1.5 volt. It should be noted that in the composition range from 1/3 to 2/3 the value of ambipolar conductivity increases with increasing of the voltage applied to the cell. As an illustration Fig.10 shows the dependence of the value of the ambipolar conductivity on the voltage for two NiO-YSZ composite electrodes with NiO volume content 82.5% and 35%.

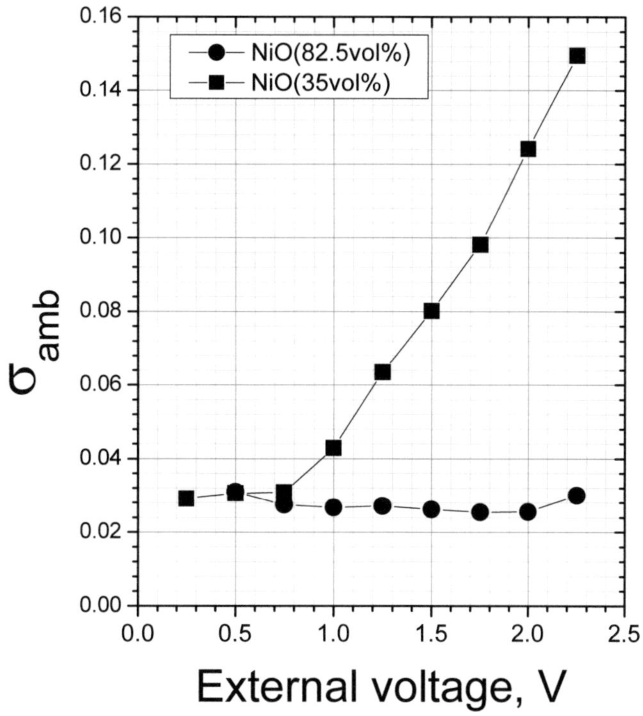

Fig. 10. The dependence of the value of the ambipolar conductivity on the cell operating voltage for two NiO-YSZ composite electrodes with NiO volume content 82.5% and 35%.

From Figs.9 and 10 it is seen that for the composition range from 1/3 to 2/3 the application of external voltage leads to 6-10 times increase of the value of ambipolar conductivity. Such unusual dependence of the value of conductivity on the applied voltage shows that change of the chemical composition of NiO-YSZ electro-catalytic electrode takes place under the cell operation.

b. Reduction-oxidation processes and the structure of the electro-catalytic electrode

The distinguishing feature of such reactors is the artificial nano-structure formed in the NiO/YSZ interface of the electro-catalytic electrode under operation. The essential changes occur in the interfacial boundary region between NiO and YSZ grains. After the electrochemical cell operation for 22 hours at a cell voltage lower than 2.2 Volts they are as follows (Aronin et al., 2005).

1. New grains nucleate and grow in the pore zone of NiO near-boundary regions. The zone of new grains spreads into the depth of the NiO grain (Fig. 11). In some cases this zone of new grains devours "old" NiO grain completely (Figs. 12a and 12b). The electron diffraction patterns from these regions contain ring reflections of Ni in addition to NiO reflections (Aronin et al., 2005). The size of Ni grains is 5-20 nm and they are seen in the dark field electron microscopy image (Fig. 12b). This image was obtained in the reflection marked by the arrow in the electron diffraction pattern (Fig. 12a, insert). The new grains are both Ni and NiO phases.

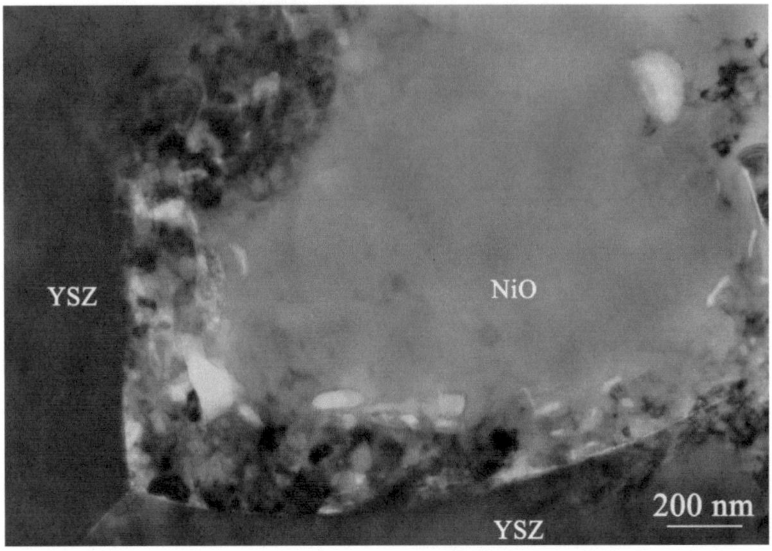

Fig. 11. Interfacial boundary between YSZ and NiO grains of the sample after cell operation.

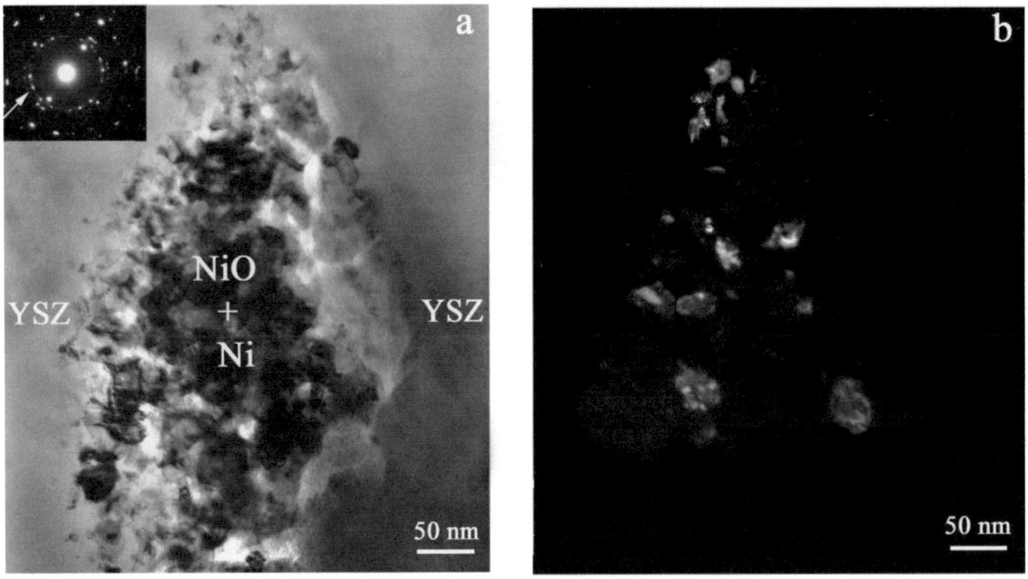

Fig. 12. Bright field TEM image (a) and dark field TEM images (b) of new NiO and Ni grains zone devoured "old" NiO grain.

2. In the region with new NiO grains the pores are located both in the interfacial vicinity and before the front of new growing grains. Sometimes small NiO grains are surrounded by the pores on all sides. Separate pores are also observed in the region of new grains, located between the new grains (Fig. 11). The high resolution electron microscopy images of the near-boundary region of YSZ grain and small NiO grain formed during the cell operation at different magnifications are shown in Fig. 13. The size of the new NiO grains varies from 10 to 100 nm depending on the location.

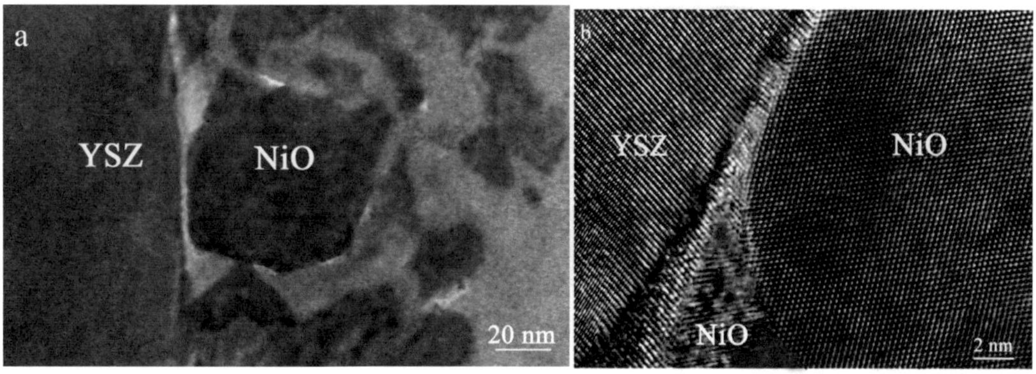

Fig. 13. HREM images of a near-boundary region of an YSZ grain and small NiO grains formed during the cell operation (a and b different magnification).

3. An orientation relationship has been found to exist between the lattices of YSZ and new NiO grains:

$$(310)_{YSZ} \parallel (110)_{NiO}, [001]_{YSZ} \parallel [1\bar{1}1]_{NiO}$$

The presence of the orientation relationship between the lattices of YSZ and NiO grains indicates that the new NiO grains may nucleate on the YSZ grains as on the substrate and in this case their surface energy will decrease. This result is a direct evidence of the process of oxygen spillover from YSZ to Ni grains even in the presence of oxygen in the surrounding gas.

The distinguishing feature of the microstructure of the YSZ/(Ni-NiO) interface is the 10-50 nm Ni grains reversibly produced during the reactor operation. Schematically the microstructure of the YSZ/NiO interface reversibly produced during the cell operation is represented in Fig.14. It is well known that adsorption and decomposition of NOx gas molecules occurs in preference to oxygen gas molecules on Ni grain surfaces (Lindsay et al., 1998; Miura et al., 2001; Rickardsson et al., 1998). In addition, we should mention that rough surfaces and nano-size Ni grains are much more active for breaking of NO chemical bonds than smooth, flat surfaces (Garin, 2001; Lindsay et al., 1998). Based on the above results, the following reaction mechanism was proposed for NO decomposition on the nano-size Ni grains produced during the reactor operation.

$$NO + Ni \rightarrow Ni\text{–}NO \tag{14}$$

$$2Ni\text{–}NO \rightarrow 2NiO + N_2 \tag{15}$$

NO gas molecules are first chemisorbed on Ni. As a second step the chemisorbed NO decomposes to form N_2, oxidizing Ni to NiO.

Oxygen ionic current passed though the network of YSZ particles surrounding the Ni grains. This process removed oxygen species from the electrode and permitted the reactions (14) and (15) to reoccur. The regeneration reaction of the reduction of NiO to Ni takes place at the NiO/YSZ interface under the reactor operation

$$NiO + V_O(ZrO_2) + 2e \rightarrow Ni + O^{2-}(ZrO_2) \quad (16)$$

Therefore, the reduction of NiO grains into Ni grains and the oxidation of Ni grains into NiO take place continuously during reactor operation. As a result the catalytic activity for NO decomposition is independent of the operation time.

At the same time oxygen gas molecules have a preference for adsorption by F-type centers on the surface of YSZ.

$$O_2 + 4e^- + 2\,V_O(ZrO_2) \rightarrow 2\,O^{2-}(ZrO_2) \quad (17)$$

From this consideration it follows that under the reactor operation adsorption and decomposition of NO and O_2 gas molecules occur on the surface of Ni grains and by F - centers on the surface of YSZ grains, respectively. The design of an electrochemically-assembled electrode with two kinds of active sites provides a way to suppress the unwanted reaction of oxygen gas adsorption (Eq.(17)) and to increase the desirable reaction of NO gas decomposition Eqs.(14,15).

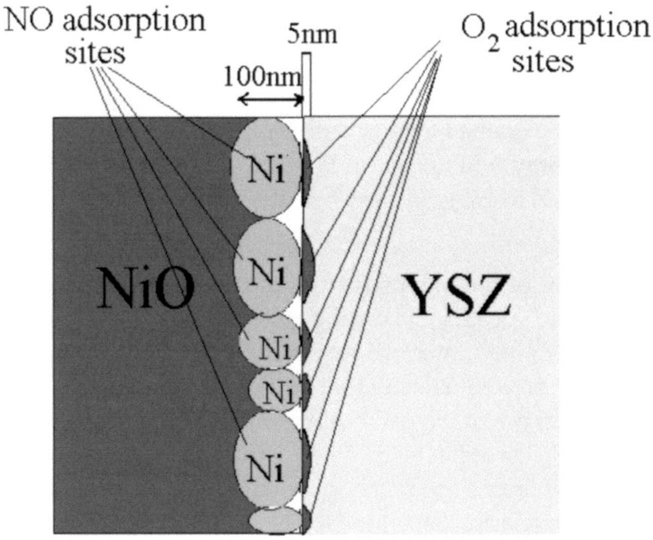

Fig. 14. Schematic representation of the microstructure of the YSZ-NiO interface and of the sites for NO and Oxygen gases adsorption in the self-assembled electro-catalytic electrode.

4.3 Covering layer

The deposition of a thin (2–3 nm) covering YSZ layer leads to a suppression of the oxygen adsorption and decomposition. Additionally, the deposition of the covering layer leads to an increase of the amount of nanosize Ni grains located in the near interface boundary porous region between the grains of NiO and YSZ. As a result, electrochemical reactors with the functional multilayer electrode show much better selectivity for NO gas decomposition even with respect to the electrochemical cells with electro-catalytic electrode but without a covering layer.

To optimize the characteristics of the electrochemical cells with multi-layer electrode, we have carried out the investigations of rate of NOx decomposition depending of the upper layer microstructure. Our investigations has shown that the best characteristics of the cells for selective NO decomposition can be reached for the electrochemical cells with the thin upper YSZ layer (2-3 μm) sintered at temperature 1450^{0}C.

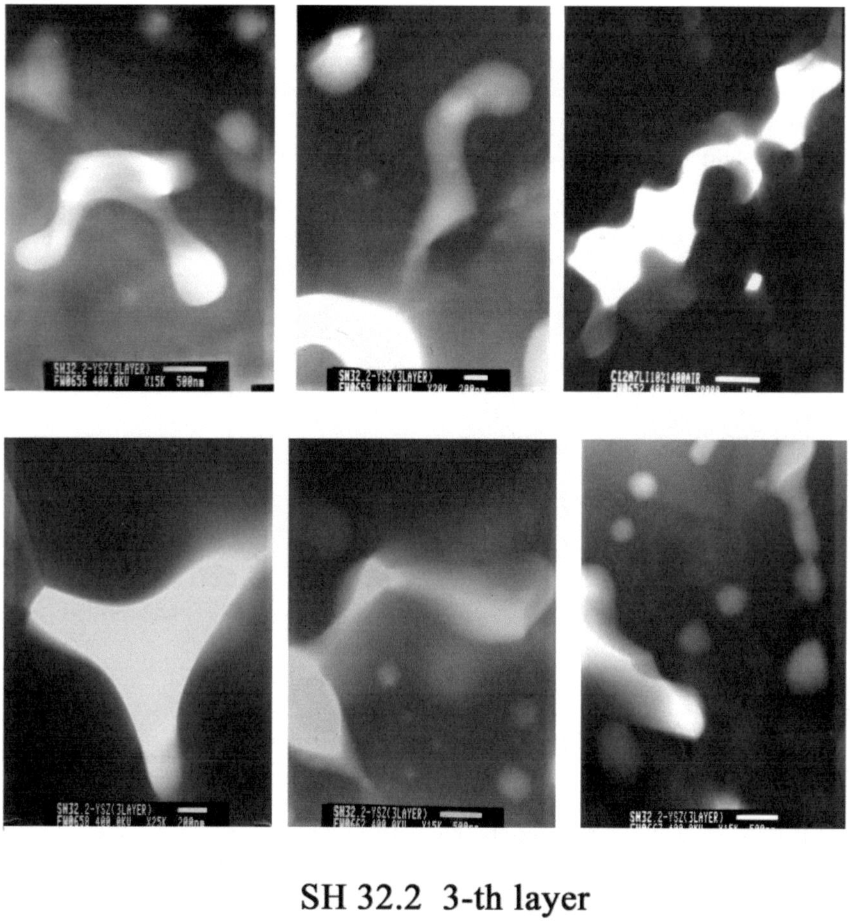

Fig. 15. TEM images of the structure of the covering YSZ layer sintered at 1450^{0}C.

The transmission electron microscopy investigations of the upper layer and of the electro-catalytic electrode were carried out on a JEOL 4000 FX microscope. Foils for the electron microscopy investigations were prepared by mechanical polishing followed by ion milling. The structure of the upper YSZ layer sintered at temperature 1450^{0}C is shown in Fig.15. This figure shows some typical microstructures of the upper layer. It is seen that upper layer is porous and that the pores form channels with a size of 200-500 nm, or they have ellipsoidal shape with a size 50-100 nm.

The structure of the electro-catalytic electrode in the electrochemical cell with multi-layer electrode is shown in Fig.16. This figure shows a typical structure of electro-catalytic electrode after the cell operation. Data of chemical composition of this structure obtained by

EDS method are also displayed in the same figure. From Fig.16 it is seen that the structure consists of NiO, Ni and YSZ grains. The Ni grins are located in the near interface boundary porous region between the grains of NiO and YSZ. From above consideration it follows that this structure is the same as a structure of the field-quenched electro-catalytic electrode in the cells without upper layer (Aronin et al., 2005; Bredikhin et al, 2006). It is obvious that deposition of the upper layer leads to suppressing of the oxygen gas adsorption and to an increase of the concentration of the oxygen vacancies in the YSZ grains. As the result both the rate of the reduction of the NiO to Ni and of the amount of new Ni grains increase in the YSZ/(Ni-NiO) interface region. Therefore the electrochemical cells with multi-layer electrode should show a much higher selectivity for NOx gas decomposition in the presence of excess oxygen than all known cells.

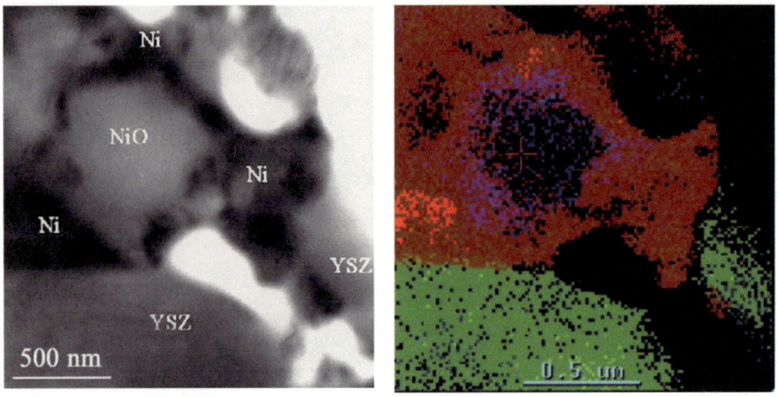

Fig. 16. TEM image of the structure of the electro-catalytic electrode (A) and the chemical composition of this structure obtained by EDS method (B).

One more important advantage of the electrochemical reactor with multi-layer electrode should be discussed. Our investigations have shown that electrochemical cells with multi-layer electro-catalytic electrode effectively operate even at low concentration of NOx (300-500 ppm) and at the high concentration of oxygen (10%) in the exhaust gas. In Fig.17 the NO conversion is plotted as a function of the current for one compartment electrochemical cell with multi-layer electrode at different concentration of NO (500 ppm or 1000 ppm) and oxygen (2% or 10%) at a gas flow rate 50 ml/min and at temperature 560^{0}C. From this figure it is seen that the decrease of the NO concentration from 1000ppm to 500ppm leads to the two times decrease of the value of the current required for 30% NO decomposition for both 2% and 10% of oxygen in the gas mixture. Direct proportion between NO concentration in the exhaust gas and the value of the current required for NO decomposition confirms our proposal that the process of NO gas adsorption and decomposition is practically independent of the oxygen gas adsorption and decomposition. Additional we should mention that increase of the oxygen content in the investigated gas from 2% to 10% at fixed NO concentration leads to the 1.5 times increase only of the value of the current required for NO decomposition. This result shows that the oxygen adsorption and decomposition in the electrochemical cells with multi-layer electrode is suppressed. In accordance with above we can conclude that new type of electrochemical reactor with multi-layer electro-catalytic electrode can be used for effective NO decomposition even in the presence of high oxygen concentration.

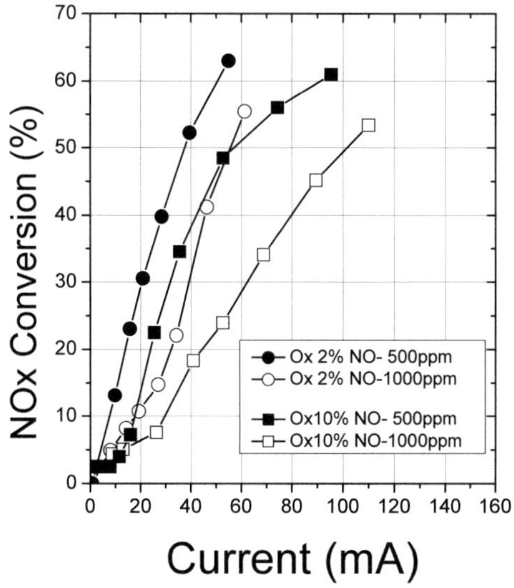

Fig. 17. The dependence of NO conversion on the value of the current for electrochemical cells with multi-layer electro-catalytic electrodes in the presence of 2% of Oxygen (-●- 500ppm and -O- 1000 ppm of NO gas) and of 10% of Oxygen (-■- 500 ppm and -□- 1000ppm of NO gas).

The design of the self-assembled electrode with two kinds of active sites provides a way to suppress the unwanted reaction of oxygen gas adsorption and to increase many times the desirable reaction of NO gas decomposition. For the first time, an electrochemical cell with multi-layer electro-catalytic electrode for selective NO decomposition in the presence of excess oxygen (10%) operating at a low value of electrical power was designed. These results indicate that electrochemical reactors with multi-layer electro-catalytic electrode can be used for practical applications.

5. Intermediate and low temperature electrochemical reactors with multilayer functional electrode

In 2006 *K.Hamamoto et.al.* (Hamamoto et.al., 2006, 2007) proposed to use an electrochemical rector with multilayer functional electrode for intermediate temperature operation. The cell performance represented on (5b) was used for intermediate temperature operation.

8YSZ (8 mol % Y_2O_3-doped ZrO_2), 10YSZ (10 mol % Y_2O_3-doped ZrO_2) and ScCeSZ (10 mol % Sc_2O_3 and 1 mol % CeO_2-doped ZrO2) were selected as solid electrolytes. The electro-catalytic electrodes with compositions NiO-8YSZ, NiO-10YSZ and NiO-ScCeSZ (with 55 mol % of NiO) were deposited on the surface of 8YSZ, 10YSZ and ScCeSZ solid electrolyte disks, respectively. Figure 18 shows the values of current efficiency (η) plotted as a function of applied voltage for such electrochemical reactors operated at 475°C. The value of selectivity (ν_{sel}) of such reactors for NO gas molecules decomposition is also displayed in Fig. 18 (Hamamoto et.al., 2006). From Fig.18 it is seen that in electrochemical reactors with electro-catalytic electrodes the probability for NO gas molecules to be adsorbed and decomposed is at least 5 times higher than for oxygen gas molecules.

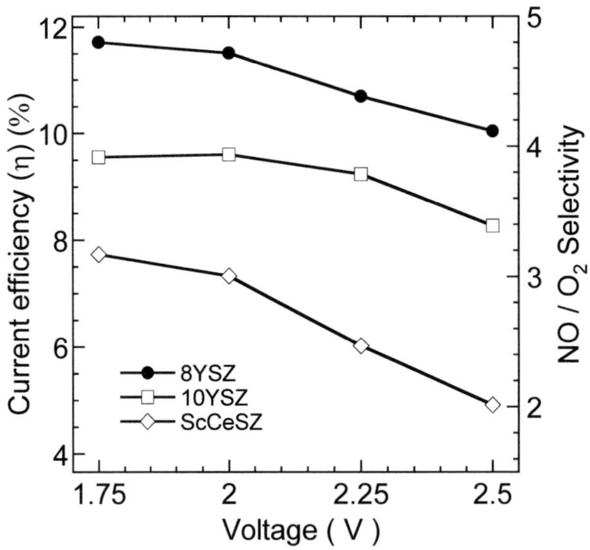

Fig. 18. Dependence of current efficiency for NO decomposition and NO/O_2 selectivity on operating voltage at 475°C.

In 2008 *K.Hamamoto et.al.* (Hamamoto et al., 2008) proposed to use an electrochemical reactor with additional NOx adsorption layer deposited on the top of the multilayer electro-catalytic electrode. Such type of cell can be represented by following cell arrangement:

NOx adsorbent | (Covering layer |Cathode |Electro-catalytic electrode) |YSZ| (Anode) (18)

The authors (Hamamoto et al., 2008) tested three systems which are generally used as NOx adsorber catalyst. The Pt/K/Al2O3, Pt/Na/Al2O3 and Pt/Cs/ Al2O3 adsorbents were prepared from a Al2O3 (Merck, p.a., specific surface area = 10 m2 g−1) support suspension in water, to which a solution containing KNO3 (NaNO3, CsNO3) and with aqueous solutions of platinum nitrate (Pt(NO3)2) was added, in order to obtain a load of 10 wt% of K (Na, Cs) and 3 wt% of Pt. The mixture was heated while being vigorously stirred until a paste was achieved, which was dried in an oven for 24 h at 200°C and crushed and calcined at 600°C for 2 h.

To clarify the capability of a NOx adsorption layer on the multilayer cathode, *K.Hamamoto et.al.* (Hamamoto et al., 2008) carried out the NOx decomposition measurements of the YSZ based cells with and without a NOx adsorbent at a fixed operating voltage, U=2.5 V, on the cells under various O2 concentrations at 500°C by passing a mixed gas with 1000 ppm of NO in He through the cell at a gas flow rate of 200 ml/min (Fig. 19). As a result, NOx adsorption layers have improved the NOX decomposition properties though the current values of each cell were almost the same.

From Fig.19 it is seen that the values of current efficiency (η) for electrochemical reactor with the Pt/K/Al2O3 adsorbent are four - five times higher compared with the reactor without adsorbent. The values of current efficiency in such reactors increase up to 20% and the values of the NO/O_2 selectivity up to 25 (Fig.20). Additionally we should mention that such electrochemical reactors effectively decompose NOx even at low temperature range (<300°C).

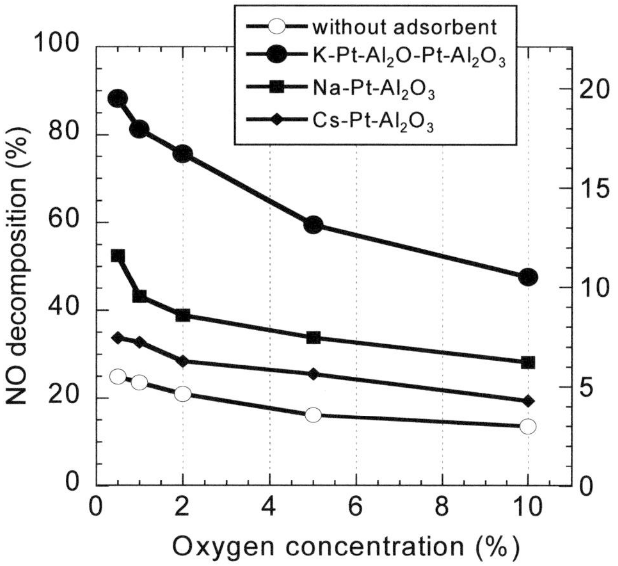

Fig. 19. Dependence of NO decomposition and of the value of current efficiency on the O_2 concentrations for the YSZ based electrochemical reactor with and without a NOx adsorbent at 500°C.

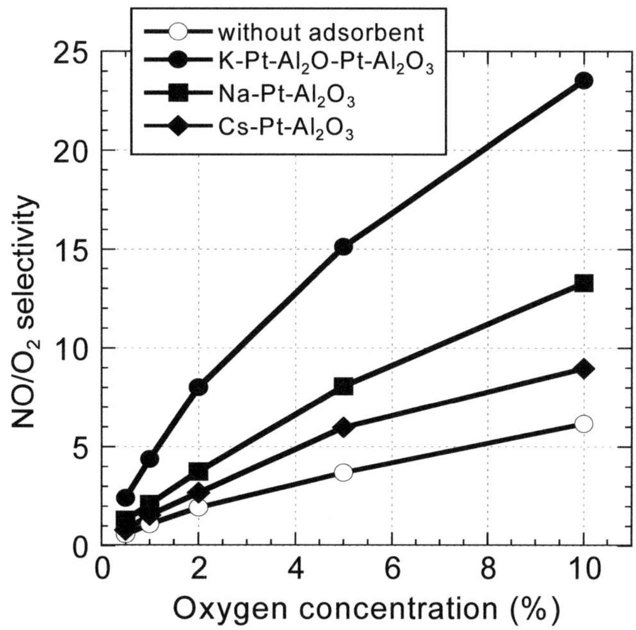

Fig. 20. Dependence of NO/O_2 selectivity on the O_2 concentrations for the YSZ based electrochemical reactor with and without a NOx adsorbent at 500°C.

6. Conclusion

It was shown that the electrochemical reduction of nitric oxide in an atmosphere containing excess oxygen is not effective due to the low selectivity of the cathode materials. To solve the problem of effective electrochemical reduction of nitric oxide in the presence of the excess oxygen *S.Bredikhin et al.* (Awano et al., 2004b; Bredikhin et al., 2001a, 2001b) proposed the concept of artificially designed multilayer structure which should operate as an electrode with high selectivity. Our investigations have shown that substitution of traditional cathodes by the multilayer electro-catalytic electrode with an electrochemically assembled nanostructure leads to a dramatic decrease in the value of the electrical power required for NO decomposition. It was shown that multilayer electro-catalytic electrode should consist at list from three main functional layers: Cathode; Electro-catalytic electrode; Covering layer, in order to operate as an electrode with high selectivity. In 2008 *K.Hamamoto et.al.* (Hamamoto et al., 2008) proposed to use an electrochemical reactor with fourth additional NOx adsorption layer. The values of current efficiency in such reactors increase up to 20% and the values of the NO/O_2 selectivity up to 25. These results indicate that this new type of electro-catalytic reactor can be used for practical applications. From our point of view, such systems should substitute traditional catalytic systems for exhaust gas purification.

7. References

Armor, J.N., "Catalytic reduction of nitrogen oxides with methane in the presence of excess oxygen: A review", *Catalysis Today*, 26, 147-158 (1995).

Aronin, A., Abrosimova, G., Bredikhin, S., Matsuda, K., Maeda, K. & Awano, M., "Structure evolution of a NiO-YSZ Electro-catalytic Electrode", *Journal of the American Ceramic Society* 88, 5, 1180-1185, (2005).

Awano, M., Bredikhin, S., Aronin, A., Abrosimova, G., Katayama, S. & Hiramatsu, T., "NOx decomposition by electrochemical reactor with electrochemically assembled multilayer electrode", *Solid State Ionics* 175, 605–608, (2004).

Awano, M., Fujishiro, Y., Hamamoto, K., Katayama, S. & Bredikhin, S., "Advances in nano-structured electrochemical reactors for NOx treatment in the presence of oxygen", *International Journal of Applied Ceramic Technology*, 1 (2), 277-286, (2004).

Baker, R.A. & Doerr, R.C., *Ind. Eng. Chem. Process Des. Dev.* 4 (1965), p. 188.

Baker, R.A. & Doerr, R.C., *J. Air Pollut. Control Assoc.* 14 (1964), p. 409.

Bosch, H. & Janssen, F., *Catalysis Today*, 2, 369 (1988).

Bredikhin, S., Abrosimova, G., Aronin, A. & Awano, M., "Electrochemical Cells with Multilayer Functional Electrodes", Part I. Reduction-oxidation reactions in a NiO-YSZ electro-catalytic electrode, *Journal of Ionics* , 12, 1, 33-39, (2006).

Bredikhin, S., Abrosimova, G., Aronin, A., Hamamoto, K., Fujishiro, Y., Katayama, S. & Awano, M., "Pt-YSZ Cathode for Electrochemical Cells with Multilayer Functional Electrode", *Journal of the Electrochemical Society* 151 (12) J95-J99 (2004).

Bredikhin, S., Maeda, K. & Awano, M., "Electrochemical cell with two layer cathode for NO decomposition", *Journal of Ionics*, 7, 109-115, (2001).

Bredikhin, S., Maeda, K. & Awano, M., "NO decomposition by an electrochemical cell with mixed oxide working electrode", *Solid State Ionics* 144, 1-9, (2001).

Bredikhin, S., Maeda, K. & Awano, M., "Peculiarity of NO decomposition by electrochemical cell with mixed oxide working electrode", *Journal of the Electrochemical Society* 148, (10), D133-D138, (2001).

Finlayson-Pitts, B.J. & Pitts, J.N., "Tropospheric Air Pollution: Ozone, Airborne Toxics, Polycyclic Aromatic Hydrocarbons, and Particles", *Science* 276 (5315), 1045-1051 (1997).

Garin, F., "Mechanism of NOx decomposition", *Applied Catalysis* A: General 222, (2001), 183.

Gur, T.M. & Huggins, R.A., "Decomposition of Nitric Oxide on Zirconia in a Solid State Electrochemical Cell", *J.Electrochem.Soc.* 126, no.6, 1067-1075, (1979).

Hamada, H., Kintaichi, Y., Sasaki, M., Ito, T. & Tabata, M., *Appl. Catal.* 64 (1990), p. L1.

Hamamoto, K., Fujishiro, Y. & Awano, M., "Low temperature NOx Decomposition using electrochemical reactor", *Journal of The Electrochemical Society*, (In print),(2008).

Hamamoto, K., Fujishiro, Y. & Awano, M., "Reduction and Reoxidation Reaction of Catalytic Layers in Electrochemical Cells for NOx Decomposition", *Journal of The Electrochemical Society*, 154 (9) F172-F175, (2007).

Hamamoto,.K., Fujishiro, Y. & Awano, M., "Intermediate Temperature Electrochemical Reactor for NO*x* Decomposition", *Journal of the Electrochemical Society*, 153 (11) D167-D170, (2006).

Hibino, T., "Electrochemical Removal of NO and CH_4 from Oxidizing Atmosphere", *Chem. Lett.* 927, (1994).

Hibino, T., Inoue, I. & Sano, M., "Electrochemical reduction of NO by alternating current electrolysis- using yttria-stabilized zirconia as the solid electrolyte; Part I. Characterizations of alternating current electrolysis of NO", *Solid State Ionics* 130, 19-29, (2000).

Hibino, T., Inoue, I. & Sano, M., "Electrochemical reduction of NO by alternating current electrolysis using yttria-stabilized zirconia as the solid electrolyte; Part II. Modification of Pd electrode by coating with Rh", *Solid State Ionics* 130, 31-39, (2000).

Hiramatsu, T., Bredikhin, S., Katayama, S., Shiono, O., Hamamoto, K., Fujishiro, Y. & Awano, M., "High selective deNO x electrochemical cell with self-assembled electro-catalytic electrode", *Journal of Electroceramics*, 13 (1-3), 865-870, (2004).

Hwang, H.J., Towata, A., Awano & Maeda, K., "Sol-gel route to perovskite-type Sr-substituted $LaCoO_3$ thin films and effects of polyethylene glycol on microstructure evolution", *Scripta Materialia*, 44, 2173, (2001).

Iwamoto, M., Proceedings of Meeting of Catalytic Technology for Removal of Nitrogen Monoxide, Tokyo, Japan, 1990, p. 17

Iwayama, K. & Wang, X., "Selective decomposition of nitrogen monoxide to nitrogen in the presence of oxygen on RuO_2/Ag(cathode)/yttria-stabilized zirconia/Pd(anode)", *Applied Catalysis B*: Environmental, 19, 137 (1998).

Janssen, F. & Meijer, R., "Quality control of $DeNO_x$ catalysts Performance testing, surface analysis and characterization of $DeNO_x$ catalysts", *Catalysis Today*, 16, 157 (1993).

Kammer K., "Electrochemical DeNOx in solid electrolyte cells—an overview" *Applied Catalysis B*: Environmental 58, 33–39 (2005).

Klimisch, R.L. & Barnes, G.J., *Environ. Sci. Technol.* 6 (1972), p. 543.

Kobayashi, K., Yamaguchi, Sh., Higuchi, T., Shin, Sh. & Iguchi, Y., *Solid State Ionics* 135, 643, (2000).

Lerdau, M.T., Munger, J.W. & Jacob, D.J., "Enhanced: The NO_2 Flux Conundrum", *Science* 289 (5488), 2291-2293 (2000).

Libby, W.F., "Promising catalyst for auto exhaust", *Science* 171, 499-500 (1971).

Lindsay, R., Theobald, A., Gießel, T., Schaff, O., Bradshaw, A.M., Booth, N.A. & Woodruff, D.P., "The structure of NO on Ni(111) at low coverage", *Surface Science* 405, *L566-L572*, (1998).

Marwood, M. & Vayenas, C.G., "Electrochemical Promotion of the Catalytic Reduction of NO by CO on Palladium", *Journal of Catalysis*, 170, 275 (1997).

Matsuda, K., Kanai, T., Awano, M. & Maeda, K., "NO decomposition Properties of Lanthanum Manganite Porous Electrode", MRS Proceedings, Vol. 658, G9, 36, 1-5 (2001).

Miura, K., Nakagawa, H., Kitaura, R. & Satoh, T., "Low-temperature conversion of NO to N_2 by use of a novel Ni loaded porous carbone", *Chemical Engineering Science* 56, 1623-1629, (2001).

Nakatani, J., Ozeki, Y., Sakamoto, K. & Iwayama, K., "NO Decomposition in the Presence of Excess O_2 Using the Electrochemical Cells with Pd Electrodes Treated at High Temperature and Coated with $La_{1-x}Sr_xCoO_3$", *Chem. Lett.*, no.4, 315-319, (1996).

Nishihata, Y., Mizuki, J., Akao, T., Tanaka, H., Uenishi, M., Kimura, M., Okamoto, T. & Hamada, N., "Self-regeneration of a Pd-perovskite catalyst for automotive emissions control", *Nature* 418, 164-167 (2002).

Pancharatnam, S., Huggins, R.A. & Mason, D.M., "Catalytic Decomposition of Nitric Oxide on Zirconia by Electrolytic Removal of Oxygen", J. Electrochem. Soc. 122, no.7, 869-875, (1975).

Parvulescu, V.I., Grange P. & Delmon, B., "Catalytic removal of NO", *Catalysis Today*, 46, 233-316, (1998).

Rickardsson, I., Jönsson, L. & Nyberg, C., "Influence of surface topology on NO adsorption: NO on Ni(100) and Ni(510)", *Surface Science* 414, 389-395, (1998).

Roth, J.F. & Doerr , R.C., *Ind. Eng. Chem.* 53 (1961), p. 293.

Sato S., Hirabayashi, H., Yahiro, H., Mizuno, N. & Iwamoto, M., *Catal. Lett.* 12 (1992), p. 193.

Schoonman, J., in: Chowdari, B.V.R. & Radakrishna, S. (Eds.), *Proceedings of the 3rd International Symposium on Solid State Ionics Devices*, World Scientific Singapore, 697, (1988).

Simonsen, V.L.E., Find, D., Lilliedal, M., Petersen, R., & Kammer, K., "Spinels as cathodes for the electrochemical reduction of O_2 and NO", *Topics in Catalysis* Vol. 45, 143-148, (2007).

Wagner, C., in: Delahay, P. & Tobias, C.W. (Eds.), Advances in Electrochemistry and Electrochemical Engineering, Interscience Publishers, New York, (1966).

Walsh, K.J. & Fedkiw, P.S., "Nitric oxide reduction using platinum electrodes on yttria-stabilized zirconia", *Solid State Ionics* 93, 17 (1997).

Washman, Eric D., Jayaweera, P., Krishnan, G. & Sanjurjo, A., "Electrocatalytic reduction of NO_x on $La_{1-x}A_xB_{1-y}B'_yO_{3-}$: evidence of electrically enhanced activity", *Solid State Ionics* 136-137, 775-782, (2000).

9

Sequential Injection Anodic Stripping Voltammetry at Tubular Gold Electrodes for Inorganic Arsenic Speciation

José A. Rodríguez[1], Enrique Barrado[2], Marisol Vega[2],
Yolanda Castrillejo[2] and José L.F.C. Lima[3]
[1]Universidad Autónoma del Estado de Hidalgo
[2]Universidad de Valladolid
[3]Universidade do Porto
[1]Mexico
[2]Spain
[3]Portugal

1. Introduction

Arsenic is one of the most feared contaminants because of its high toxicity at low concentrations. Exposure to high levels of arsenic can cause problems in humans ranging from gastrointestinal symptoms to arsenicosis. Once this element is dissolved in water and is ingested, it is accumulated in the body. Contamination of groundwater with arsenic is one of the major environmental and public health problems on a global scale (NRC, 1999).

The World Health Organization guideline has established a concentration of 10 $\mu g\ l^{-1}$ as the maximum residue limit for arsenic in drinking water (WHO, 2004). Well-known arsenic contaminated regions include Bangladesh, India and other countries. In Mexico, sources of drinking water exceeding 10 $\mu g\ l^{-1}$ have been found in Baja California Sur, Chihuahua, Coahuila, Durango, Zacatecas, Hidalgo, Morelos, Guanajuato, Sonora and San Luis Potosí (Camacho et al., 2011). In Spain, the presence of naturally occurring arsenic in groundwater has been reported in the sedimentary Duero and Tajo Cenozoic basins, located in central Spain (García-Sánchez et al., 2005; Gómez et al., 2006; Vega et al., 2008) .

Arsenic is usually distributed as water soluble species, colloids, suspended forms and sedimentary phases. Mobilized arsenic is most likely transported by water and accumulated in downstream river sediments as a result of the great affinity of arsenic to iron rich phases. The most common sources of non-naturally arsenic worldwide arise from the presence of alloys used in manufacture of transistors, laser, semi-conductors and mining industry (Smedley & Kinniburgh, 2002).

There are more than 20 arsenic compounds identified in environmental and biological systems (Gong et al., 2002). The dominant forms of arsenic present in the environment are As(III) (arsenite) and As(V) (arsenate) (Mondal, 2006). As(III) binds to sulfhydryl groups impairing the function of many proteins and affects respiration by binding to the vicinal thiols in pyruvate dehydrogenase and 2-oxoglutarate dehydrogenase. As(V) is a molecular analogue of phosphate and inhibits oxidative phosphorylation, the main energy generation system. As(V) is most frequently present in surface water while As(III) is commonly found in anaerobic groundwaters. Redox potential, pH and organic matter control the species present in water.

To determine the potential transformation and risk of arsenic in the environment, the analysis of arsenic should include identifying and quantifying both, the total quantity of arsenic present and the specific chemical forms, a procedure known as speciation (Bednar et al., 2004; Burguera & Burguera, 1997; Gong et al., 2002).

2. Arsenic speciation

In-house laboratory assays are generally required to accurately measure arsenic in environmental samples at the $\mu g\ l^{-1}$ level in waters. The preferred laboratory methods for measurement of arsenic involve sample pre-treatment, either with acid addition or acidic digestion of the sample. Pre-treatment transfers all the arsenic in the sample into an arsenic acid solution, which is subsequently measured using techniques such as graphite furnace atomic absorption spectroscopy (ETAAS), atomic fluorescence spectrometry (AFS), inductively coupled plasma atomic emission spectrometry (ICP-AES), inductively coupled plasma-mass spectrometry (ICP-MS), high performance liquid chromatography (HPLC) coupled to ICP-MS, X-ray fluorescence (XRF), neutron activation analysis (NAA) and capillary electrophoresis (CE) (B'Hymer & Caruso, 2004; Burguera & Burguera, 1997; EPA, 1999; Gong et al., 2002; Melamed, 2005).

The most commonly used speciation techniques often involve a combination of chromatographic separation with spectroscopic detection. HPLC is the most used in the ion-pairing and ion exchange modes (B'Hymer & Caruso, 2004; Gong et al., 2002). Such techniques are expensive to operate and maintain and require fully equipped and staffed laboratories.

Anion- and cation-pairing chromatography techniques have been developed for separation of arsenic species. Tetrabutylammonium is the common pairing cation for separating As(III) and As(V) using reverse phase columns for the separation. The resolution depends on the concentration of ion-pair reagent, the flow rate, ionic strength and pH of the mobile phase (Guerin et al., 1999). Anion-exchange chromatographic techniques have been used for inorganic arsenic speciation analysis. A gradient elution using ammonium phosphate as mobile phase allows the resolution of As(III) and As(V) from organoarsenic species (Terlecka, 2005). Speciation of trace levels of arsenic in environmental samples requires high sensitivity, then the use of HPLC-MS with electrospray ionization, or HPLC-ICP-MS are often needed (B'Hymer & Caruso, 2004).

On the other hand, electrochemical assays, in particular stripping analysis, have demonstrated to be useful for detection of arsenic traces in water samples. Cathodic stripping voltammetry (CSV) or adsorptive cathodic stripping voltammetry (AdCSV) using hanging mercury drop electrodes (HMDE) was used in the past for arsenic analysis (Ferreira & Barros, 2002; Sadana, 1983). In the last years the analytical use of mercury has been discouraged due to its toxicity. Different materials have been reported for the determination of arsenic, including platinum (Williams & Johnson, 1992), gold (Forsberg et al, 1975), bismuth (Long & Nagaosa, 2008), carbon substrates (Sun et al., 1997) and boron doped diamond (Ivandini, et al. 2006).

Anodic stripping voltammetry (ASV) provides an alternative technique for measuring inorganic arsenic in water samples. ASV at gold film electrodes (Sun et al. 1997) or solid gold electrodes (Kopanica & Novotny, 1998) have been extensively used for inorganic arsenic speciation as they allow to determine separately As(III) and total As. The analysis by ASV involves three major steps (Figure 1). First, the electrode surface is conditioned for analysis (cleaning the surface of the solid electrode and/or plating a gold film). The As(III) is then deposited as elemental arsenic on the working gold electrode by electrochemical reduction. After the deposition step, the elemental arsenic is electrochemically oxidized (stripped) back to As(III). As(V), the most stable form of the element in oxidizing environments, is determined after chemical or electrochemical (Muñoz & Palmero, 2005) reduction to As(III), total As is then determined and As(V) is calculated by difference between total As and As(III). The Environmental Protection Agency (EPA) has approved an analytical method (EPA, 1999) for arsenic determination in water samples based on the use of ASV at gold film electrodes.

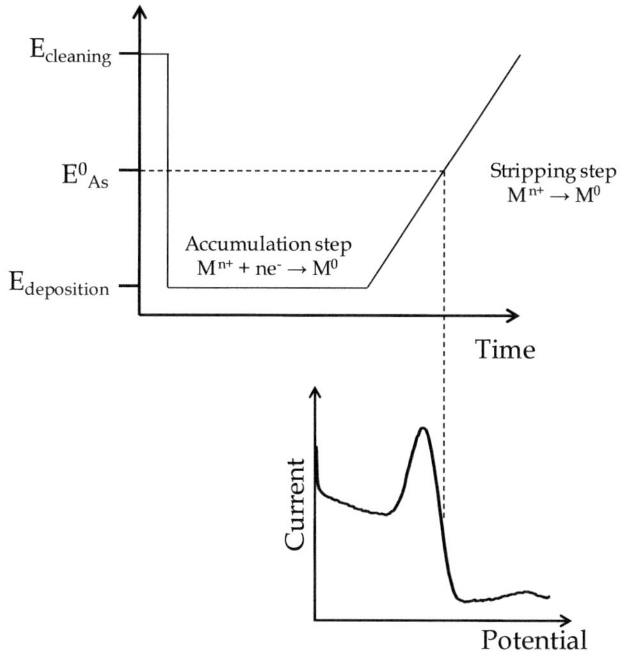

Fig. 1. Anodic stripping voltammetry: the potential-time waveform with the resulting voltammogram.

The remarkable sensitivity, broad scope and low cost of stripping analysis have led to its application in the determination of arsenic in water, soils and food samples. From early years of stripping analysis two main different research areas have been considered. The use of microelectrodes and disposable electrodes (Gibbon et al., 2010), and the development of hyphenated techniques using flow manifolds (Economou, 2010). On-line stripping analysis using flow analysis has demonstrated the viability and potentialities of this coupling such as: a) lower consumption of sample and reagents, b) higher precision and accuracy and c) higher degree of automation.

3. On-line stripping analysis

Flow analysis methodologies are based on the measuring of a transient non-steady signal, allowing a high sampling rate without the need for segmentation to limit analyte dispersion. This concept has simplified the measuring systems, and has resulted in a rapid increase in the interest of these techniques (Ruzicka & Hansen, 1988).

Continuous flow methodologies are the common approach for analysis by flow systems coupled to stripping analysis. This coupling mode uses a selection valve (SV), different streams of solutions are selected by the valve and pumped unidirectionally through the electrochemical flow cell for electrode modification/conditioning, pre-concentration and stripping (Fig.2). On-line medium exchange is an alternative to minimize the interference of the analytical matrix. The continuous flow mode is based on simple instrumentation, but its greatest drawback is the high consumption of sample and reagent solutions (Muñoz & Palmero, 2004).

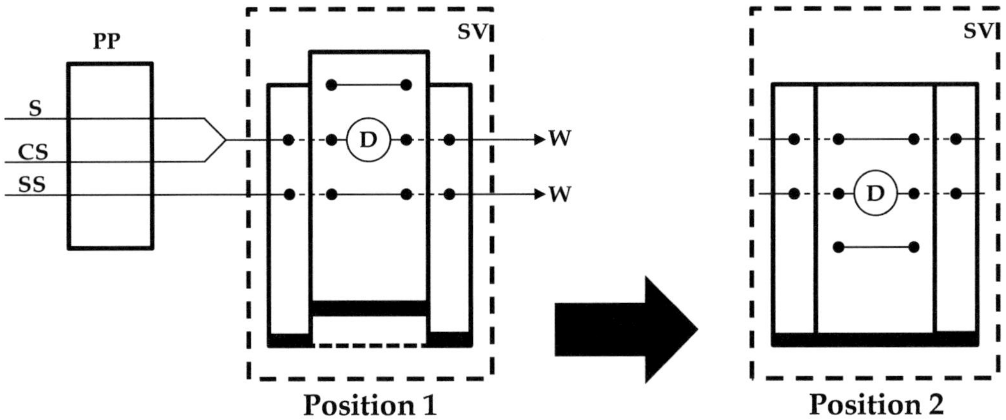

Fig. 2. Continuous flow system. S, sample; CS, conditioning solution; SS, stripping solution; PP, peristaltic pump; SV, selection valve; D, electrochemical flow cell; W, waste.

Flow injection analysis (FIA) is based on the injection of a known amount of sample in a flowing carrier solution stream via an injection valve (IV); the flowing carrier transports the sample to the detector. The main advantages of FIA compared with continuous flow systems are the operational simplicity and the lower consumption of sample and carrier (Bryce et al., 1995). However, the reduction in the amount of analyte deposited on the electrode surface as a result of the decrease in the contact time between the sample and the working electrode and the dispersion of the sample in the flow manifold results in a decrease of the analytical signal. A typical FIA system is illustrated in Fig. 3.

Sequential injection analysis (SIA, Fig. 4.) is other flow methodology that has been coupled to on-line stripping analysis. The heart of SIA manifold is the multiport selection valve; solutions are aspired and transported as zones using a bidirectional pump. SIA advantages are the low consumption of sample and reagents, the flexibility and the potential for automated sample manipulation (Ivaska & Kubiak, 1997). The sample volumes used in SIA are smaller than those employed for continuous flow systems and FIA, the amount of analyte deposited is lower, thus yielding a decreased signal.

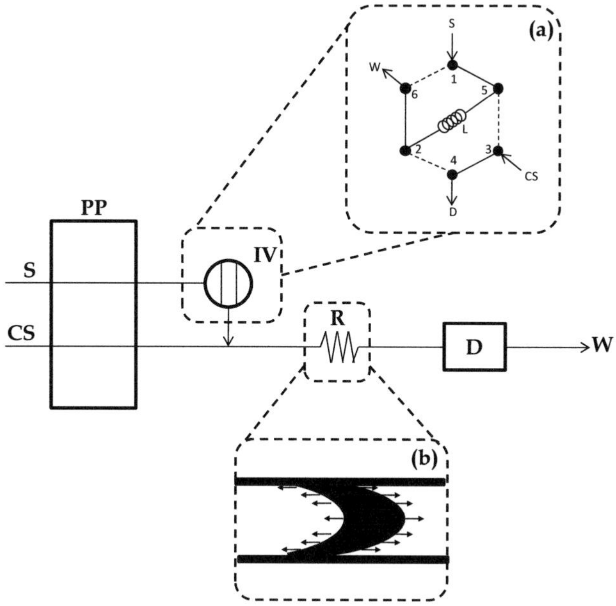

Fig. 3. Flow injection analysis system. a) insertion of S into CS, b) dispersion phenomena. S, sample; CS, carrier solution; PP, peristaltic pump; IV, injection valve; R, reactor; D, electrochemical flow cell; W, waste.

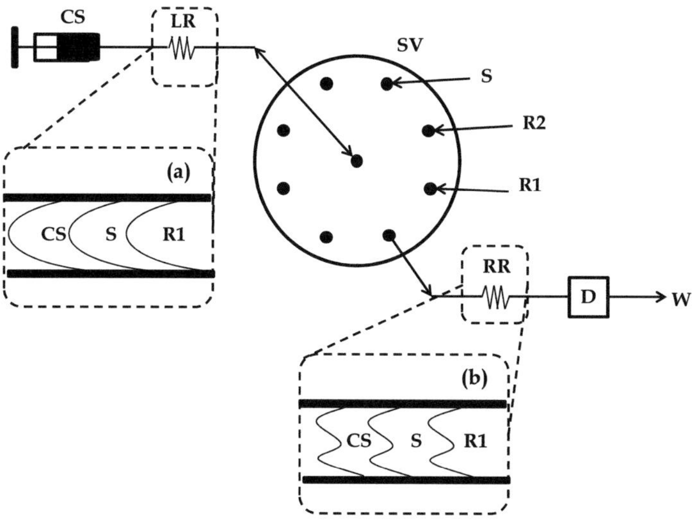

Fig. 4. Sequential injection analysis system. a) sample and reagents aspiration, b) mixture dispense. CS, carrier solution; LR, loading reactor; SV, selection valve; S, sample; R1 and R2, reagents; RR, reaction coil; D, electrochemical flow cell; W, waste.

Another critical part of flow methods is the detector (electrochemical flow cell). An electrochemical detector uses the electrochemical properties of analytes for determination in the flowing stream. Electrochemical detection is usually performed by controlling the potential of the working electrode at a fixed value and monitoring the current as a function of time. The current response thus generated is proportional to the concentration of the analyte. During on-line stripping analysis the analyte is pre-concentrated on the surface electrode in flowing conditions, whilst the stripping step can be done in flowing conditions or stopped flow (Economou, 2010).

Different cell designs have been used for electrochemical detection. The cell design must fulfil the requirements of high signal-to-noise ratio, low dead volume, well defined hydrodynamics, small ohmnic drop, high contact area and easy of construction and maintenance. In addition, the reference and counter electrodes should be located downstream next to the working electrode, so that reaction products at counter electrode or leakage from the reference electrode do not interfere with the working electrode detection.

The most widely used detectors are based on the wall-jet, thin-layer, and tubular configurations (Fig. 5.) (Trojanowicz, 2009):

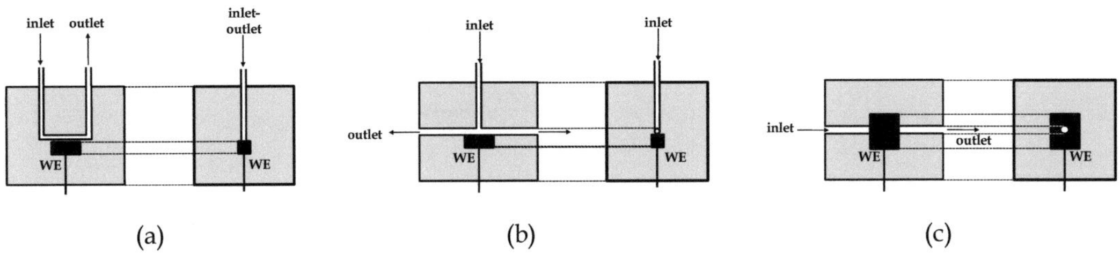

Fig. 5. Schematic representation of front and side view of electrochemical flow cells configurations. a) thin-layer, b) wall-jet and c) tubular. WE, working electrode.

In the wall-jet design, the stream flows perpendicularly to the working electrode surface, and then spreads radially over it improving the contact between the analyte and the electrode. The thin-layer cell consists on a thin layer of solution that flows parallel to the planar electrode surface, the main disadvantage is the small contact area. Tubular configuration provides minimal flow disturbance and a higher contact area, compared with thin-layer configuration. This feature has enabled the application of tubular configuration in flow injection systems with sequential determination in which the detector is relocated inside the flow manifold (Catarino et al., 2002).

4. Experimental conditions

4.1 Reagents, equipment and analysis

All solutions were prepared by dissolving the respective analytical grade reagent in ultrapure water (Milli Q, Millipore) with a specific conductivity lower than 0.1 μS cm^{-1}, and used without further purification.

A stock solution of 1000 mg l^{-1} As(V) was prepared by dissolution of $Na_2HAsO_4{\cdot}7H_2O$ (Panreac, Spain) in water and acidified with concentrated HCl (0.1% v/v). A stock solution of 1000 mg l^{-1} As(III) was prepared by dissolving As_2O_3 (Sigma, St Louis, MO, USA) with NaOH 1x10^{-2} M (Panreac) and then acidified to pH 1 with concentrated HCl (Panreac). Stock solutions were renewed weekly. Standard solutions of As(V) and As(III) of concentrations ranging from 0 to 50 μg l^{-1} were prepared daily by dilution of the respective stock solution.

L-cysteine 1x10^{-3} M (HOOC-CH(NH_2)-CH_2-SH, Sigma) was used as reducing agent. A supporting electrolyte solution of 2.0 M HCl was used for all the experiments.

Figure 6 shows a scheme of the sequential injection anodic stripping voltammetry system (SI-ASV) used for inorganic arsenic speciation in water samples. The system consisted of a MicroBu 2030 multisyringe burette with programmable speed (CS, Crison, Spain) used to aspire and dispense the reagent solutions, an eight-way selection valve (SV, Crison), a home-made tubular electrochemical cell (D), and a mixer chamber (MC).

The tubular electrochemical detection cell was made up of a Perspex body in which working and auxiliary electrodes were placed. These electrodes were built from gold and carbon discs (7.0 mm diameter) with length of 1.0 and 2.0 mm, respectively. Both have a tubular channel (0.8 mm diameter) in the centre of the electrode. These electrodes were used in connection with a saturated Ag/AgCl reference electrode (Metrohm, Switzerland). The instrumental devices were controlled by means of the Autoanalysis 5.0 software (Sciware, Spain). All tubing connecting the different components of the flow system was made of Omnifit PTFE with 0.8 mm (i.d.). Electrochemical experiments were performed with an Autolab PGSTAT10 potentiostat/galvanostat (EcoChemie) equipped with GPES 4.6 software. Unless otherwise stated, a frequency (f) of 25 Hz, pulse amplitude (E_{sw}) of 50 mV, step height (ΔE_s) of 8 mV, and deposition potential (E_d) and time (t_d) of −0.4 V for 40 s were chosen as the square wave anodic stripping voltammetry (SWASV) parameters. A conditioning potential and time (2 s at 2.0 V) was added to increase the reproducibility (Kopanica & Novotny, 1998).

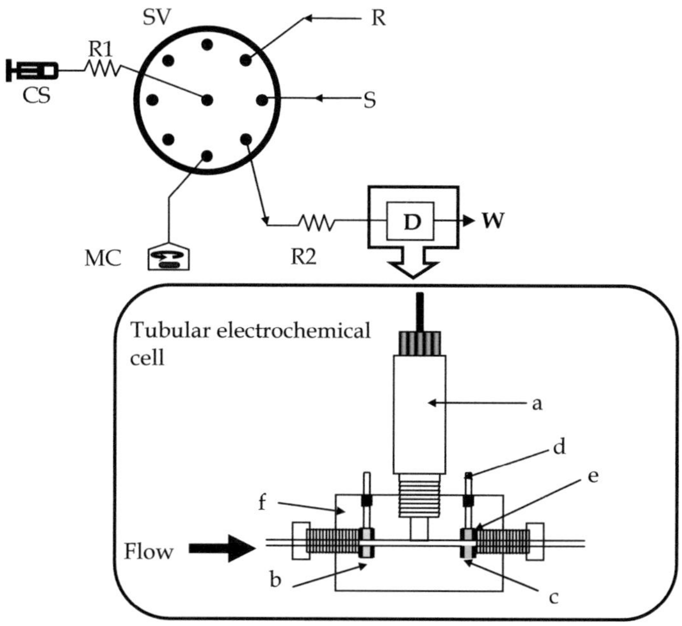

Fig. 6. Schematic set up of the SI-ASV flow system: CS, carrier solution; R1, holding coil; R2, reaction coil; SV, selection valve; R, reductant ; S, sample; MC, mixer chamber; D, detector; W, waste. Components of the electrochemical cell: a, reference electrode; b, tubular gold electrode; c, glassy carbon counter electrode; d, connector, e, O-ring.

4.1.1 Sampling

500 ml polyethylene bottles were conditioned by filling them with 2% v/v HNO_3 for at least three days. Once in the sampling site, bottles were rinsed several times with the water to be collected. Groundwater samples were obtained from deep and shallow wells in the area of Tierra de Pinares (Segovia, Spain), affected by arsenic contamination of aquifers. The well was pumped for at least 10 min before a water sample was collected. Bottles were completely filled with the sample to minimize the oxidation of As(III) by air. Immediately after sampling, the bottled water samples were acidified with HCl (pH<2), wrapped in hermetic plastic bags, and transported to the laboratory in iceboxes. Water samples were stored not longer than one week at 4°C prior analysis.

4.1.2 Analytical cycle

Initially, a 0.4 ml sample aliquot and a 0.2 ml of the reductant (L-cysteine, R-SH) are aspirated sequentially to the holding coil (R1). The resulting solution is directed towards the mixer chamber MC at 1.2 ml min^{-1} (50 s), and is allowed to stand for 2 minutes to reduce chemically As(V) to As(III) according to the following reaction:

$$2\,\text{R-SH} + \text{As(V)} \rightarrow \text{R-S-S-R} + \text{As(III)} + 2\text{H}^+ \qquad (1)$$

Simultaneous to the reduction of As(V) to determine in a second step total As, the measurement of As(III) was carried out. The sample was introduced in the system using a binary sampling strategy consisting on intercalation of multiple small sample segments of 50 µl (aspirated to S channel) with small segments (50 µl) of carrier solution (dispensed by CS to R2). The total aspirated volume was therefore 400 µl of sample and 250 µl of carrier solution. The binary sampling strategy creates multiple reaction interfaces which contribute to a faster homogenization of the sample media. The sample mixture is then propelled towards the electrochemical cell by the CS at a flow rate of 0.6 ml min^{-1} to be electrochemically deposited (40 s at -0.4 V).

$$\text{As(III)} + 3\,e^- \rightarrow \text{As}^0 \qquad (2)$$

After deposition of elemental arsenic, 2.0 ml of the carrier solution are pumped through the detection cell at a flow rate of 30.0 ml min^{-1}. The elemental arsenic is then stripped off in stop flow mode using the CS as clean medium under the SWASV parameters mentioned above. The exchange of the sample solution by the CS electrolyte solution minimizes the interference produced by the chlorine generated at the auxiliary electrode (Billing et al., 2002).

$$\text{As}^0 \rightarrow \text{As(III)} + 3e^- \qquad (3)$$

For total As determination, the solution contained in the mixing chamber (port 4) is now aspirated to R1 and then propelled through the detection cell at a flow rate of 1.2 ml min^{-1} (40 s). The resulting As(III) (sum of As(III) plus reduced As(V)) is then electrochemically reduced and stripped in a clean medium as described for As(III). The As(V) is then determined as the difference between total As and As(III).

5. Results and discussion

5.1 Optimization of the SI-ASV system for inorganic arsenic speciation

The optimization of the variables is a critical step in the design of new analytical methods. Optimization involves the selection of the chemical and instrumental factors which may affect the analytical signal, and the choice of the values of the variables to obtain the best response from the chemical system. For this purpose, two different strategies can be used. In the traditional univariate optimization, all values of the different factors except one are constant, and this one is the object of the examination. The alternative to this strategy is the use of chemometric techniques based mainly on the use of experimental designs (Tarley, et al. 2009).

Factorial designs are used to identify the significant variables (factors) affecting the selected response and as a tool to explore and model the responses as a function of these significant experimental factors. Two-level full factorial designs are a powerful alternative to find the adequate experimental conditions to produce the best response of the chemical system. This type of design fits the response to a linear model. For a two-factor case, the response surface is given by the linear model:

$$\hat{y} = b_0 + b_1 x_1 + b_2 x_2 + b_{12} x_1 x_2 \tag{4}$$

If the interaction term $b_{12}x_1x_2$ is negligible, then the response surface is planar. The more important the interaction term, the greater is the degree of twisting that the planar response surface experiences. Chemometrical optimization commonly uses the following procedure: a) choose a statistical design to investigate the experimental region of interest, b) perform the experiments in random chronological order, c) perform analysis of variance (ANOVA) on the regression results so that the most appropriate model with no evidence of lack of fit can be used to represent the data.

The factors investigated that could affect the response for As(III) and total As determination are listed in Table 1. Two levels for each factor were selected for a complete factorial design (replicates, i.e., 2^{3+1}) allowing to identifying the critical factors. Sixteen experiments were performed in random order for each of As(III) and total As optimization. The optimal values obtained for factors B and C from the optimization of SI-ASV for As(III) were fixed for total As determination, and new factors related to the prior reduction of As(V) were included in the second design of experiments. The current of the stripping peak of arsenic (in μA) using a 10.0 μg l^{-1} standard of arsenic was selected as the response to be optimized.

The flow rate is critical in on-line stripping methods, as it controls the dispersion between the sample and the carrier solution. This factor and the deposition time contribute to the amount of analyte deposited on the electrode surface. With respect to the reaction coil length, it has to be sufficiently long for loading and pre-treatment of sample previous to the deposition step. In addition, for total As determination the chemical parameters such as reduction time and reducer concentration must guarantee the complete reduction of the As(V) contained in the sample to As(III).

Analyte	Variable	Level	
		-	+
As(III)	A. Flow rate (ml min^{-1})	0.6	1.2
	B. Deposition time (s)	20	40
	C. Reaction coil length (cm)	40.0	80.0
Total As	A. Flow rate (ml min^{-1})	0.6	1.2
	B. Reduction time (s)	120	300
	C. [L-cysteine] (M)	$1x10^{-3}$	$1x10^{-2}$

Table 1. Selected levels of each variable for the analysis of As(III) and total As.

The design matrix and mean values obtained for the peak height of the 10.0 μg l^{-1} arsenic solution are provided in Table 2.

Exp	A	B	C	Mean peak height (μA) and %RSD	
				As(III)	Total As
1	-	-	-	2.14(5.29)	0.63(1.13)
2	-	-	+	13.35(1.59)	0.66(9.72)
3	-	+	-	5.58(7.35)	0.70(5.09)
4	+	-	-	7.26(2.53)	3.99(2.84)
5	-	+	+	12.20(1.16)	0.73(6.83)
6	+	+	-	4.41(3.05)	4.01(1.06)
7	+	-	+	9.53(2.67)	3.30(7.29)
8	+	+	+	8.89(1.03)	3.06(2.08)

Table 2. Design matrix and experimental results (mean and RSD, in brackets; n=2). A, B and C as in Table 1.

An Analysis of Variance of the results of the experimental design revealed the mean effect of each factor on the stripping signal (Table 3). In turn, these values enabled to calculate the variance of each factor using the Yates algorithm (column 3) (Massart, et al. 1997). By comparing the variance shown by each factor with the variance of the residuals, a Fischer F-test was then performed for each source of variation.

The F-test indicated that, at a significance level of p=0.05, the critical factors for As(III) determination were the deposition time, the length of the reactor R2 and the binary interactions flow rate-deposition time and flow rate-reaction coil length. The significance of the binary interactions is a consequence of the influence of both factors in the correct mixture of the sample solution and the electrolyte support solution. On the other hand, for total As determination, the variable with major contribution was the flow rate.

Figure 7 shows the effect of the control factors on the peak height of the stripping signal, among which the reaction coil length and the flow rate are the most important factors for As(III) and total As determination, respectively. Based on the results shown in Fig. 7, the combination of settings that generates the highest peak height was selected and is shown in Table 4.

Factor	As(III)			Total As		
	Effect	Variance	$F_{calculated}$	Effect	Variance	$F_{calculated}$
A	-0.80	2.54	6.86	2.91	33.96	**403.08***
B	2.92	34.13	**91.98***	-0.21	0.17	2.07
C	6.15	151.04	**407.02***	-0.40	0.63	7.45
AB	-1.45	8.38	**22.58***	-0.09	0.03	0.40
AC	-2.77	30.69	**82.70***	-0.43	0.73	**8.63***
BC	-0.60	1.42	3.82	-0.07	0.02	0.21
ABC	1.70	11.56	**31.15***	-0.07	0.02	0.21
Residual		0.37			0.08	

Table 3. ANOVA of the results of the experimental design showing the factors and/or interactions affecting significantly the peak height ($F_{calculated}$>7.57 at 95% confidence level, in bold). A, B and C as in Table 1.

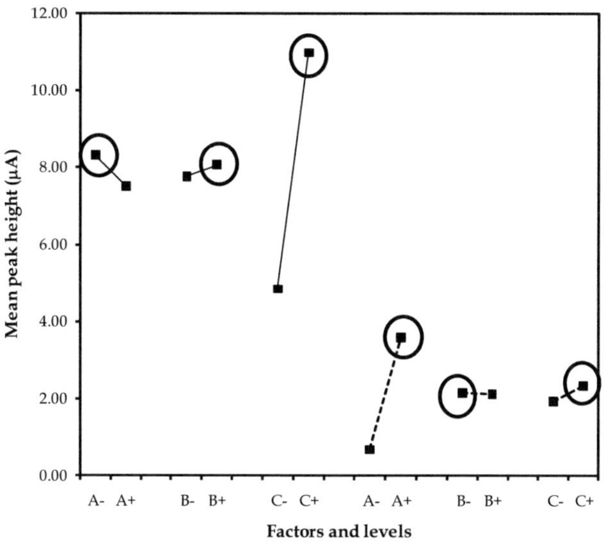

Fig. 7. Effect of control factors on the mean peak height of the stripping signal. (——) As(III) and (- - -) total As variables. A, B and C meaning as in Table 1. In circles the optimal level for each factor.

Analyte	Variable	Value
As(III)	A. Flow rate (ml min^{-1})	0.6
	B. Deposition time (s)	40
	C. Reaction coil length (cm)	80.0
Total As	A. Flow rate (ml min^{-1})	1.2
	B. Reduction time (s)	120
	C. [L-cysteine] (M)	$1x10^{-2}$

Table 4. Optimized experimental conditions of SI-ASV system for inorganic arsenic speciation in water samples.

5.2 Analytical properties of the optimized flow system

Calibration plots for As(III) and total As (added as As(V)) were obtained in the experimental conditions described in Table 4. Three replicate measurements of each standard As solution were made and average values were used for calculations. A linear dependence of the height of the stripping signal with the injected concentration of arsenic was found in the concentration range 2-40 μg l^{-1} for both total As and As(III) (Fig. 8.).

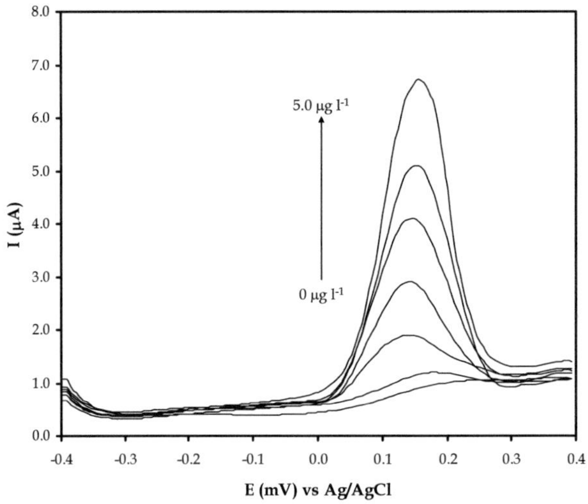

Fig. 8. SW-voltammograms for As(III) standards obtained by SI-ASV in 2M HCl.

The limit of detection calculated according to the IUPAC criterion as $3s_e/b_1$, where s_e is the square root of the residual variance of the calibration plot and b_1 is the slope, resulted in values of 1 μg l^{-1} for As(III) and 2 μg l^{-1} for total As. These limits of detection are adequate to assess the compliance of the method with the maximum tolerable levels for As in drinking water (WHO, 2004).

The reproducibility of the procedure, expressed as relative standard deviation of six replicate determinations of a water sample containing 5.0 and 30.0 μg l^{-1} of As(III) and total As, were 3.6 and 7.6% respectively. The regression parameters of both regression lines are tabulated in Table 5.

Parameter	As(III)	Total As
Square root of residual variance, s_e	0.60	0.63
Determination coefficient, R^2	0.99	0.99
Intercept confidence interval, $b_0 \pm t\ s(b_0)$	0.43±0.46	0.50±0.52
Slope confidence interval, $b_1 \pm t\ s(b_1)$	1.27±0.03	1.24±0.04
Linear range (μg l^{-1})	3-40	6-40
Limit of detection (μg l^{-1})	1	2

Table 5. Regression parameters of the calibration plots of peak current (in μA) vs. arsenic concentration (in μg l^{-1})

Technique	Electrode geometry	Working electrode	E_d (V), t_d (s)	LOD (μg l^{-1})	Reference
Continuous flow	Wall-jet	Gold	-0.30, 300	0.15	Kopanica & Novotny, 1998
Continuous flow	Thin layer	Gold film	-0.2, 80	0.5	Huang & Dasgupta, 1999
Continuous flow	Wall jet	Gold film	-0.65, 60	100	Billing et al. 2002
Continuous flow	Wall jet	Gold film	-0.1, 120	0.55	Muñoz & Palmero, 2004
Continuous flow	Thin layer	Gold nanoparticles	-0.4, 600	0.25	Majid et al. 2006
SIA	Tubular	Gold	-0.4, 40	1.0	This work

Table 6. Compilation of on-line stripping techniques for As(III) determination.

Listed in Table 6 there are some methods designed for on-line ASV determination of As(III). Sample introduction by continuous flow is the main strategy used for on-line ASV. This tendency is related with the effort to increase sensitivity by increasing the amount of sample used. Tubular electrodes have higher contact area; this characteristic, in combination with the use of binary sampling strategy, generates a robust and sensitive flow system which is competitive with other flow systems proposed, but using less sample volume and shorter deposition time.

The developed flow system was applied to the determination of the inorganic species of arsenic in groundwater samples originating from Tierra de Pinares (Segovia, Spain) as described in the Experimental section. Five replicate determinations of both total As and As(III) were carried out on each sample by the standard additions method. Results are displayed in Table 7. Total As was also determined in samples by electrothermal atomic absorption spectrometry (ETAAS) for comparison.

Sample	SI-ASV system			ETAAS
	As(III)	As(V)	total As	total As
1	12	24	36	30
2	37	16	53	57
3	18	19	37	39
4	41	96	137	127
5	21	17	38	38
6	44	120	164	174
7	14	4	18	24
8	116	104	220	218
9	75	85	160	164
10	42	84	126	129

Table 7. Contents (mean, n=5) of As(III), As(V) and total As, determined in groundwater samples by the proposed SI-ASV system, and comparison between total arsenic concentrations determined by SI-ASV and ETAAS. Concentration units, μg l^{-1}.

For each groundwater sample, average total As concentrations determined by both methods were compared by means of a paired t-test. Calculated t value was compared with the tabulated t value for 9 degrees of freedom and a significance level of p=0.05 (t=2.26). The calculated t value (1.00) is lower than the tabulated one, thus the null hypothesis that the methods do not give significantly different values for the mean total As concentration is accepted.

6. Conclusions

A sequential injection system with anodic stripping voltammetric detection for speciation of inorganic arsenic has been exploited. The reagents consumption is minimal. Limits of detection of 1 and 2 µg l^{-1} for As(III) and total As can be achieved. The tubular configuration of the working electrode, in combination with binary sampling strategy, was successfully applied to determine As(III) and total As in groundwater samples without any pre-treatment.

The LODs obtained for As(III) and total As were similar to those reported using continuous flow as sample introduction method. The SI-ASV system uses 0.4 ml of water sample for each determination, which is noticeably smaller than the sample amount reported using other flow methodologies without loss of sensitivity.

The proposed SI voltammetric system is an alternative for cost-effective higher degree of automation. The linearity and response achieved with the system makes it very suitable for arsenic measurements in natural waters with usual concentrations. Samples with higher or lower concentration can also be easily measured by varying the amount of sample aspirated and/or the deposition time.

7. Acknowledgments

The authors wish to thank the economical support provided by Consejería de Educación, Junta de Castilla y León (projects VA023A10-2 y GR170) and CONACyT (project 61310).

8. References

B'Hymer, C., Caruso, J.A. (2004). Arsenic and its speciation analysis using high-performance liquid chromatography and inductively coupled plasma mass spectrometry. *Journal of Chromatography A*. Vol. 1045, pp. 1-13

Bednar, A.J., Garbarino, J.R., Burkhardt, M.R., Ranville, J.F., Wildeman, T.R. (2004). Field and laboratory arsenic speciation methods and their application to natural water analysis. *Water Research*. Vol. 38, pp. 355-364

Billing, C., Groot, D.R., van Staden, J.F. (2002). Determination of arsenic in gold samples using matrix exchange differential pulse stripping voltammetry. *Analytica Chimica Acta*. Vol. 453, pp. 201-208

Bryce, D.W., Izquierdo, A., Luque de Castro, M.D. (1995). Flow injection anodic stripping voltammetry at a golg electrode for selenium (IV) determination. *Analytica Chimica Acta*. Vol. 308, pp. 96-101

Burguera, M. & Burguera, J.L. (1997). Analytical methodology for speciation of arsenic in environmental and biological samples. *Talanta*. Vol 44. pp. 1581-1604

Camacho, L.M., Gutierrez, M., Alarcón-Herrera, M.T., Villalba, M.L., Deng, S. (2011). Occurrence and treatment of arsenic in groudwater and soil in northern Mexico and southwestern USA. *Chemosphere,* Vol. 83, pp. 211-225

Catarino, R.I.L., Garcia, M.B.Q., Lima, J.L.F.C., Barrado, E., Vega, M. (2002) Relocation of a Tubular Voltammetric Detector for Standard Addition in FIA. *Electroanalysis.* Vol. 14, pp. 741-746

Economou, A. (2010). Recent developments in on-line electrochemical analysis-An overview of the last 12 years. *Analytica Chimica Acta.* Vol. 683, pp. 38-51

Environmental Protection Agency (EPA). (1999). *Analytical methods support document for arsenic in drinking water EPA -815-R-00-010.* Washington, DC. pp.57

Ferreira, M.A., Barros. A.A. (2002) Determination of As(III) and arsenic(V) in natural waters by cathodic stripping voltammetry at a hanging mercury drop electrode. *Analytica Chimica Acta,* Vol. 459, pp. 151-159

Forsberg, G., O'Laughlin, J.W., Megargle, R.G., Koirtyohann, S.R. (1975). Determination of arsenic by anodic stripping voltammetry and differential pulse anodic stripping voltammetry. *Analytical Chemistry.* Vol. 47, pp. 1586-1593

García-Sánchez, A., Moyano, A., Mayorga, P. (2005). High arsenic contents in groundwater of central Spain. *Environmental Geology,* Vol. 47, pp. 847-854.

Gibbon-Walsh, K., Salaun, P., van der Berg, C.M.G. Arsenic speciation in natural waters by cathodic stripping voltammetry. Analytica Chimica Acta. Vol. 662, pp 1-8

Gómez, J.J., Lillo, J., Sahun. B. (2006). Naturally ocurring arsenic in groundwater and identification of the geochemical surces in the Duero Cenozoic Basin, Spain. *Environmental Geology,* Vol. 50, pp. 1151-1170

Gong, Z., Lu, Z., Ma, M., Watt, C., Le, C. (2002). Arsenic speciation analysis. *Talanta.* Vol. 58, 99. 77-96

Guerin, T., Astruc, A., Astruc, M. (1999). Speciation of arsenic and selenium compounds by HPLC hyphenated to specific detectors: a review of the main separation techniques. *Talanta.* Vol. 50, pp. 1-24

Huang, H., Dasgupta, P.K. (1999). A Field-deployable instrument for the measurement and speciation of arsenic in potable water. *Analytica Chimica Acta.* Vol. 380, pp. 27-37

Ivandini,T.A., Sato, R., Makide, Y., Fujishima, A., Einaga, Y. (2006). Electrochemical detection of arsenic(III) using iridium-implanted boron-doped diamond electrodes. *Analytical Chemistry.* Vol. 78, pp. 6291-6298

Ivaska, A. & Kubiakb, W.W. (1997). Application of sequential injection analysis to anodic stripping voltammetry. *Talanta.* Vol. 44, pp. 713-723

Kopanica, M. & Novotny, L. (1998). Determination of traces of arsenic (III) by anodic stripping voltammetry in solutions, naturals waters and biological material. *Analytica Chimica Acta.* V ol. 38, pp. 211-218

Long, J. & Nagaosa, Y. (2008). Determination of trace arsenic(III) by differential-pulse anodic stripping voltammetry with in-situ plated bismuth-film electrode. *International Journal of Environmental Analytical Chemistry.* Vol. 88, pp. 51-60

Majid, E., Hrapovic, S., Liu, Y., Male, K.B., Luong, J.H.T. (2006). Electrochemical determination of arsenite using a gold nanoparticle modified glassy carbon electrode and flow analysis. *Analytical Chemistry.* Vol. 78, pp. 762-769

Massart, D.L., Vandegiste, B.M.G., Buydens, L.M.C., de Jong, S., Lewi, P.J., Smeyers-Verbeke, *Handbook of Chemometrics and Qualimetrics*: Part A, Elsevier, Amsterdam, 1997.

Melamed, D. (2005) Monitoring arsenic in the environment: a review of science and technologies with the potential for field measurements. *Analytica Chimica Acta*. Vol. 532, pp. 1-13

Mondal, P., Majumber, C.B., Mahanty, B. (2006). Laboratory based approches for arsenic remediation from contaminated water: recent developments. *Journal of Hazardous Materials*. Vol. B137, pp. 464-479

Muñoz, E., Palmero, S. (2004) Potentiometric stripping of arsenic(III) using a wall-jet flow cell and gold(III) solution as chemical reoxidant. *Electroanalysis*. Vol. 16, pp. 1956-1962

Muñoz, E., Palmero, S. (2005). Analysis and speciation of arsenic by stripping potentiometry: a review. *Talanta*. Vol. 65, pp. 613-620

National Research Council (NRC). (1999). *Arsenic in dinking water*. The National Academies Press, Washinton, DC, 330pp.

Ruzicka, J. & Hansen, E.H. (1988). *Flow Injection Analysis*. Nueva York: Wiley&Sons

Sadana, R.S. (1983). Determination of arsenic in the presence of copper by differential pulse cathodic stripping voltammetry at a hanging mercury drop electrode. *Analytical Chemistry*. Vol. 55, pp. 304-307

Smedley P.L. & Kinniburgh D.G. (2002). A review of the source, behaviour and distribution of arsenic in natural waters. *Applied Geochemistry*. Vol. 17, pp. 517-568

Sun, Y., Mierzwa, J., Yang, M. (1997) New method of gold-film electrode preparation for anodic stripping voltammetric determination of arsenic (III and V) in seawater. *Talanta*. Vol. 44, pp. 1379-1387

Tarley, C.R.T., Silveira, G., dos Santos, W.N.L., Matos, G.D., da Silva, E.G.P., Bezerra, M.A., Miro, M., Ferreira, S.L.C. (2009). Chemometric tools in electroanalytical chemistry: Methods for optimization based on factorial design and response surface methodology. *Microchemical Journal*. Vol. 92, pp. 58-67

Terlecka, E. (2005). Arsenic speciation analysis in water samples: a review of the hyphenated techniques. *Environmental Monitoring and Assessment*. Vol. 107, pp. 259-284

Trojanowicz, M. (2009). Recent developments in electrochemical flow detections-A review. Part I. Flow analysis and capillary electrophoresis. *Analytica Chimica Acta*. Vol. 653, pp- 36-58

Vega, M., Carretero, C., Fernandez, L., Pardo, R., Barrado, E., Deban, L. (2008). Hydrochemistry of groundwater from the Tierra de Pinares region (Douro basin, Spain) affected by high levels of arsenic. *Proceedings of Arsenic 2008, 2nd International Congress on arsenic in the environment: Arsenic from nature to humans*. pp. 83-84, Valencia, Spain

Williams, D.G. & Johnson, D.C. (1992). Pulsed voltammetric detection of arsenic(III) at platinum electrodes in acidic media. *Analytical Chemistry*. Vol. 64, pp. 1785-1789

World Health Organization (WHO). (2004). *Guidelines for drinking water quality recommendations*. Vol. 1, 3r ed., World Health Organization, Geneva, pp. 306-308.

10

Electrode Materials a Key Factor to Improve Soil Electroremediation

Erika Méndez[1], Erika Bustos[1], Rossy Feria[2], Guadalupe García[2] and Margarita Teutli[3]*

[1]*Laboratorio de Tratamiento de Suelo, Centro de Investigación y Desarrollo Tecnológico en Electroquímica, Sanfandila, Pedro Escobedo, Querétaro,*

[2]*Departamento de Química, Universidad de Guanajuato, Centro Guanajuato, Guanajuato,*

[3]*Facultad de Ingeniería, Benemérita Universidad Autónoma de Puebla, Ciudad Universitaria, Puebla, Pue;*

México

1. Introduction

Pollution of top and subsurface soil, as well as underneath groundwater has been one of the consequences of industrial activities; actually environmental professionals are facing a time consuming issue when they look for potential solutions for heavy metals and organic compounds removal. Although many soil remediation technologies are available, electrokinetic processing has been an emerging technology offering advantages for a wide variety of pollutants being either organic or inorganic; as well as its versatility of being applied in soil wetting conditions ranging from unsaturated to saturated; one of the main advantages of this technology is the fact that this process can be applied to low permeability soils, like clays.

Initially electrokinetics was applied for soil consolidation, in this process water flux is forced by an electrical field action, an approach to explain how it works is based on setting up a soil structural change analysis based on modification of soil matrix, plasticity index and crystalline state (Gray, 1970). For clayey soils it has been accepted that they behave like an osmotic membrane, therefore it is important to understand how physicochemical factors affect its response in regulating osmotic pressure into the soil matrix (Fritz, 1986). Another report (Darmawan, 2002) reinforce the necessity of knowing how the solid matrix response to the electric field, since obtained electrical current is function of electrolyte concentration, buffering capacity, and chemical form of involved metals, these can be in either soluble, electrostatically adsorbed, or surface complexed forms; also, metal migration is favored when soil is dominated by clay minerals, otherwise migration is lowered when soil has a high buffering capacity and/or high humic content, the last one acts like an additional resistance to the current transference throughout the soil.

Electrokinetics as a remediation experimental procedure requires having a wetted soil in which electrodes are inserted and terminals are connected to a power source. As soon as an electric field is generated, electrode reactions take place producing protons (H^+) at the anode and hydroxide (^{-}OH) at the cathode; concentration of these ions increases exponentially creating and acid front which moves from anode to cathode, and a basic front moving from cathode to anode; during its passage through the soil protons and hydroxides interact with sorbed pollutants releasing them into solution. Soluble ion transport occurs by three mechanisms: 1) Diffusion due to concentration gradients, 2) Convection due to fluid movement and 3) Migration due to the electric field.

A sample of initial published results (Acar et al, 1994; Hamed et al, 1991; Khan & Alam, 1994; Kim et al, 2002; Pamucku et al 1990; Pamucku & White, 1992; Reddy et al, 1999) is enough to claim that this method is highly efficient on restoration actions for clayey soils having very low heavy metal concentrations, for which regular mining procedures would result very expensive; although, for this method one of the minuses is the time required to get metal removals above 90%. Most of these studies report soil characterization providing information about: sand, clay, silt content; organic matter as well as hydraulic permeability. Although, few reports have covered soil electrical resistance, which according to Vázquez et al (2004), it could be used as a method for analyzing soil behavior in presence or absence of an electrical field.

In order to improve the process and get shortening of experimental times, applied efforts have covered a wide set of conditions. Some examples of reported research have addressed for modifying pH and current density (Hamed& Bhadra, 1997), chemical conditioning of electrode wells (Reed et al, 1995; Murillo-Rivera et al, 2009), cation inclusion (Colleta et al, 1997), as well as addition of complexing (Yeung et al, 1996) and lixiviant agents (Cox et al 1996); another approach has been the inclusion of reactive barriers into the soil matrix (Cundy & Hopkinson, 2005; Ruiz et al, 2011).

When treating PAHs (Polyaromatic Hydrocarbons) soil pollution it is important to care about lateral effects like lowering the electroosmotic flow rate (EOFR), which can be consequence of controlling pH at the electrode wells, by doing this an affectation of soil and/or solution chemistry can be induced producing an accumulation at the neutral or alkaline soil regions (Saicheck & Reddy, 2003). Another factor affecting EOFR derives from surfactant inclusion, then it becomes necessary to evaluate if it is worthy lowering EOFR for increasing PAHs removal (Kolosov et al, 2001). Also, pollutant mobilization should be evaluated as a function of pH control and surfactant addition, since flow can occurs in anodic direction (Ribeiro et al, 2005), or be enhanced in the cathodic direction.

Finally, organics and metal removal can be affected by geometry cell and flow direction, in this sense a report about an upward electrokinetic soil remediation (Wang et al, 2007) points out that removal efficiency is increased for organics when electroremediation cell is smaller in diameter, or larger in height; otherwise for metals, removal is improved when the cell is smaller in diameter or shorter in height.

Electrode materials are a key parameter to assure that electron transference takes place at fast rates. Selection of materials should be based on thermodynamic and kinetic response, so interface molecular interactions are fast enough to release the oxidant species. Also, it must be considered aspects like mechanical, thermic and corrosion resistance; as well as the procedures and solutions used in surface cleaning, pretreatment, and surface activation. In this sense, materials which satisfy these requirements are: vitreous carbon, titanium, stainless steel, platinum, gold and silver; but it must not be discarded that all materials can reduce its activity as experimental conditions favor metal deposition and so far electrode passivation. Then, in order to increase active sites number, reduce passivation and increase useful lifetime; some efforts had addressed electrode modification using oxide materials such as Carbon|TiO_2, Ti|SnO_2-Sb, Ti|IrO_2-Ta_2O_5, Ti|RuO_2. Electrodes prepared in this way have been named Dimensionally Stable Anodes (DSA) and they have proven be effective in organic degradation (Comminellis, 1994), because in addition to their high capacity to generate hydroxyl radicals, they also are mechanically resistant to abrupt pH changes.

Titanium dioxide exists in three crystalline forms namely anatase, rutile, and brookite, from which the first is the one exhibiting higher catalytic activity; which is highly dependent on the surface specific area of the particulate form, a phenomena which is exemplified by Baiju et al (2009), in their work they used anatase-TiO_2 in dye removal; these authors provide evidence on how switching from a particulate form to nanotubes it results in a concurrent change in the mechanism of dye removal from an aqueous solution. Although, other authors (Yigit & Inan, 2009) have provided evidence on how a mixture anatase-rutile provides a higher efficiency in humic acid mineralization.

In wastewater treatment use of Dimensionally Stable Anodes (DSA) made of a Ti mesh covered with a film of iridium or tin has probed being an effective tool for organics degradation while keeping its mechanical resistance (León et al, 2009); also, there is a theory proposed by Comminellis (1991) in which it is shown that the iridium DSA posses a highly reactive surface which directly oxidizes the substrate, at this respect an improvement in oxidant activity is obtained when titanium mesh is covered with a film including a mixture of oxides like Ir-Ta (Hu et al, 2002). Use of modified electrodes favors electrocatalytic processes by supporting higher current densities and so far reducing the corresponding oxidation potentials, because inner sphere mechanisms allow for hydroxyl radicals generation at an interfacial level, by which there it is possible to increase organic molecules decomposition reaching total mineralization levels. About the electrode matrix it can be affirmed that DSA usually exhibit higher mechanical resistance because they are prepared onto a rigid matrix, while the ones based on Reticulated Vitreous Carbon (RVC) have higher reactive surface but a lower mechanical resistance.

Considering that few attention have been dedicated to the role that electrode material plays on soil electroremediation, since a higher electrode activity would allow developing conditions which could enhance pollutant removal, specially when the pollutant is an hydrocarbon product. In this sense, this contribution presents results from two research approaches in which it is tested how inclusion of a catalytic specie, like anatase (TiO_2) in a reticulated vitreous carbon (RVC) or a titanium anode covered with IrO_2-Ta_2O_5 film will help to improve occurrence of water electrolysis reactions; also, each experimental approach allowed to elucidate the role played by materials used as anode and cathode, as well as the influence exerted by additional resistance factors like electrode position in respect to the soil interphase. Finally, it is an opportunity to test the theory by which it is expected an enhancement of electrokinetic hydrocarbon removal as consequence of the higher electrode activity.

Experimental work is organized as follows: 1) Electrodes made of Reticulated Vitreous Carbon (RVC), in which anode was modified by inclusion of an anatase deposit (TiO_2), and keeping constant a RVC cathode, this provides two combinations bare RVC electrodes and RVC-TiO_2 anode with bare RVC cathode; 2) Anode made of a titanium plate covered with

IrO_2-Ta_2O_5, and two cathode materials carbon felt (CF) and titanium plate (Ti), plus a variation of its position respect to the soil matrix. These experiments are described in the following paragraphs.

2. Methodology

Results correspond to two independent experimental sets, so far methodology for each one will be discussed in an independent section.

2.1 Anode modification

For this set of experiments soil was collected at 50 cm depth in an undisturbed site of Guanajuato, México. Applying the ASTM D4318-10 methodology soil was characterized for liquid (LL) and plastic (PL) limits, the water content difference between LL and PL provides the plasticity index (PI), which is an indicator of soil response, because greater PI values correspond to a soil which is more plastic and compressible and so far it exhibits greater volume changes using LL and PI values on the plasticity chart (Helwany, 2007), can be obtained a fast classification of soil type, providing an insight on the possible soil response, before any electrical perturbation is applied. Also, textural classification was done following the USCS-P13-B-2 procedure. This soil was artificially polluted by mixing it with a phenanthrene solution, let stand overnight, and room temperature dried. Later on, it was characterized for sorbed phenanthrene, which resulted in 12 mg Kg^{-1}. For experiments polluted soil was rewetted with deionized water.

This set of experiments considered electrodes made from 100 ppi (pores per inch) reticulated vitreous carbon (RVC), impregnated with TiO_2 by a sol-gel method. Impregnating solution was prepared dissolving metallic Ti in concentrated HCl; later on it was precipitated with NH_4OH, obtained product was filtered and washed before dispersing it into a 10% ethanol solution; in this solution RVC pieces were immersed for 24 hours; after that, they were dried and subjected to a 3 hours calcination, this procedure pursues formation of the anatase phase in the deposit; confirmation of anatase presence was done by Raman spectroscopy and some micrographs were obtained with a Leica S8APO stereomicroscope. The electrochemical cell was a rectangular one (14 cm length, 10 cm width, 9 cm high).

Electrokinetic experiments were run for 24 hours, a 25 mA cm^{-2} electrical current density was imposed with an electrophoresis power supply FR500-125 BIOELEC. Electrode arrangement considered keeping constant a bare RVC cathode, and switching the anode from bare RVC to RVC-TiO_2, recorded parameters correspond to: pH and electric conductivity measured with a Multipurpose Lab Interphase Vernier Software; electroosmotic flow was registered with an illuminated multitester MUL-270 Steren.

In order to quantify residual phenanthrene concentrations, at the end of each experiment, soil sample was cut in slices, room temperature dried, and phenanthrene was Soxhlet extracted with ethylic ether. After that, the solvent was evaporated and phenanthrene was solubilized in 5 mL of HPLC grade acetonitrile, this sample was centrifuged and injected into an inverse phase Hypersil chromatographic column C-18 ODS (100 mmx4.6 mm, and 3 mm particle size); detection was done using a mobile phase made of a mixture $CH_3CN/H_2O/CH_3OH$ in proportions 30:15:55 % v/v; using a flow rate of 0.4 mL min^{-1} and 254 nm detection wavelength.

2.2 Cathode modification

Soil sample corresponds to a hydrocarbon polluted weathered soil, for which a physical characterization is done by using the ASTM D4318 as well as the USCS-P13-B-2 procedure, by which it was established soil type and textural composition. Since this soil has hydrocarbon pollutants, its concentration was determined as oil and grease by the Soxhlet technique. According to published results (Murillo-Rivera et al, 2009), 0.1M NaOH solution is an electrolyte that works fine for hydrocarbon polluted soils, so this one was chosen as the electrolyte for soil wetting, and wells replenishment.

Experimental cell was a rectangular one (10 cm length, 2 cm width, 4 cm high), a current density of 20 mA cm^{-2} was imposed with a PDC-GP 4303DU Power Source, in a galvanostatic mode during 4 hours.

Considering that a DSA anode provides oxidant species at higher rates, then for this hydrocarbon polluted soil electroremediation it was chosen an electrode arrangement, considering a modified DSA made of a titanium plate covered with an iridium-tantalum film ($Ti|IrO_2$-Ta_2O_5), and two types of cathode: carbon felt (CF) and a titanium plate (Ti). Also, in these set of experiments it was considered two electrode positions: in the first one, a physical barrier of filter paper was included between soil and electrode, while in the second one the electrode was set in direct contact with soil sample.

Registered experimental parameters were: applied electrical current, developed electrical potential, with these it was possible to calculate cell resistance and energy consumption. At the end of each experiment soil was cut in 3 sections (anode, middle, cathode), hydrocarbon removal was estimated from soil residual concentrations, which were extracted by the Soxhlet technique, and later analyzed by UV-Vis (XLS, Perkin-Elmer) gas chromatography coupled to mass spectroscopy (CG-EM, Agilent GC 19091-413).

3. Results

3.1 Anode modification

Soil characterization results are reported in Table 1. As it can be observed this soil is classified as a Low plasticity Clay (CL), then it will no exhibit a great volume change during experimentation; also, clay and silt content indicate that this is a low permeability soil.

Parameter	Value	Methodology
Liquid Limit (LL) %	36	ASTM D4318-10
Plastic Limit (PL) %	24	ASTM D4318-10
Plasticity Index (PI) %	12	ASTM D4318-10
Classification	Low plasticity clay (CL)	Plasticity Chart (Helwany, 2007, page 13)
Sand %	34	USCS-P13-B-2
Silt %	53	USCS-P13-B-2
Clay %	13	USCS-P13-B-2

Table 1. Physical and textural properties of Guanajuato soil.

Based on the above described soil characteristics, it was considered important to determine how this type of soil responses to the action of an electric field. To accomplish this step, clean soil was wetted with deionized water, and later on tested with the bare RVC electrodes. Measurements of pH were done at the electrode interface (anode, cathode) and two middle points (4 and 7 cm). Experimental pH profiles are presented in Figure 1. As it can be observed, natural pH is slightly alkaline (pH=8); also, even though protons are generated at the anode, its penetration is slow, and their amount is not enough to get a high pH depletion at this position; otherwise intermediate points (4 and 7 cm) show an alkalinization since its values are increased by up to 2 units; also, at the cathodic position, pH response satisfies expectations of high alkaline values, since final pH is closer to 13. Considering that alkaline pH favors organic pollutants removal (Murillo-Rivera et al, 2009), then this soil is considered adequate to evaluate hydrocarbon removal.

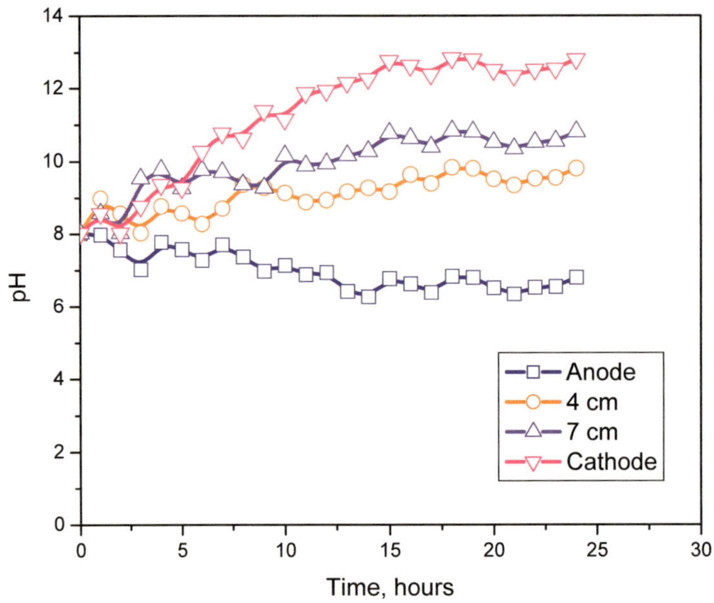

Fig. 1. Graph of pH evolution at anodic and cathodic wells during the electrokinetic experiments for clean soil, applying a current density of 25 mA cm^{-2}, and using bare RVC anode and cathode electrodes.

Artificial polluted sample was prepared as it is described in the methodology section. Next step is to use this sample with different anode materials, and establish if these are useful on improving electrokinetic process performance for removal of phenanthrene from polluted soil. In Figure 2 it is shown the pH evolution at the anodic and cathodic wells, for a 24 hours soil electroremediation experiment. It can be observed that using bare RVC electrodes (Anode I and Cathode I) makes pH at the anode be slightly depleted during the first hours, but later on occurs an increase of its value, which keeps it around 8 (the initial value) during the rest of the experiment; otherwise, at the cathodic well a fast alkalinization is observed, this remains around 10.5 during the whole experiment, but this value is lower than the one obtained with clean soil.

Otherwise, anode replacement by the RVC-TiO_2 option (Anode II, Cathode II), it enhances proton production, so in short time pH goes to acidic values, and after 5 hours it stabilizes around 5. Keeping the cathode as bare RVC makes that the cathodic well pH go to slightly higher alkaline values, than those registered when the anode was bare RVC; but still pH values are lower than the ones obtained with the clean soil. It seems that pollutant inclusion makes a more resistive system by which hydroxyl production is lowered respect of the rates achieved with clean soil.

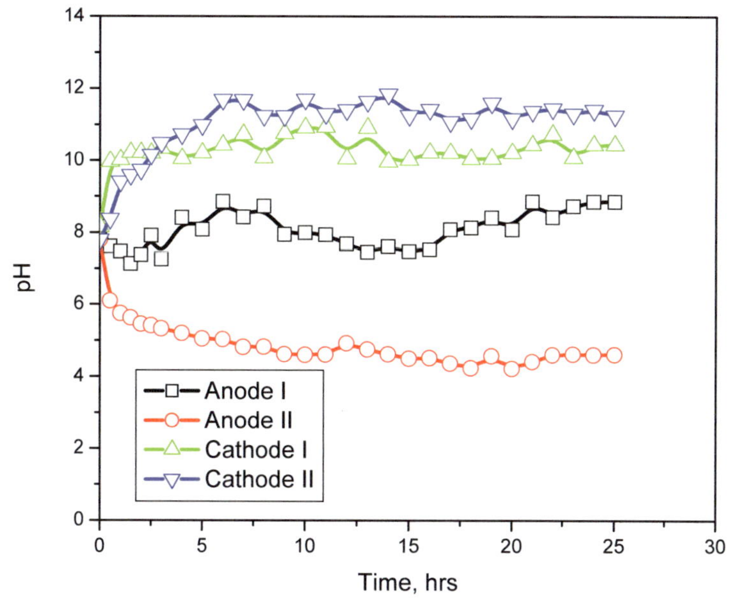

Fig. 2. Graph of pH evolution at anodic and cathodic wells during the electrokinetic experiments for clay soil contaminated with 12 mg Kg^{-1} of phenanthrene, applying a current density of 25 mA cm^{-2}, and using different electrode materials: (I) Bare RVC anode and cathode; and (II) RVC-TiO_2 anode with RVC cathode.

For these experiments drained volume was collected, results are presented in Figure 3. As it can be observed keeping a bare RVC cathode, and switching from (I) bare RVC to (II) RVC-TiO_2 anodes, exhibit similar water transport during the first 3 hours; but after that water transport is increased for the RVC-TiO_2 anode; therefore this modification it allows enhancing the amount of water being displaced from anode to cathode. After 10 hours, water transport reach and steady rate of transport, corresponding regression lines are described as follows: for the RVC (system I) y= 72.31x-54.995, R^2=0.995, while the RVC-TiO_2 (system II) is described by y= 88.15x-2.0373, R^2= 0.999.

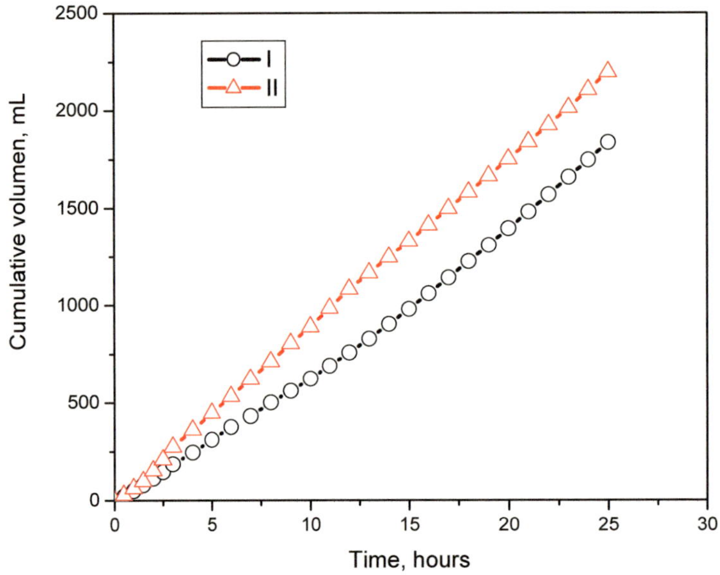

Fig. 3. Graph of collected cumulative volume at the cathodic well during electrokinetic experiments for clay soil contaminated with 12 mg Kg^{-1} of phenanthrene, applying a current density of 25 mA cm^{-2}, having a bare RVC cathode and using different anode materials: (I) Bare RVC anode; and (II) RVC-TiO_2 anode.

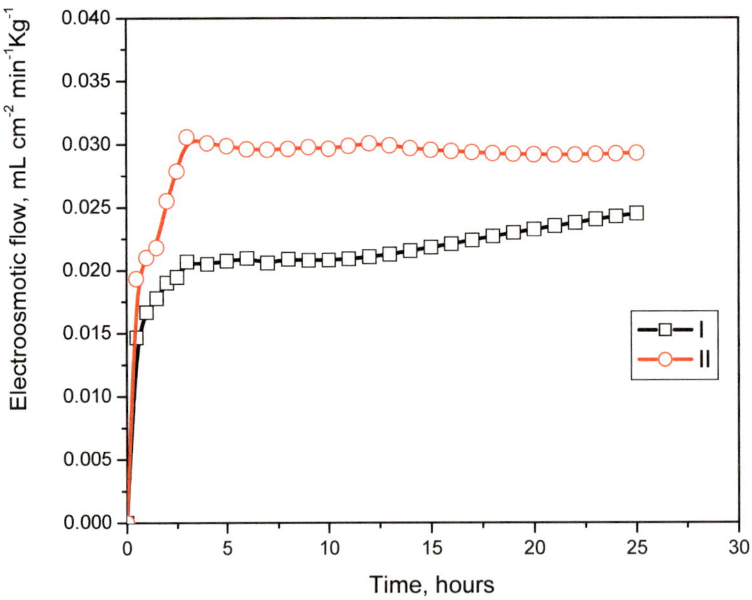

Fig. 4. Comparison of cumulative electroosmotic flow registered at the cathodic well during the electrokinetic experiments for clay soil contaminated with phenanthrene using different anode materials: (I) Bare RVC anode and cathode; and (II) RVC-TiO_2 anode with RVC cathode.

Collected drained volume values, cell characteristics and soil mass were used to mathematically obtain cumulative electroosmotic flow (mL cm^{-2} min^{-1} Kg^{-1}), results are shown in Figure 4. From the plot can be established that effectively catalytic activity of TiO_2 allows for getting higher electroosmotic flow. As it can be observed in three hours the RVC-TiO_2 anode (II) reached steady response, while the bare RVC anode (I) provides a much lower electroosmotic flow which seems to smoothly reach steady response at similar times, but after 10 hours, a new perturbation takes place and it goes to a transient response, the last taking place at an slower rate in respect to the initial one.

As it was mentioned in the methodology section, electroremediated soil sample was cut in slices and recovered residual phenantrene was injected in an inverse phase Hypersil chromatographic column. Elution time for phenanthrene was 5.8 min, while humic and fulvic acids appear at about 2 min of elution time (Chongsan et al, 2006; Xing & Kang, 2005; Yanzheng et al, 2007). In Figure 5 it is shown a chromatogram for the soil extracted phenanthrene before any electrokinetic experiment, this reference signal is about 1 arbitrary unit (A.U.).

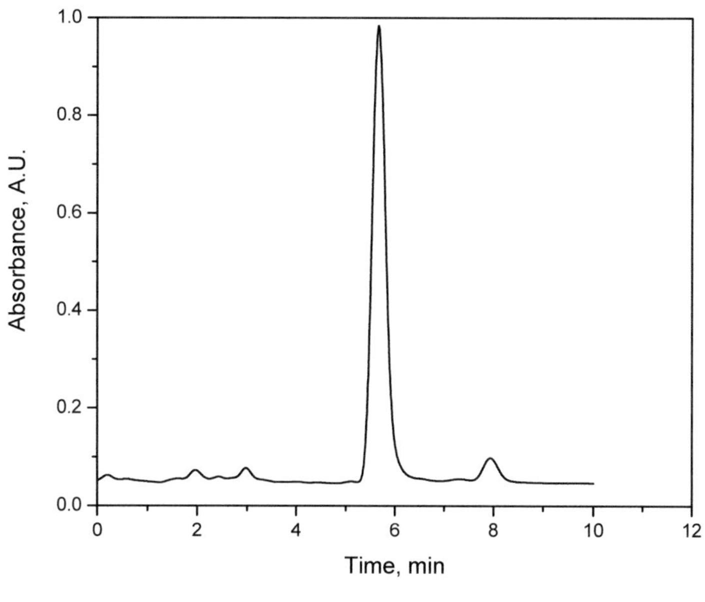

Fig. 5. Chromatogram of the Phenanthrene standard, reference signal obtained from the artificially polluted soil.

Obtained chromatograms for each soil section, after the electrokinetic experiment with bare RVC anode and cathode are shown in Figure 6, and those for the RVC-TiO_2 anode with bare RVC cathode are shown in Figure 7.

As it can be observed in Figure 6, none of the positions amount concentrations higher than 0.25 A. U., also there are several peaks between the humic and fulvic acids (2 min) and the phenanthrene (5.8 min), these peaks are smaller than other signals, and they can be associated to a phenanthrene decomposition by products from lateral reactions, which take place as electrolyte moves through the soil during the electrokinetic experiment; the higher residual phenanthrene concentration for bare RVC electrodes was about 0.25 A.U. and it occurs at the 0.7 dimensionless position, that is the section before to the one closer to the cathode.

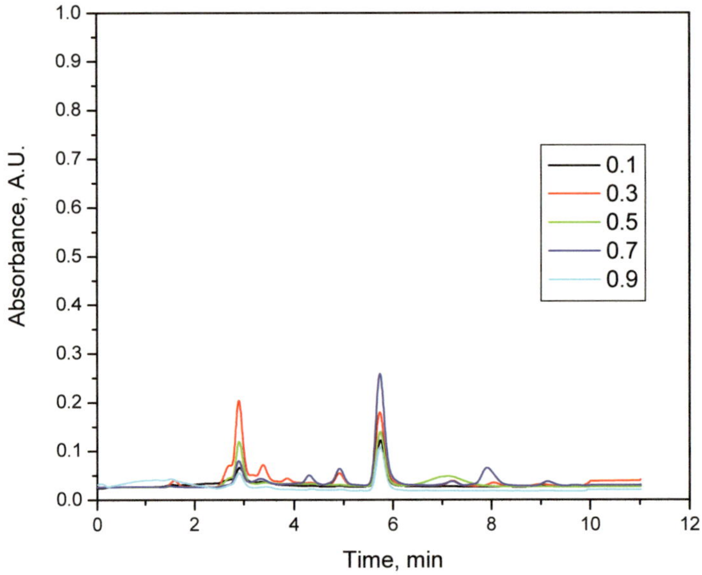

Fig. 6. Chromatograms of residual phenanthrene concentrations in each soil slice for the system of bare RVC electrodes.

Otherwise, when the experiment was run with the RVC-TiO_2 anode (Figure 7) the presence of smaller peaks it is practically null; also, an opposite phenomena is observed since in this case the higher residual concentration was about 0.9 A.U., while the lower one is not less than 0.6 A. U., this last takes place at the 0.3 cm position (near the anode), In general, with this option phenanthrene removal was lower than the one attained with the bare RVC electrodes.

In order to make more explicit the above expressed, concentration was calculated from each soil slice chromatogram, this was done by an integration of the area under phenanthrene peak; in this way, its residual concentration was estimated. Results are reported in Figure 8 as percentage of the original concentration in soil (12 mg Kg^{-1}=100%), it can be observed that effectively higher removal was obtained with the bare RVC electrodes, an average of 80%; and even though replacing the anode by the RVC-TiO_2 provides higher oxidation conditions and a faster water transport; this fact does not allowed for getting a right residence time for solubilizing and transporting phenanthrene, since removal amounts an average of 20%. It is noticeable that for the bare RVC electrodes, higher residual phenanthrene concentration took place at the same position where it is the lower one when the anode was RVC-TiO_2.

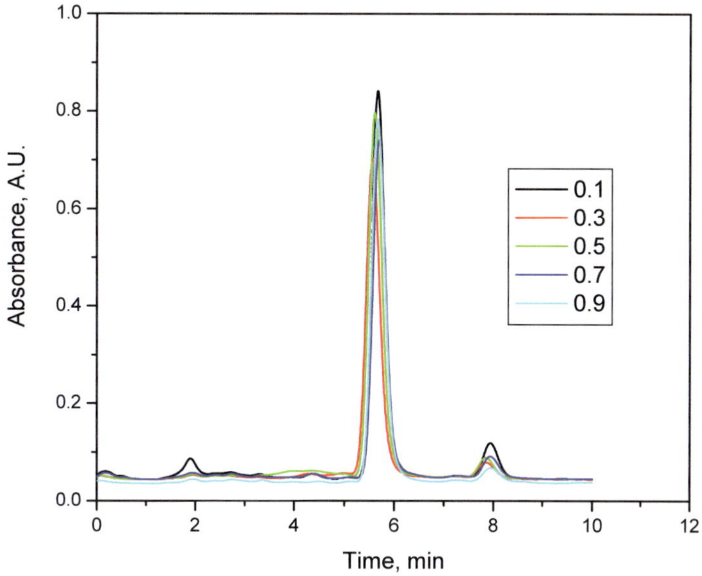

Fig. 7. Chromatograms of residual phenanthrene concentrations in each soil slice for the system of TiO_2-RVC anode and bare RVC cathode.

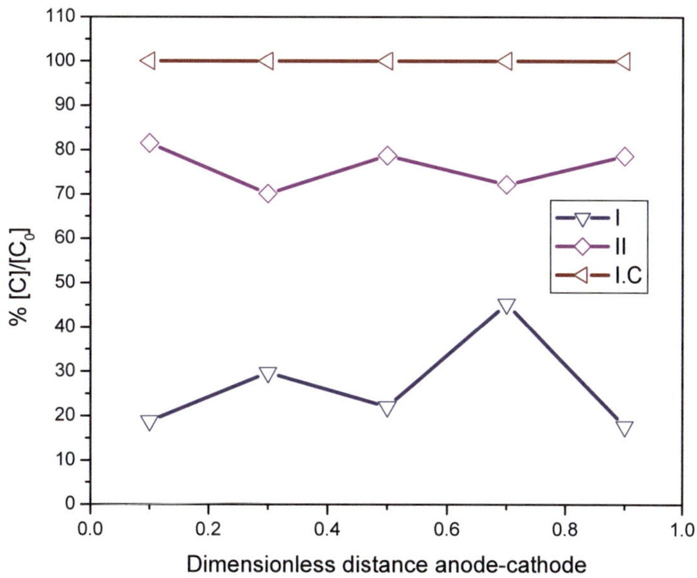

Fig. 8. Comparison of residual phenanthrene concentrations in each soil slice for the system of (I) bare RVC electrodes, and the (II) RVC-TiO_2 anode, RVC cathode, versus initial concentration (IC).

Correlating these results with corresponding pH data it can be affirmed that keeping a shorter difference in pH wells, as it happens with bare RVC electrodes, it favors a soil alkaline condition which, even though produces a slower liquid movement, so far this is good enough for phenanthrene removal since it provides a higher residence time.

Up to here, it was shown that inclusion of one catalytic specie, like anatase, in the anode allowed increasing the oxidant specie production and the electroosmotic flow rate, but obtained phenanthrene removal was lowered. So next step is to analyze what happen if the catalytic activity is maintained at the anode, but cathode is chosen between different materials. Experimental set-up objective was data collection for two different cathode materials, and also to clarify how much the system becomes affected by inclusion of additional physical barriers like a thick filter paper.

3.2 Cathode modification

For this set of experiments soil sample was collected at an industrial area located in Nuevo Teapa, Veracruz, México, this is a highly polluted and weathered area. Physical characterization is reported in Table 2.

Parameter	Value	Methodology
Liquid Limit (LL) %	42	ASTM D4318-10
Plastic Limit (PL) %	28	ASTM D4318-10
Plasticity Index (PI) %	14	ASTM D4318-10
Classification	Medium plasticity clay (CI)	Plasticity Chart (Helwany, 2007, page 13)
Sand %	56	USCS-P13-B-2
Silt %	24	USCS-P13-B-2
Clay %	20	USCS-P13-B-2

Table 2. Physical and textural properties of Nuevo Teapa soil

For this set of experiments it was chosen a dimensionally stable anode (DSA) made of a titanium plate with an iridium-tantalum film ($Ti|IrO_2\text{-}Ta_2O_5$) which was maintained constant; considered cathode materials were: carbon felt (CF) and a titanium plate (Ti). Physical barrier inclusion was evaluated considering two electrode positions: Array I: placing the physical barrier at the soil interphase and the electrode after it; Array II: electrode placed at the soil interphase.

An initial test of soil response was done with the combination DSA-Ti, pH profiles for the option including a physical barrier between electrode and soil (Array I) are presented in Figure 9. As it can be observed soil tends to be acidified, and after 4 hours it is reached a stable pH condition which at the anode is about 2 units lower, and at the cathode is one unit lower, in respect to the initial value. It seems that barrier favors a buffering effect by which the system does not get a drastic pH drop. Although, having a pH variation is evidence of getting a high rate for proton generation and transport throughout soil. It is important to point out that after 3 hours, pH at the middle and cathode sections have similar values, fact which reflects a strong neutralization of transported protons.

Results for Array II are presented in Figure 10, pH profiles for the combination DSA-Ti without barrier clearly show a fast pH drop, and after 1 hour, at the middle section seems to occur an hydroxide accumulation, which could be a factor to accelerate proton penetration, such that at two hours the anode section starts to lowering its pH, and even though it does not reach an acidic condition pH drop is about 5 units at this section, but this pH does not exert a strong impact over the other sections; since it seems that protons penetration displaced hydroxyls to the middle and cathode sections, which suffer a temporary raise in concentration. This behavior corresponds to a pulsed function being displaced from the middle to the cathode section, since when pH starts to decay in the middle, the cathode one start to raise its pH, which it gets a higher value than the initial one. At the final time, the middle section has decreased its pH in 1 unit, while the cathode section has reached stability at pH 12.

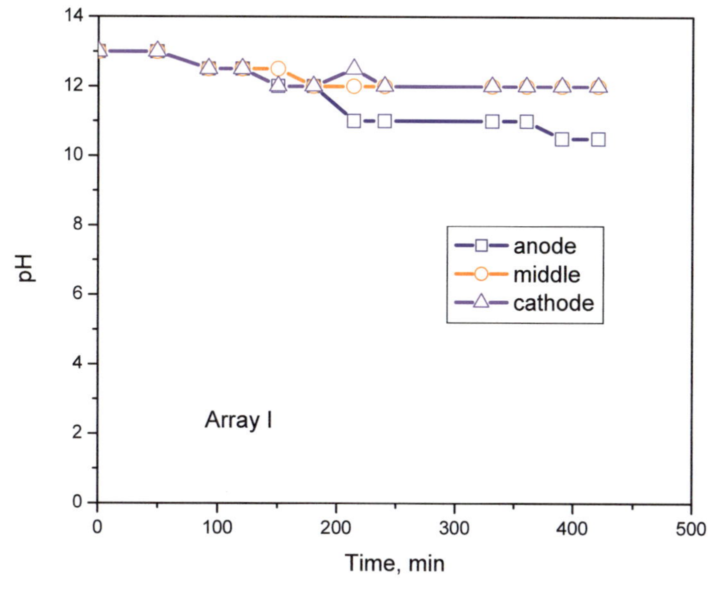

Fig. 9. Graph of pH profiles for array I of the DSA anode and Ti cathode, experimental conditions: 0.1 M NaOH wetting electrolyte, current density 20 mA cm^{-2}.

Experimental approach considered a follow up through the global electrical resistance (R, Ohms), which was indirectly calculated from experimental values of electrical potential (E, Volts) and applied current (I, Amperes), parameters related by Ohm's law ($R=E\ I^{-1}$). Results are shown in Figure 11.

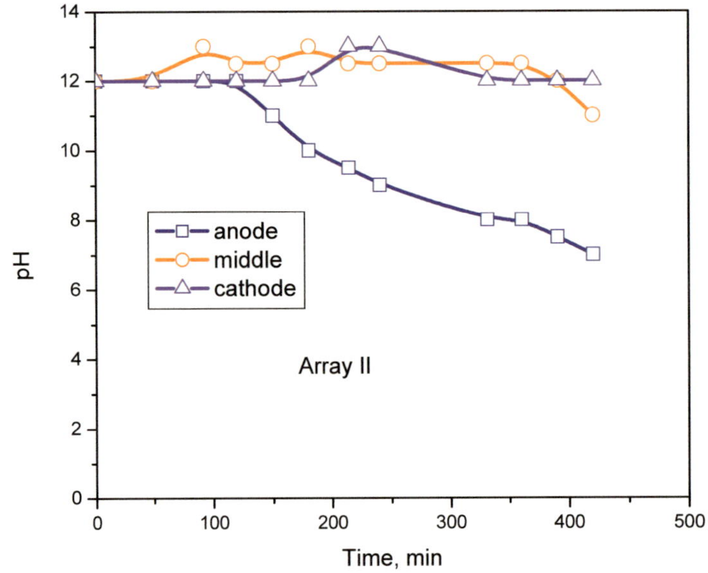

Fig. 10. Graph of pH profiles for array II of the DSA anode and Ti cathode, experimental conditions: 0.1 M NaOH wetting electrolyte, current density 20 mA cm^{-2}.

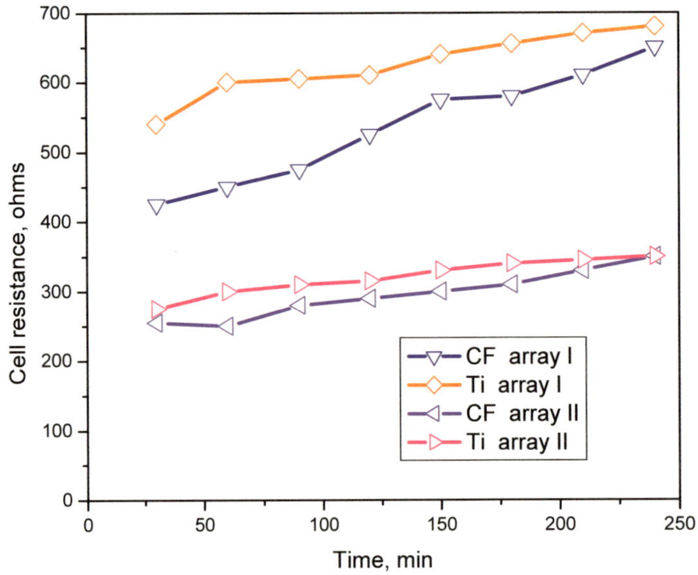

Fig. 11. Calculated resistance for soil electroremediation experiments using a modified IrO_2-Ta_2O_5 anode with either Carbon Felt (CF) or Titanium (Ti) cathode. Array I physical barrier inclusion, Array II soil contact experimental conditions: 0.1 M NaOH wetting electrolyte, current density 20 mA cm^{-2}.

Observing Figure 11 it is evident that allowing electrodes make contact with the soil (array II) provides a less resistive system; and also there is not a clear advantage between using CF or Ti as cathode, since both systems provide similar initial values at 30 min, R=255 ohms for CF, and R=275 ohms for Ti; final resistance values are identical for both systems R=350 ohms. Although, CF exhibit a slightly higher raise (95 ohms) in respect to Ti (75 ohms). Applying a linear regression analysis to CF data, its slope corresponds to 0.462 ohm-min^{-1}, while the Ti slope is 0.339 ohm-min^{-1} then, even though initial resistance value is higher for the Ti electrode, this electrode provides a more stable system.

As it can be observed inclusion of the physical barrier makes the experimental system to be more resistive than that where electrodes make contact with the soil, also resistance trends in this system are similar to those of the previous one; since at initial times system with a CF cathode seems to be less resistive at 30 min, R=425 ohms, than that with a Ti cathode at 30 min, R=540 ohms, but at the end of the experiment (240 minutes), the resistance of the system with CF was increased by about 185 ohms, while the one with Ti by only 140 ohms. Applying a regression analysis the slope for CF is 1.087 ohm-min^{-1}, while the one with Ti has a slope of 0.599 ohm-min^{-1}; these values are higher than the ones observed when the electrode is placed at the soil interphase; but again it is confirmed that using Ti as cathode provides a more stable system. The failure of the CF electrode can be attributed to a poisoning effect since there is a possibility that desorbed hydrocarbons get retained at the cathode.

At the end of each experiment, residual hydrocarbon content in soil was estimated by a Soxhlet technique at 3 points: 0.25, 0.5, 0.75 anode to cathode dimensionless distance; concentration values are normalized respect to the initial concentration condition an presented as percentage. Results are shown in Figure 12 for carbon felt (CF) cathode, and in Figure 13 for titanium (Ti) cathode.

As it can be observed from both Figures (12 and 13) switching the cathode position provides opposite trends in residual hydrocarbon concentrations, since when the physical barrier is between soil and electrode, residual concentration goes from higher to lower in the anode-cathode direction, it seems that transported hydrocarbons are no allowed to accumulate near the cathode; also, CF cathode provides the best conditions for hydrocarbon transport since in the anode-cathode direction hydrocarbon removal goes from 36% to 65% (27% difference), while with the Ti cathode goes from 30 to 42% (12% difference).

In opposite way, when electrodes are at the soil interphase it happens that residual hydrocarbon concentrations increase from anode to cathode, and the higher ones are registered near the cathode; in this case again CF cathode provides the best removal since hydrocarbon removal goes from 60% to 40% (20 % difference), while the Ti cathode removal goes from 40% to 30% (10 % difference).

Based on the cell resistance results, there is an assumption about carbon felt being passivated due to adsorption of those transported hydrocarbon molecules. In order to assess which type of hydrocarbons migrated, and accumulated at the cathode, CF cathodes were washed with a dichloromethane solution, and eluted samples were used for PAHs estimation.

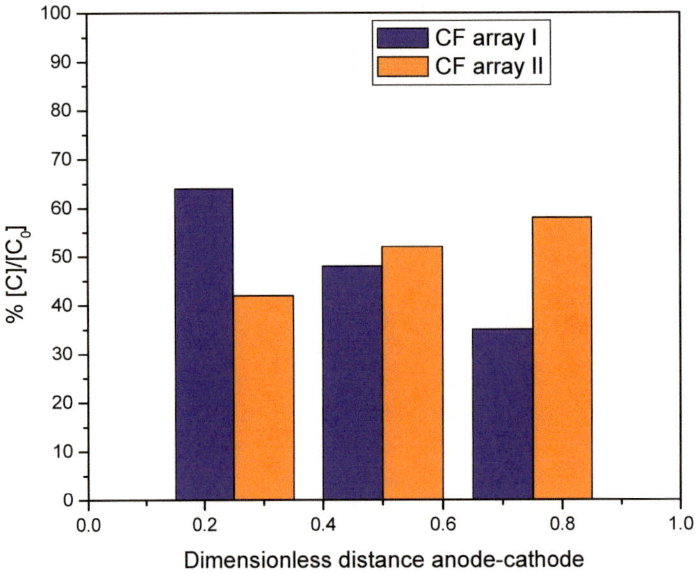

Fig. 12. Graph of normalized residual concentrations for carbon felt (CF) cathode. Array I physical barrier inclusion, Array II soil contact, experimental conditions: 0.1 M NaOH wetting electrolyte, current density 20 mA cm^{-2}.

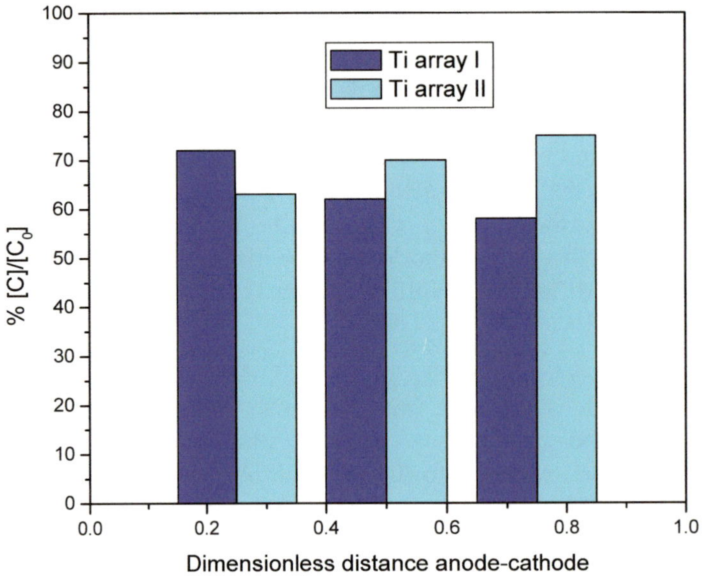

Fig. 13. Graph of normalized residual concentrations for Titanium (Ti) cathode. Array I physical barrier inclusion, Array II soil contact, experimental conditions: 0.1 M NaOH wetting electrolyte, current density 20 mA cm^{-2}.

In order to assess risk exposition levels for oil exploration and production sites, it takes relevance to detect the EPA's 16 priority Polycyclic Aromatic Hydrocarbons (PAHs) (Bojes & Pope, 2007). Analytical techniques that can be applied to PAHs detection consider HPLC coupled with UV-Vis detection, with this technique the 16 priority PAHs can be detected at wavelengths between 227 and 297 nm (Maureen, 2011); another useful technique is gas chromatography (GC) coupled with mass spectroscopy (MS) (Amzad Hossain & Salehuddin, 2011).

Based on this information, a first approach to PAHs detection was done with UV-Vis, in Figure 14 are shown obtained results for CF cathode in array I (physical barrier included), and array II (soil contact), from these spectra it is obvious that physical barrier presence has enhanced PAHs partition at the interphase soil-water, so keeps hydrocarbon accumulation low in the region nearby; also, the electrode behind the physical barrier has acted like a sink for transported PAHs, so far it is logical to get lower concentrations in the position near the cathode. Otherwise, having the cathode in contact with soil makes PAHs partition to occur at slower rate since PAHs face the hydroxide production at the soil boundary.

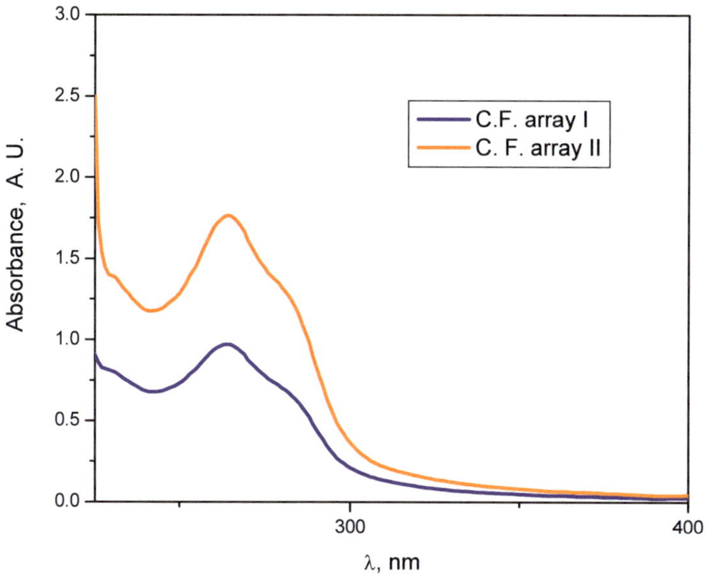

Fig. 14. UV-Vis spectra of sorbed hydrocarbons onto carbon felt cathode in Array I and Array II. Experimental conditions 0.1 M NaOH, current density 20 mA cm^{-2}, experimental time 4 hours.

Otherwise, observing the spectra, it can be notice that there is not a clear and unique peak; which means there is a possibility of having more than one PAH in the desorbed material. Therefore, a more refined technique should be used for PAHs detection and quantification, requirements widely covered by GC coupled to MS; analytical detection was limited to three of the 16 PAHs in EPA's priority list, the ones chosen were one having three aromatic rings (phenanthrene) and two with 4 rings (fluoranthene and pyrene).

In order to asses soil electroremediation efficiency in removal of phenanthrene, fluoranthene and pyrene, GC-MS was applied to the soxhlet extracted samples including both original and electroremediated soil. Concentration values were calculated from the area under the curve, and these were converted to percentage taking as reference concentration the one for each PAH registered in the extract from the original weathered soil. Results for the Ti cathode are shown in Figure 15 for the array I (physical barrier inclusion), and Figure 16 for array II (soil contact).

As it can be observed in Figure 15, (array with the physical barrier) there is a higher to lower trend from anode to cathode for the three PAHs which were analyzed, and removals are low; it seems that molecule size exerts an influence on their movement through the soil, since the 3 rings molecule (phenanthrene) has reached removals between 80 and 90%, while those with 4 rings (fluoranthene, pyrene) get similar removals between 60 and 85%. Otherwise, in Figure 16 it can be observed that allowing the electrode to make contact with soil enhances PAHs removal, getting similar residual concentrations for all, in this experiment removals are above 90%.

Correlating these residual concentrations with pH observations (Figures 9 and 10) it seems that the fact of having a physical barrier between soil and electrode, which produces a more resistive system, does not allow for getting a high concentration gradient between electrodes, resulting in lower removals than those obtained when the electrode make contact with the soil; since the last arrangement produces a higher pH gradient between anode and cathode, so far a higher driving force for PAHs transport.

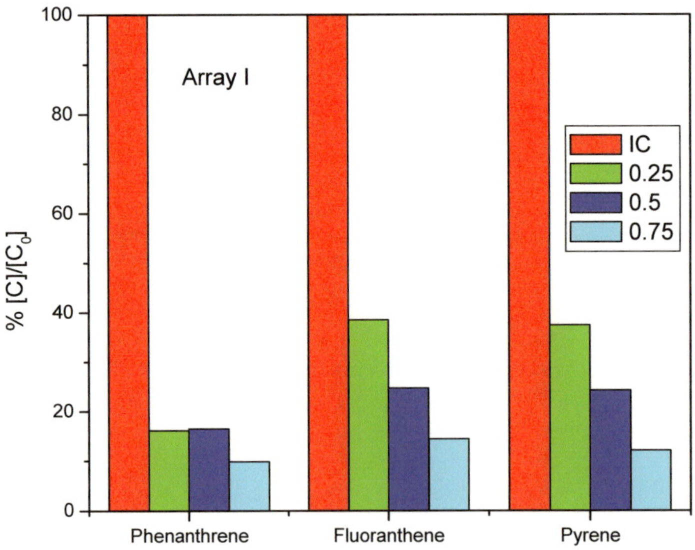

Fig. 15. Residual concentrations of representative PAHs in Array I physical barrier included, experimental conditions 0.1 M NaOH, current density 20 mA cm^{-2}, experimental time 4 hours.

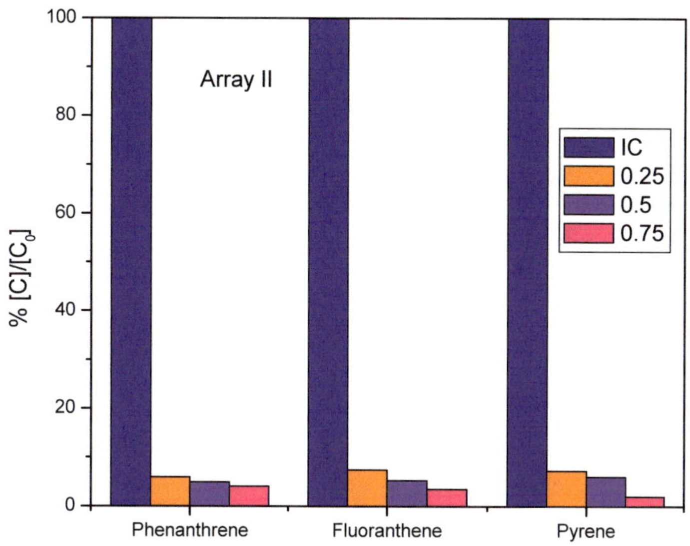

Fig. 16. Residual concentrations of representative PAHs in Array II, experimental conditions 0.1 M NaOH, current density 20 mA cm^{-2}, experimental time 4 hours.

4. Conclusions

For anode modification obtained results allows to claim that, effectively inclusion of anatase into the RVC matrix makes electrode reaction being more efficient. Also, by using the modified RVC-TiO_2 electrode it is possible to increase the rate at which protons are generated and transported throughout the soil, and so far this influences pH at both electrode wells: anodic and cathodic. Also, it provides a higher electroosmotic flow, but this fast water transport does not allow for an adequate residence time, lowering phenanthrene removal. So far, the bare RVC electrodes provided a lower pH gradient between anode-cathode, as well as a lower electroosmotic flow, both parameters are providing a better environment for phenanthrene removal, since with this option it was obtained up to 80% lowering in soil phenanthrene concentration.

For cathode modification obtained results have shown that cell resistance is lower when electrodes are in contact with soil sample, and this allowed for higher hydrocarbon mobility, so residual concentration profile exhibits an increasing trend from anode to cathode. Otherwise, physical barrier inclusion increased soil resistance and so far, hydrocarbon mobility is lowered, this fact resulted in a decreasing concentration trend from anode to cathode. From oil and grease extractions it was determined that CF provides higher hydrocarbon removal, although this option is not the best because transported hydrocarbons get adsorbed in the electrode, being difficult its recovery.

Even though Ti cathode provided lower hydrocarbon removal as it was estimated from Soxhlet extractions, when extracted samples were tested by GC-MS for quantification of three priority hydrocarbon pollutants, it happens that phenanthrene, fluoranthene and pyrene concentrations have been lowered in values higher than 60% when the physical barrier was included; and higher removal was obtained more than 95% when cathode was placed at the soil interface.

5. Acknowledgments

The authors would like to thank to Ciencia Básica 2007-Consejo Nacional de Ciencia y Tecnología (CONACyT) No. 84955 and to Fondo Mixto (FOMIX)-Veracruz-CONACyT, No 9631 for their financial support of this research

6. References

Acar, Y. B., Hamed, T., Alshawabkeb, A. N., & Gale, R. J. (1994). Removal of cadmium (II) from saturated kaolinite by the application of electrical current. *Géotechnique*, Vol.44, No.2, 239-254.

Amzad Hossain, M., Salehuddin, S. M. (2011). Polycyclic aromatic hydrocarbons (PAHs) in edible oils by gas chromatography coupled with mass spectroscopy. *Arabian Journal of Chemistry*, DOI: 10.1016/j.arabjc.2010.09.012

Baiju, K.V., Shukla, S., Biju,S., Reddy, M. L. P., & Warrier, K. G. K. (2009). Morphology-dependent dye-removal mechanism as observed for anatase-titania photocatalyst. *Catalysis Letters*, Vol.131, 663-671.

Bojes, H. K, Pope, P. G. (2007). Characterization of EPA's 16 priority pollutant polycyclic aromatic hydrocarbons (PAHs) in tank bottom solids and associated contaminated soils at oil exploration and production sites in Texas. *Regulatory Toxicology and Pharmacology*, Vol.47, 288-295.

Chongshan, L., Baohua, X., Huang, W., & Congquiang, L. (2006). Equilibrium sorption of phenanthrene by soil humic acids. *Chemosphere, Vol.*63, 1961-1968.

Colleta, Th. F., Bruell, C. J., Ryan, D. K, & Inyang, H. I., (1997). Cation-enhanced removal of lead from kaolinite by electrokinetics. *Journal of Environmental Engineering*, Vol.123, No.12, 1227-1233.

Comninellis, Ch., C. Pulgarin (1991). Anodic oxidation of phenol for wastewater treatment, *Journal of Applied Electrochemistry*, Vol.21, 703-708.

Comninellis, Ch. (1994) Electrocatalysis in the Electrochemical cConversion-Combustion of Organic Pollutants for Wastewater Treatment, *Electrochimica Acta*, Vol.39, 1857-1862.

Cox, C. D., Shoesmith, M. A. & Ghosh, M. M. (1996). Electrokinetic remediation of mercury-contaminated soils using iodine/iodide lixiviant. *Environmental Science Technology*, Vol.30, No. 6, 1933-1938.

Cundy, A. B., Hopkinson, L., (2005). Electrokinetic iron pan generation in unconsolidated sediments: implications for contaminated land remediation and soil engineering. *Applied Geochemistry*, Vol.20, 841-848.

Darmawan, Wada, S. I. (2002). Effect of clay mineralogy on the feasibility of electrokinetic soil decontamination technology. *Applied Clay Science*, Vol.20, 283-293.

Fritz, S. J. (1986). Ideality of clay membranes in osmotic processes. *Clay and Clay Minerals*, Vol.34, No.2, 214-223.

Gray, D. H. (1970). Electrochemical hardening of clay soils. *Géotechnique*, Vol.20, No.1, 81-93.

Hamed, J., Acar, Y.B., Gale, R. J. (1991). Pb(II) removal from kaolinite by electrokinetics. *Journal of Geotechnical Engineering*, Vol.117, 241- 271.

Hamed, J. T., Bhadra, A. (1997). Influence of current density and pH on electrokinetics. *Journal of Hazardous materials*, Vol.55, 279-294.

Helwany, S. (2007). *Applied soil mechanics with ABAQUS applications. Chapter 1.* John Wiley &Sons, Inc. ISBN 978-0471-79107-2, pp 11-13

Hu, J.M., Meng, H. M., Zhang, J. Q., & Cao, C. N. (2002). Degradation mechanism of long service life Ti/IrO2-Ta2O5 oxide anodes in sulphuric acid. *Corrosion Science*, Vol.44, No.8, 1655-1668.

Khan, L. I., Alam, S. (1994). Heavy metal removal from soil by coupled electric-hydraulic gradient. *Journal of Environmental Engineering*, Vol.120, No.6, 1524-1543.

Kim, S-O., Kim, K-W., Stüben, D. (2002). Evaluation of electrokinetic removal of heavy metals from tailing soils. *Journal of Environmental Engineering* , Vol.128, No.8, 705-715.

Kolosov, A. Yu., Popov, K. I., Shabanova, N. A., Arter'eva, A. A., Kogut, B. M., Frid, A. S., Zel'venskii, V. Yu., & Urinovich, E. M. (2001). Electrokinetic removal of hydrophobic organic compounds from soil. *Russian Journal of Applied Chemistry*, Vol.74, No.4, 631-635.

Leon, M. T., Pomposo, G.G., Suárez, G.J., & Vega, S. S. (2009). Treatment of acid orange 24 solutions with dimensionally stable anodes. *Portugalia Electrochimica Acta*, Vol.27, No.3, 227-236

Maureen, J. (2011). HPLC detector options for the determination of polynuclear aromatic hydrocarbons LC 7. Varian application notes. In: *www.chem.agilent.com/library/application/lc07.pdf*

Murillo-Rivera, B., Labastida, I., Barón, J., Oropeza-Guzmán, M. T., Gonzalez, I., & Teutli-León, M. M. M. (2009). Influence of anolyte and catholyte composition on TPHs removal from low permeability soil by electrokinetic reclamation. *Electrochimica Acta*, Vol.54, 2119-2114.

Pamucku, S., Khan, L. I., & Fang, H-Y., (1990). Zinc detoxification of soils by electro-osmosis. Transportation Research Record No. 1288, Soils, Geology and Foundations. *Geotechnical Engineering*, 41-46.

Pamucku, S., Wittle, J. K. Electrokinectic removal of selected heavy metals from soil. *Environmental Progress*, Vol.11, No.3, 241 250.

Reddy, K. R., Donahue, M., Saicheck, R. E., & Sasaoka, R. (1999). Preliminary assessment of electrokinetic remediation of soil and sludge contaminated with mixed waste. *Journal of Air & Waste Management Association*, Vol.49, 823-830.

Reddy, K. R., Ala, P. R., Sharma, S., & Kumar, S. N. (2006). Enhanced electrokinetic remediation of lead contaminated soil. *Sci. Technol.*, Vol.85, 123-132.

Reed, B. E., Berg, T. E., Thompson, J. C., & Hatfield, J. H. (1995). Chemical conditioning of electrode reservoirs during electrokinetic soil flushing of Pb-contaminated silt loam. *Journal of Environmental Engineering*, Vol.121, No.11, 805-815.

Ribeiro, A. B., Rodriguez-Maroto, J. M., Mateus, E. P., & Gomes, H. (2005). Removal of organic contaminants from soils by an electrokinetic process: the case of atrazine. Experimental and modeling. *Chemosphere*, Vol.59, No.9, 1229-39.

Ruiz, C., Anaya, J. M., Ramírez, V., Alba, G. I., García, M. G., Carrillo-Chávez, A., Teutli, M. M., & Bustos, E. (2011). Soil arsenic removal by a permeable reactive barrier of iron coupled to an electrochemical process. *International Journal of Electrochemical Science*, Vol.6, 548-560.

Xing, B., Kang, S., (2005). Phenanthrene sorption to sequentially extracted soil humic acids and humin. *Environ. Sci. Technol.*, Vol.39, 134-140.

Saichek, R. E., Reddy, K. R. (2003). Effect of pH control at the anode for the electrokinetic removal of phenathrene from kaolin soil, *Chemosphere,* Vol.51, No.4, 273-287.

Vázquez, M. V.,Hernández-Luis, F., Grandoso, D., & Arbelo, C. D. (2004). Study of electrical resistance of andisols subjected to electro-remediation treatment. *Portugalia Electrochimica Acta* Vol.22, 399-410.

Wang, Jing-Yuan; Huang, Xiang-Jun; Kao, J. M. C., & Stabnikova, Olena. (2007). Simultaneous removal of organic contaminants and heavy metals from kaolin using an upward electrokinetic soil remediation process. *Journal of Hazardous Materials*, Vol.144, No.1-2, 292-299.

Yanzheng, G., Wanting, L., Wang, X., & Li, Q. (2007). Impact of exotic and inherent dissolved organic matter on sorption of phenanthrene by soils. *Journal of Hazardous Materials*, Vol.140, 138-144.

Yeung, A. T., Hsu, Cheng-non, & Menon, R. M. (1996). EDTA-enhanced electrokinetic extraction of lead. *Journal of Geotechnical Engineering*, Vol.122, 8, 666-673.

Yigit, Z., Inan, H. (2009). A study of photocatalytic oxidation of humic acid on anatase and mixed-phase anatase-rutile TiO_2 nanoparticles. *Water Air Soil Pollution*: *Focus*, Vol.9, 237-243